Probability Theory and Stochastic Modelling

Volume 110

Probability Theory and Stochastic Modelling publishes cutting-edge research monographs in probability and its applications, as well as postgraduate-level textbooks that either introduce the reader to new developments in the field, or present a fresh perspective on fundamental topics.

Books in this series are expected to follow rigorous mathematical standards and all titles will be thoroughly peer-reviewed before being considered for publication.

Probability Theory and Stochastic Modelling covers all aspects of modern probability theory including:

- Gaussian processes
- Markov processes
- Random fields, point processes, and random sets
- Random matrices
- Statistical mechanics and random media
- Stochastic analysis
- High-dimensional probability

as well as applications that include (but are not restricted to):

- Branching processes and other models of population growth
- Communications and processing networks
- Computational methods in probability theory and stochastic processes, including simulation
- Genetics and other stochastic models in biology and the life sciences
- Information theory, signal processing, and image synthesis
- Mathematical economics and finance
- Statistical methods (e.g., empirical processes and MCMC)
- Statistics for stochastic processes
- Stochastic control and stochastic differential games
- Stochastic models in operations research and stochastic optimization
- Stochastic models in the physical sciences

Probability Theory and Stochastic Modelling is a merger and continuation of Springer's Stochastic Modelling and Applied Probability and Probability and Its Applications series.

Richard Durrett

Dynamics on Graphs

Second Edition

Springer

Richard Durrett
Duke University
Durham, NC, USA

ISSN 2199-3130 ISSN 2199-3149 (electronic)
Probability Theory and Stochastic Modelling
ISBN 978-3-032-21907-7 ISBN 978-3-032-21908-4 (eBook)
https://doi.org/10.1007/978-3-032-21908-4

This Springer imprint is published by the registered company Springer Nature Switzerland AG
The registered company address is: Gewerbestrasse 11, 6330 Cham, Switzerland

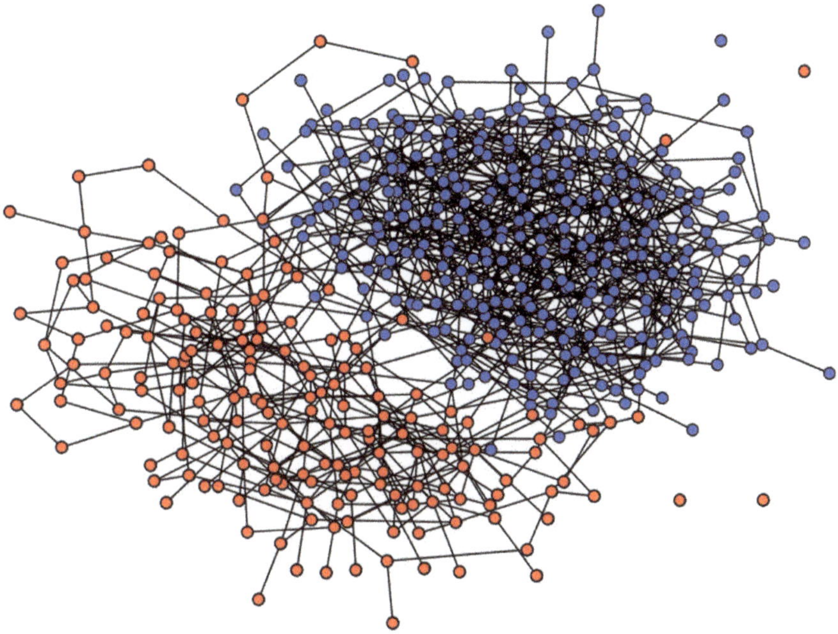

Graph fission in an evolving voter model, see Sect. 8.1.1

For Tom Liggett, and Harry Kesten
Mentors, colleagues, and friends, now departed

Harry Kesten (1931–2019): A Personal and Scientific Tribute. G.R. Grimmett and G.F. Lawler. Notices of the A.M.S. vol. ***67****, issue 6 (2020), pp. 822–831.*

The life and mathematical legacy of Thomas M. Liggett. Paul Jung, Amber Puha, et al. Notices of the A.M.S. vol. ***68****, issue 1 (2021), pp. 67–79.*

Preface

Mathematicians are like Frenchmen: whatever you say to them they translate into their own language and forthwith it is something entirely different,

Johann Wolfgang von Goethe

This volume is an extensive rewrite of the 2007 book *Random Graph Dynamics*. Frankly I wrote that book to learn the subject. The previous version concentrated on the geometry of random graphs and covered a wide variety of examples: Erdős-Rényi random graphs, the configuration model, preferential attachment, small worlds, the Chung-Lu model, and the CHKNS model named in honor of the five authors: Duncan Callaway, John Hopcroft, Jon Kleinberg, Mark Newman, and Steve Strogatz.

The 2007 book surveryed the then well-developed theory of random walks on graphs and touched on the theory that was just beginning to be developed for the voter model and contact process. Nineteen years has brought a lot of progress on the study of intereacting particle systems on graphs. For this reason, the new version studies a wide variety of dynamics on a small number of classes of random graphs, primarily the Erdős-Rényi model, configuration model, and inhomogeneous random graphs. These are covered in the first three chapters. After that we turn to the processes that we study on random graphs: epidemics, the contact process, random walks and the voter model/coalescing random walks. The final chapter explores the difficult and still poorly understood subject of coevolving networks, where the states of the vertices and the edges in the graph both evolve.

I have tried (and likely failed) to write a book that has the style of a good seminar talk. My goal was to focus on the ideas that go into the proofs, and not get bogged down in technical details. In most cases this means we will not give full proofs of the most exciting results we discussing. As my postdoctoral mentor Tom Liggett once said: "a book should not just be a union of papers."

I would like to thank my coauthors and others who helped me write the material included here: Shankar Bhamidi, Tom Britton, Shirshendu Chatterjee, Ted Cox, Alice Ding, Zoe Huang, Nicholas Lanchier, Laurent Ménard, Peter Mucha, Claudia

Neuhauser, David Sivakoff, Dong Yao, and the Duke DOMath teams I supervised in 2018, 2020, and 2022. Like most mathematicians, I have been inspired by the writings of others. That list is very long but in addition to Harry Kesten and Tom Liggett to whom this book is dedicated I would particularly like to mention David Aldous, Goeff Grimmett, Svante Janson, Yuval Peres, Allan Sly, and Remco van der Hofstad.

The publication of this book has been a long and winding road starting with it submission to another publisher in early 2024. I'll spare the reader the unhappy details except to mention that in April 2025 a hacker did me the favor of breaking into the math department server at Duke and relieving me of more than 100,000 emails that I had accumulated over 15 years. I would like to thank the editorial board of the Probability Theory and Stochastic Modeling for accepting the book to appear in the series, and Habiba Kalantarova for guiding me through the publication process.

It is inevitable that there will be typos and errors in this book. One review of the 2007 version of this book said "You should not trust every proof in this book." Though I am now retired from Duke, mail sent to rtd@math.duke.edu will for the foreseeable future reach me. While I appreciate lists of typos, I am more enthusiastic to hear suggestions for improvements and to learn about new results and better proofs.

Durham, USA Richard Durrett

Competing Interests The author has no competing interests to declare that are relevant to the content of this manuscript.

Contents

Notation, Etc.

In general we follow the notation of the paper whose results we are describing. We do this to make it easier for the reader find the missing details. A consequence of this is that the notation can change from section to section.

For concepts from probability theory, the notation will in most cases follow that used in the fifth edition of my graduate probability book, *Probability: Theory and Examples*, referred to here as PTE5. Some basic examples:

If a sequence of random variables *converges in distribution* to X we write $X_n \Rightarrow X$.

We use χ for the *standard normal distribution*. This notation derives from the fact that the square of a normal has a chi-squared (χ^2) distribution.

We will use many comparisons or "couplings" of random variables, so the concept of stochastic order is important: $X \le_d Y$ means that $P(X \le z) \ge P(Y \le z)$ for all z, or equivalently $Eh(X) \le Eh(Y)$ for all bounded nondecreasing h. In this case we can define random variables $X' =_d X$ and $Y' =_d Y$ so that $X' \le Y'$ (almost surely).

Being a mathematician, when log is written, we mean the natural logarithm. When another base b is needed we will write $\log_b$.

You can't find the phrase in PTE5 but with high probability (or whp) means that the statement holds with a probability that $\to 1$ as some parameter $n \to \infty$.

If $f(t), g(t) > 0$ then

$$f(t) = o(g(t)) \quad or \quad f \ll g \quad \text{if } f(t)/g(t) \quad \to \quad 0 \text{ as } t \to \infty$$

$$f(t) \sim g(t) \quad \text{if } f(t)/g(t) \quad \to \quad 1 \text{ as } t \to \infty$$

$$f(t) = O(g(t)) \quad \text{if } f(t)/g(t) \;\leq\; C \text{ for all } t$$

$$f(t) = \Omega(g(t)) \quad \text{if } f(t)/g(t) \;\geq\; c \;> 0 \text{ for all } t$$

$$f(t) = \Theta(g(t)) \quad \text{if } c \;\leq\; f(t)/g(t) \;\leq\; C \text{ for all } t\,.$$

We use the name and year system. For example: Mountford, Valkesin, and Yao (2013). After the first citation in a section later references will have the name shortened to initials, e.g. MVY (2013), unless that is ambiguous, in which case the year is added.

Chapter 1
Erdős-Rényi Random Graphs

In this chapter we will analyze the random graph model introduced by Erdős and Rényi in the 1950s. This example has been extensively studied. A very nice account of many of the results can be found in the classic book of Bollobás (2001), where the analysis has been done primarily by combinatorial methods. In contrast to this and many other treatments, we mainly rely on methods from probability.

To define the model, we begin with the set of vertices $V = \{1, 2, \dots, n\}$. For $1 \le x < y \le n$ let $\eta_{x,y}$ be independent, $= 1$ with probability p and $= 0$ otherwise, and then set $\eta_{y,x} = \eta_{x,y}$. If $\eta_{x,y} = 1$ there is an edge connecting x and y. Here, we will be primarily concerned with the situation when $p = \lambda/n$ and in particular with showing that there is a phase transition. Let $\mathcal{C}_x$ be the component containing x, i.e., the set of vertices that can be reached from x.

- When $\lambda < 1$ all of the components are small, i.e., there is a constant $a(\lambda)$ so that $P(\max_{1\le x\le n} |\mathcal{C}_x| \ge a(\lambda) \log n) \to 0$ as $n \to \infty$
- When $\lambda > 1$ there is a **giant component** with $\sim g(\lambda)n$ vertices, and the second largest component is of size $O(\log n)$. There is also a central limit theorem for the size of the giant component.
- At the critical value, $\lambda = 1$, the largest component has size $n^{2/3}$, and there are many other components of this size
- As the reader will learn in Sects. 1.7 and 1.8, the critical value is surrounded by a critical regime $\lambda = 1 + tn^{-1/3}$, $t \in (-\infty, \infty)$, in which cluster growth as a function of t can be approximated by a **multiplicative coalescent**, which gives an intuitive picture of the emergence of the giant component.

The first three results show that there is a **double jump transition**: the size of the largest component goes from $\log n$ to $n^{2/3}$ to n.

In Sect. 1.1 we define **branching processes** and describe results that will be useful in understanding the phase transition in the Erdős-Rényi random graph. To compute the size of $\mathcal{C}_x$, we set $I_0 = \{x\}$, let I_1 be the neighbors of x, and for $t \ge 2$ let I_t be the vertices y for which the distance $d(x, y) = t$. In Sect. 1.2 we will show that when t is not too large $Z_t = |I_t|$ is approximately a branching process in which each individual

R. Durrett, *Dynamics on Graphs*, Probability Theory and Stochastic Modelling 110,
https://doi.org/10.1007/978-3-032-21908-4_1

has an average of λ children. If $\lambda < 1$ the branching process dies out exponentially fast and all components are small. When $\lambda > 1$, the branching process survives with probability $g(\lambda)$, and all sites with surviving branching processes combine to make the giant component.

In a branching process, a large number of individuals are born at one time, so in Sect. 1.3 we introduce the **exploration process** in which we reveal the neighbors of sites in the cluster one site at a time. In this process the number of active sites is a random walk run until it hits 0, so the large body of results about random walk leads to a detailed understanding of cluster sizes, much more precise than from the branching process viewpoint.

In Sect. 1.4 we compute the threshold for the graph to be connected, i.e., all vertices in one components. It is trivial that if there is a vertex with degree 0 then the graph is not connected. A result of Erdős and Rényi from 1959 states that if $p = (c_n + \log n)/n)$, $H(n, p)$ is the probability that Erdős-Rényi(n, p) is connected, and $c_n \to c$ then

$$P(H(n, p)) \to \exp(-e^{-c}).$$

Most of the results in this chapter are well-known but Sect. 1.5 is an exception. In 1981 Ajtai, Komlós, and Szemerédi showed that when $\lambda > 1$ not only is there a connected set of vertices of size cn but there is also a self-avoiding path with length $c'n$. This is useful for studying the contact process on Erdős-Rényi graphs because it is known that the survival time of the contact process on $[0, m]$ is $\geq \exp(cm)$ with high probability as $m \to \infty$.

Returning to well-known results, in Sect. 1.6 we prove a central limit theorem for the size of the giant component by using a clever argument of Martin-Löf based on the exploration process. In Sect. 1.7 we describe a combinatorial approach that leads to explicit formulas for the distribution of the cluster size in the critical regime. A useful aspect of this approach is a duality between the supercritical and subcritical regimes which makes it easy to prove that the second largest component has size $O(\log n)$ when $\lambda > 1$.

In Sect. 1.8 we introduce the probabilistic approach to the critical regime. The most important results are two theorems of David Aldous that (i) gives an interesting description of the cluster sizes in terms of the excursions of a reflection Brownian with linear drift and (ii) show that on the time scale introduced in the fourth bullet point, the cluster sizes as a function of t are a multiplicative coalescent.

Finally in Sect. 1.9 we compute the critical exponents of percolation for the phase transition in the Erdős-Rényi random graph. These computations are relevant to percolation theory in $\mathbb{Z}^d$ since they are the mean-field values of the critical exponents, which are supposed to hold above (>) the critical dimension $d_c = 6$.

1.1 Branching Processes

In this section we define branching processes and describe their basic properties. Since n will be the number of vertices in our graph, we will use t and s for our discrete time parameter. Let ξ_i^t, $i, t \geq 0$, be i.i.d. nonnegative integer-valued random variables. Define a sequence Z_t, $t \geq 0$ by $Z_0 = 1$ and

$$Z_{t+1} = \begin{cases} \xi_1^{t+1} + \cdots + \xi_{Z_t}^{t+1} & \text{if } Z_t > 0, \\ 0 & \text{if } Z_t = 0. \end{cases}$$

Z_t is called a **Galton-Watson process**. The idea behind the definition is that Z_t is the number of individuals in the tth generation, and each member of the tth generation gives birth to an independent and identically distributed number of children. $p_k = P(\xi_i^t = k)$ is called the **offspring distribution**.

Lemma 1.1.1 *Let* $\mathcal{F}_t = \sigma(\xi_i^s : i \geq 1, 1 \leq s \leq t)$ *and* $\mu = E\xi_i^t \in (0, \infty)$. *Then* Z_t/μ^t *is a martingale w.r.t.* $\mathcal{F}_t$.

Proof Clearly, Z_t is measurable with respect to $\mathcal{F}_t$, or $Z_t \in \mathcal{F}_t$. Recall, see Theorem 4.1.2 in PTE5, that if $X = Y$ on $B \in \mathcal{F}$ then $E(X|\mathcal{F}) = E(Y|\mathcal{F})$ on B. On $\{Z_t = k\}$,

$$E(Z_{t+1}|\mathcal{F}_t) = E(\xi_1^{t+1} + \cdots + \xi_k^{t+1}|\mathcal{F}_t) = k\mu = \mu Z_t.$$

Dividing both sides by μ^{t+1} now gives the desired result. □

Theorem 1.1.2 *If* $\mu < 1$ *then* $Z_t = 0$ *for all* t *sufficiently large.*

Proof $E(Z_t/\mu^t) = E(Z_0) = 1$, so $E(Z_t) = \mu^t$. Now $Z_t \geq 1$ on $\{Z_t > 0\}$ so

$$P(Z_t > 0) \leq E(Z_t; Z_t > 0) = E(Z_t) = \mu^t \to 0$$

exponentially fast if $\mu < 1$. □

The last answer should be intuitive. If each individual on the average gives birth to less than one child, the species will die out. The next result shows that after we exclude the trivial case in which each individual has exactly one child, the same result holds when $\mu = 1$.

Theorem 1.1.3 *If* $\mu = 1$ *and* $P(\xi_i^t = 1) < 1$ *then* $Z_t = 0$ *for all* t *sufficiently large.*

Proof When $\mu = 1$, Z_t is itself a nonnegative martingale, so the martingale convergence theorem, Theorem 4.2.12 in PTE5, implies that Z_t converges to an a.s. finite limit Z_∞. Since Z_t is integer-valued, we must have $Z_t = Z_\infty$ for large t. If $P(\xi_i^t = 1) < 1$ and $k > 0$ then $P(Z_t = k \text{ for all } t \geq T) = 0$ for any T, so we must have $Z_\infty \equiv 0$. □

To state the next result we need a definition. For $\theta \in [0, 1]$, let $\phi(\theta) = \sum_{k \geq 0} p_k \theta^k$ be the **generating function** for the offspring distribution p_k.

Theorem 1.1.4 *If $\mu > 1$ then $\sigma = P(Z_t > 0$ for all $t) > 0$. The* **extinction probability** *$\rho = 1 - \sigma$ is the unique solution of $\phi(x) = x$ with $x \in [0, 1)$.*

Proof Differentiating gives for $\theta < 1$

$$\phi'(\theta) = \sum_{k=1}^{\infty} k\, p_k \theta^{k-1} \geq 0,$$

$$\phi''(\theta) = \sum_{k=2}^{\infty} k(k-1) p_k \theta^{k-2} \geq 0.$$

So ϕ is increasing and convex, and $\lim_{\theta \uparrow 1} \phi'(\theta) = \sum_{k=1}^{\infty} k p_k = \mu$. Our interest in ϕ stems from the following facts.

(a) *If $\theta_t = P(Z_t = 0)$ then $\theta_t = \sum_{k=0}^{\infty} p_k \theta_{t-1}^k = \phi(\theta_{t-1})$.*

Proof of (*a*). If $Z_1 = k$, an event with probability p_k, then $Z_t = 0$ if and only if all k families die out in the remaining $t - 1$ units of time, independent events with probability θ_{t-1}^k. Summing over the disjoint possibilities for each k gives the desired result. □

(b) *If $\phi'(1) = \mu > 1$ there is a unique $\rho < 1$ so that $\phi(\rho) = \rho$.*

Proof of (*b*). $\phi(0) \geq 0$, $\phi(1) = 1$, and $\phi'(1) > 1$, so $\phi(1 - \epsilon) < 1 - \epsilon$ for small ϵ. The last two observations imply the existence of a fixed point. To see it is unique, observe that $\mu > 1$ implies $p_k > 0$ for some $k > 1$, so $\phi''(\theta) > 0$ for $\theta > 0$. Since ϕ is strictly convex, it follows that if $\rho < 1$ is the smallest fixed point, then $\phi(x) < x$ for $x \in (\rho, 1)$. □

(c) *As $t \uparrow \infty$, $\theta_t \uparrow \rho$.*

Proof of (*c*). $\theta_0 = 0$, $\phi(\rho) = \rho$, and ϕ is increasing, so induction implies θ_t is increasing and $\theta_t \leq \rho$. Let $\theta_\infty = \lim \theta_t$. Taking limits in $\theta_t = \phi(\theta_{t-1})$, we see $\theta_\infty = \phi(\theta_\infty)$. Since $\theta_\infty \leq \rho$, it follows that $\theta_\infty = \rho$. For a picture see Fig. 1.1.

Combining (a)–(c) shows $P(Z_t = 0$ for some $t) = \lim_{t \to \infty} \theta_t = \rho < 1$. □

Example 1.1.5 (Poisson distribution with mean λ)

$$P(\xi = k) = e^{-\lambda} \frac{\lambda^k}{k!}.$$

In this case $\phi(s) = \sum_{k=0}^{\infty} e^{-\lambda} s^k \lambda^k / k! = \exp(\lambda(s - 1))$ so the fixed point equation is

$$\rho = \exp(\lambda(\rho - 1)). \tag{1.1.1}$$

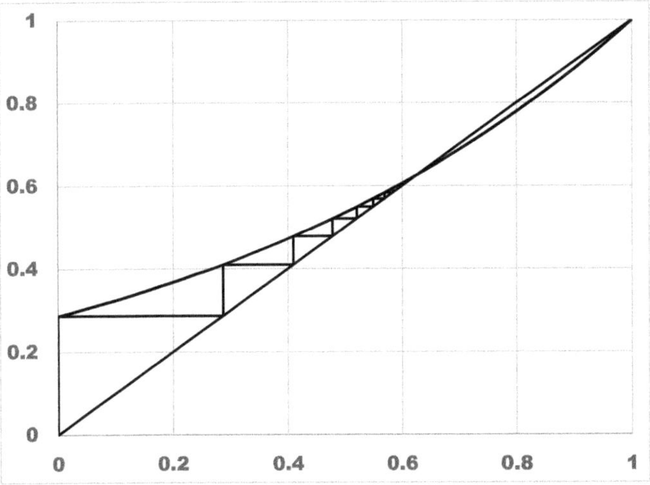

Fig. 1.1 Generating function for the Poisson distribution with mean 1.25. The alternating vertical and horizontal lines illustrate the iteration of the generating function: $(0, 0) \to (0, \theta_1) \to (\theta_1, \theta_1) \to (\theta_1, \theta_2) \ldots$ Fixed point $\rho = 0.62861$

Theorem 1.1.4 shows that when $\mu > 1$, the limit of Z_t/μ^t has a chance of being nonzero. The next theorem is not the best possible but it suffices for our purposes. For a proof see Example 4.4.9 in PTE5.

Theorem 1.1.6 *If* $\sum_k kp_k > 1$ *then* $W = \lim Z_t/\mu^t$ *exists. If* $\sum_k k^2 p_k < \infty$ *then* $W \not\equiv 0$.

The ultimate result about this question is due to Kesten and Stigum (1966). For a "conceptual proof" see Lyons et al. (1995).

Theorem 1.1.7 *Suppose that* $\mu = \sum_k kp_k > 1$, *then* $W = \lim Z_t/\mu^t \not\equiv 0$ *if and only if* $\sum_{k=1}^{\infty} p_k k \log k < \infty$.

Our next result shows that when $W \not\equiv 0$, it is positive on the set where the branching process does not die out.

Theorem 1.1.8 *If* $P(W = 0) < 1$ *then* $\{W > 0\} = \{Z_n > 0$ *for all* $n\}$, *i.e., the symmetric difference of the two sets has probability 0.*

Proof Let $r = P(W = 0)$. In order for Z_t/μ^t to converge to 0 the limit must be 0 for the branching process started by each of the children in the first generation. Breaking things down according to the number of children in the first generation

$$r = \sum_{k=0}^{\infty} p_k r^k = \phi(r).$$

This shows $r < 1$ is a fixed point of the generating function and hence $r = \rho = P(Z_t = 0 \text{ for some } t)$. Clearly, $\{W > 0\} \subset \{Z_t > 0 \text{ for all } t\}$. Since the two sets have the same probability $P(\{Z_t > 0 \text{ for all } t\} - \{W > 0\}) = 0$, which is the desired result. □

1.1.1 Conditioning

The limit theorems above describe the growth of the process when it does not die out. Our next question is: what happens in a supercritical branching process when we condition it to die out?

Theorem 1.1.9 *A supercritical branching process conditioned to become extinct is a subcritical branching process. If the original offspring distribution has generating function $\phi(z)$ then the conditioned process has generating function*

$$\bar{\phi}(z) = \rho^{-1}\phi(\rho z), \tag{1.1.2}$$

where ρ is the extinction probability. Thus if the original offspring distribution is Poisson(λ) with $\lambda > 1$ then the conditioned one is Poisson($\lambda\rho$).

Formula (1.1.2) shows the generating function of a supercritical branching process conditioned to die out is obtained by taking the graph of ϕ over $[0, \rho]$ and rescaling to make the domain and range $[0, 1]$. It is clear from Fig. 1.2 that $\phi'(\rho) < 1$ so the conditioned branching process is subcritical.

Proof Let $T_0 = \inf\{t : Z_t = 0\}$ and consider $\bar{Z}_t = (Z_t | T_0 < \infty)$. To check the Markov property for $\bar{Z}_t$ note that the Markov property for Z_t implies:

$$P(Z_{t+1} = z_{t+1}, T_0 < \infty | Z_t = z_t, \ldots Z_0 = z_0) = P(Z_{t+1} = z_{t+1}, T_0 < \infty | Z_t = z_t).$$

To compute the transition probability for $\bar{Z}_t$, observe that if ρ is the extinction probability then $P_x(T_0 < \infty) = \rho^x$. Let $p(x, y)$ be the transition probability for Z_t. The Markov property implies

$$\begin{aligned} \bar{p}(x, y) &= \frac{P_x(Z_1 = y, T_0 < \infty)}{P_x(T_0 < \infty)} \\ &= \frac{P_x(Z_1 = y) P_y(T_0 < \infty)}{P_x(T_0 < \infty)} \\ &= \frac{p(x, y)\rho^y}{\rho^x}. \end{aligned}$$

Taking the starting state $x = 1$ and computing the generating function

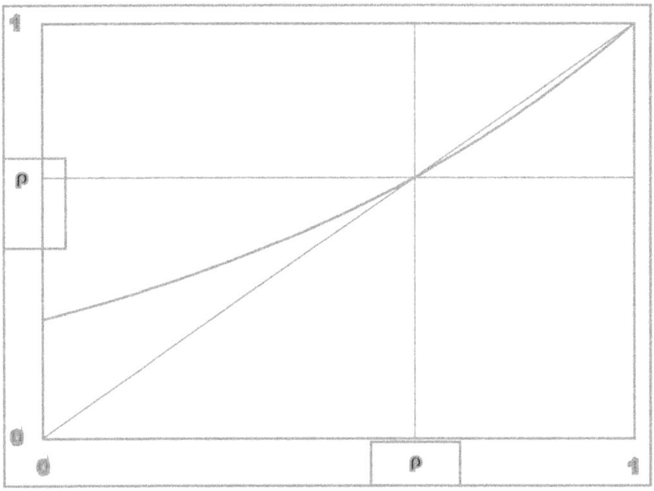

Fig. 1.2 Generating function for the Poisson distribution with mean 1.25, illustrating Theorem 1.1.9 (g.f. in lower left quadrant) and Theorem 1.1.10 (g.f. in upper right). See formulas (1.1.2) and (1.1.3)

$$\bar{\phi}(\theta) = \sum_{y=0}^{\infty} \bar{p}(1, y)\theta^y = \rho^{-1} \sum_{y=0}^{\infty} p(1, y)(\theta\rho)^y = \rho^{-1}\phi(\theta\rho).$$

The last detail is to check that the conditioned branching process is indeed a branching process. If we start with x individuals then each individual in Z_1 gives rise to an independent family. In $\bar{Z}_1$ each family must die out, so in $\bar{Z}_n$ each family of a first generation individual is an independent copy of the branching process conditioned to die out. □

Theorem 1.1.10 *Consider a supercritical branching process with offspring distribution p_k and generating function ϕ. If we condition on nonextinction and look only at the individuals that have an infinite line of descent then the number of individuals in generation t, $\hat{Z}_t$, is a branching process with offspring generating function*

$$\hat{\phi}(\theta) = \frac{\phi((1-\rho)\theta + \rho) - \rho}{1-\rho}, \tag{1.1.3}$$

where ρ is the extinction probability, i.e., the smallest solution of $\phi(\rho) = \rho$ in $[0, 1]$.

In words, we take the graph of ϕ over $[\rho, 1]$ and rescale to make the domain and range $[0, 1]$.

Proof There is nothing to prove if $\rho = 0$ so suppose $0 < \rho < 1$. If $Z_0 = 1$ and we condition on survival of the branching process, then the number of individuals in the first generation who have an infinite line of descent has distribution

$$\hat{p}_j = \frac{1}{1-\rho} \sum_{k=j}^{\infty} p_k \binom{k}{j} (1-\rho)^j \rho^{k-j}.$$

Multiplying by θ^j, summing, and interchanging the order of summation

$$\begin{aligned} \sum_{j=1}^{\infty} \bar{p}_j \theta^j &= \frac{1}{1-\rho} \sum_{j=1}^{\infty} \sum_{k=j}^{\infty} p_k \binom{k}{j} (1-\rho)^j \rho^{k-j} \theta^j \\ &= \frac{1}{1-\rho} \sum_{k=1}^{\infty} p_k \sum_{j=1}^{k} \binom{k}{j} (1-\rho)^j \theta^j \rho^{k-j}. \end{aligned}$$

Using the binomial theorem and noticing that the $j = 0$ term is missing the above

$$= \frac{1}{1-\rho} \sum_{k=1}^{\infty} p_k \{((1-\rho)\theta + \rho)^k - \rho^k\}.$$

We can add the $k = 0$ term to the sum since its value is $1 - 1 = 0$. Doing this the result is

$$\hat{\phi}(\theta) = \frac{\phi((1-\rho)\theta + \rho) - \phi(\rho)}{1-\rho}.$$

Since $\phi(\rho) = \rho$ the result follows. □

In Sect. 1.5.2 we will need a slightly different result for a supercritical branching process with $Z_0 = 1$ conditioned on $\Omega_\infty = \{n : Z_n \geq 1 \text{ for all } n \geq 1\}$. Let D_i be the event that the lineage of the ith child in the first generation dies out, and let $F = \min\{i \leq Z_1 : D_i^c \text{ occurs}\} - 1$ be the number of failures, children whose lineages die out, before we have our first success. Our goal to compute

$$P(F = k|\Omega_\infty) \quad \text{and} \quad E(F|\Omega_\infty).$$

We will eventually specialize to a Poisson offspring distribution, but we begin by considering the general supercritical situation. Let $p_j = P_1(Z_1 = j)$ be the offspring distribution and suppose that $\sigma = P(\Omega_\infty) > 0$.

$$P(F = k, \Omega_\infty) = \sum_{j=k+1}^{\infty} (1-\sigma)^k \sigma p_j.$$

Dividing by $P(\Omega_\infty) = \sigma$, we have

$$P(F = k|\Omega_\infty) = (1-\sigma)^k P_1(Z_1 \geq k+1). \tag{1.1.4}$$

To compute the expected value we multiply by k and sum.

$$E(F|\Omega_\infty) = (1-\sigma)\sum_{k=1}^{\infty} k(1-\sigma)^{k-1} P_1(Z_1 \geq k+1).$$

Let $g(k) = \sum_{j=1}^{k} j(1-\sigma)^{j-1}$ for $k \geq 1$ with $g(0) = 0$ and $g(-1) = 0$, so that $g(k) - g(k-1) = k(1-\sigma)^{k-1}$ for $k \geq 0$ and

$$\begin{aligned} E(F|\Omega_\infty) &= (1-\sigma)\sum_{k=1}^{\infty}[g(k) - g(k-1)]P_1(Z_1 \geq k+1) \\ &= (1-\sigma)\sum_{j=1}^{\infty} g(j)P_1(Z_1 \geq j+1) - (1-\sigma)\sum_{j=0}^{\infty} g(j)P_1(Z_1 \geq j+2) \\ &= (1-\sigma)\sum_{j=1}^{\infty} g(j)P_1(Z_1 = j+1) = (1-\sigma)Eg(Z_1 - 1), \end{aligned} \tag{1.1.5}$$

since we defined $g(-1) = 0$

To get a formula for $E(F|\Omega_\infty)$ we need a formula for $g(n)$. For $n \geq 1$, let

$$f(n) = \sum_{k=1}^{n} (1-\sigma)^k = \frac{(1-\sigma) - (1-\sigma)^{n+1}}{\sigma}.$$

Taking the partial derivative with respect to σ

$$\begin{aligned} g(n) = \sum_{k=1}^{n} k(1-\sigma)^{k-1} &= -\frac{\partial}{\partial\sigma} f(n) \\ &= \frac{(1-\sigma) - (1-\sigma)^{n+1}}{\sigma^2} - \frac{-1 + (n+1)(1-\sigma)^n}{\sigma} \\ &= \frac{1 - (1-\sigma)^{n+1}}{\sigma^2} - \frac{(n+1)(1-\sigma)^n}{\sigma}, \end{aligned}$$

where in the last step we have used part of the second term to cancel part of the first. Note that the last formula holds also for $n = 0$ since it gives $g(0) = 0$.

Specializing now to the case of Poisson(λ).

$$Eg(Z_1 - 1) = \sum_{k=0}^{\infty} g(k) e^{-\lambda} \frac{\lambda^{k+1}}{(k+1)!}$$
$$= \sum_{k=0}^{\infty} \left[\frac{1 - (1-\sigma)^{k+1}}{\sigma^2} - \frac{(k+1)(1-\sigma)^k}{\sigma} \right] e^{-\lambda} \frac{\lambda^{k+1}}{(k+1)!}.$$

The second sum is

$$-\lambda \sum_{k=0}^{\infty} \frac{(1-\sigma)^k}{\sigma} e^{-\lambda} \frac{\lambda^k}{k!} = -\frac{\lambda\sigma}{\sigma^2} e^{-\lambda\sigma}.$$

The first sum is

$$\frac{1}{\sigma^2} e^{-\lambda} \sum_{k=-1}^{\infty} \frac{\lambda^{k+1}}{(k+1)!} - \frac{\lambda^{k+1}(1-\sigma)^{k+1}}{(k+1)!} = \frac{1}{\sigma^2}(1 - e^{-\lambda\sigma}),$$

so when $Z_1 = \text{Poisson}(\lambda)$ we have

$$Eg(Z_1 - 1) = \frac{1}{\sigma^2}\left(1 - e^{-\lambda\sigma} - \lambda\sigma e^{-\lambda\sigma}\right).$$

To simplify this expression we use the fact that the survival probability σ satisfies $1 - \sigma = e^{-\lambda\sigma}$.

$$Eg(Z_1 - 1) = \frac{1}{\sigma^2}(1 - (1-\sigma) - \lambda\sigma(1-\sigma)) = \frac{1}{\sigma}(1 - \lambda\sigma). \qquad (1.1.6)$$

Using (1.1.5) it follows that

$$E(F|\Omega_\infty) = \frac{1-\sigma}{\sigma}(1 - \lambda\sigma). \qquad (1.1.7)$$

1.2 Cluster Growth as a Branching Process

In this section we use branching process results to study the connected component, or cluster, containing 1. To do this we introduce a process that is the same as a discrete time epidemic. To begin the process, we let $S_0 = \{2, 3, \ldots, n\}$, $I_0 = \{1\}$, and $R_0 = \emptyset$. The letters are motivated by the epidemic interpretation of the growing cluster. S_t are the susceptibles, I_t are infected, and R_t are removed (recovered and immune to further infection). In graph theory terms, we have already examined the connections of all sites in R_t, I_t are the sites to be investigated on this turn, and S_t are unexplored. These sets evolve as follows:

$$\begin{aligned} R_{t+1} &= R_t \cup I_t, \\ I_{t+1} &= \{y \in S_t : \eta_{x,y} = 1 \text{ for some } x \in I_t\}, \\ S_{t+1} &= S_t - I_{t+1}, \end{aligned}$$

where $\eta_{x,y} = \eta_{y,x} = 1$ if and only if there is an edge between x and y. The cluster containing 1, $\mathcal{C}_1 = \cup_{t=0}^{\infty} I_t = R_\infty$.

Kendall (1956) was the first to suggest a branching process approximation for epidemics. We begin with the easy result

Theorem 1.2.1 *If $\lambda < 1$ then $E|\mathcal{C}_1| \le 1/(1-\lambda) < \infty$.*

Proof To define a comparison branching process we introduce a new set of variables $\zeta^t_{x,y}$, $x, t \ge 1$, $1 \le y \le n$, that are independent, $= 1$ with probability λ/n, and 0 otherwise. Let $Z_0 = 1$, $S_t^c = \{1, 2, \dots, n\} - S_t$ and

$$Z_{t+1} = \sum_{x \in I_t, y \in S_t} \eta_{x,y} + \sum_{x \in I_t} \sum_{y \in S_t^c} \zeta^t_{x,y} + \sum_{x=n+1}^{n+Z_t-|I_t|} \sum_{y=1}^{n} \zeta^t_{x,y}. \tag{1.2.1}$$

Here, the second term is the set of extra births in the branching process due to the fact that $|S_t| < n$. The third term produces births for these individuals that are in the branching process but not in the epidemic.

It is immediate from the construction that Z_t is a branching process with offspring distribution binomial$(n, \lambda/n)$ and it provides an upper bound on the growing cluster

$$Z_t \ge |I_t|.$$

When $\lambda < 1$, $EZ_t = \lambda^t$, so the mean cluster size

$$E|\mathcal{C}_1| = E\left(\sum_{t=0}^{\infty} |I_t|\right) \le \sum_{t=0}^{\infty} \lambda^t = 1/(1-\lambda),$$

which proves the desired result. □

Theorem 1.2.2 *Suppose $\lambda > 1$ and let $\rho < 1$ be the probability the branching process Z_t dies out. Then the probability vertex 1 belongs to a giant component, one with size $\ge \epsilon n$ for some $\epsilon > 0$, tends to $1 - \rho$ as $n \to \infty$.*

In the next section, we will give a simpler proof that there is a unique giant component and it has $\sim (1-\rho)n$ members.

Proof Since Z_t is an upper bound on I_t the probability of a giant component is $\le 1 - \rho$. To prove the result in the theorem, we want to define a branching process that lower bounds the growing cluster. To be able to control the growth of the branching process we build it using a **breadth-first search**. If the word search sounds strange note that the construction gives an algorithm that can be used to compute the cluster. Modifying the notation used above,

At time 0, $A_0 = \{1\}$ is an ordered list of **active sites**, $U_0 = \{2, \ldots, n\}$ are **unexplored sites**, and $R_0 = \emptyset$ are **removed sites** whose neighbors are known.

At time 1, $R_0 = \{1\}$, $A_1 = \{y : \eta_{1,y} = 1\}$ and $U_1 = U_0 - A_1$.

At time 2, we let i_2 be the smallest member of the ordered set A_1, let $N_2 = \{y \in U_1 : \eta_{i_2,y} = 1\}$, and let $U_2 = U_1 - N_2$.

At time m, if $A_m = \emptyset$ we are done and R_m is the cluster containing 1. If $A_m \neq \emptyset$ we pick the smallest element i_m in this ordered set, let $N_m = \{y \in U_{m-1} : \eta_{i_m,y} = 1\}$, let $U_m = U_{m-1} - N_m$ and $n_m = |N_m|$. To create A_m we put the elements of N_m in increasing order and add them after the current elements of A_{m-1},

The branching process is embedded in the breadth first search. A picture (see Fig. 1.3) is useful to help explain the definitions. Let $Z_0 = 1$ and $Z_1 = n_1$. Define $Y_0 = 1$ and for $m \ge 2$ let

$$Y_{m-1} = Z_0 + \cdots + Z_{m-1} = Y_{m-2} + Z_{m-1}$$

be the total number of children up to generation $m - 1$, and

$$Z_m = n_{Y_{m-2}+1} + \cdots + n_{Y_{m-1}}.$$

To check the definitions in the example, $Y_0 + 1 = 2$ and $Y_1 = Z_0 + Z_1 = 4$, so

$$Z_2 = n_2 + n_3 + n_4 = 2 + 4 + 1 = 7.$$

This means $Y_2 = 4 + 7 = 11$ and hence $Z_3 = n_5 + n_6 + \cdots + n_{11}$.

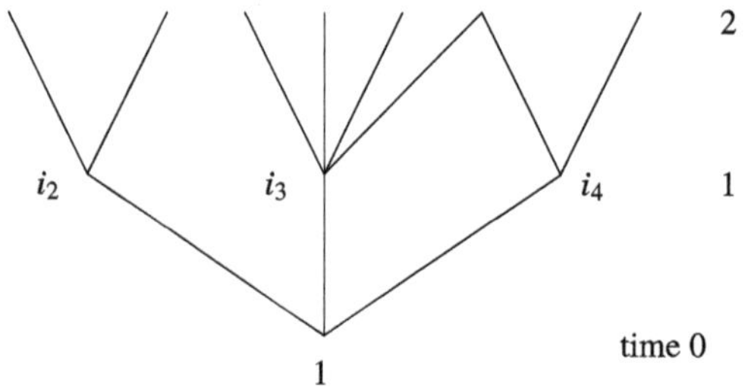

Fig. 1.3 Example of breadth-first search. The number of vertices added on the ith step. Here $n_1 = 3$, $n_2 = 2$, $n_3 = 4$, and $n_4 = 1$ since i_4 has a child that is also a child of i_3

Turning to the proof of Theorem 1.2.2, let δ be chosen so that $\bar{\lambda} = \lambda(1-\delta) > 1$. While the total number of offspring $Y_m \le \delta n$ the number of offspring n_m dominates a binomial$(n, (1-\delta)\lambda/n) \approx$ Poisson$(\lambda(1-\delta))$. The extinction probability of the Poisson branching process satisfies

$$\rho_\delta = \exp(\lambda(1-\delta)(\rho_\delta - 1)).$$

It is easy to see that $\rho_\delta \to \rho$ as $\delta \to 0$. Completing the proof is a little slippery because we need to interchange the limits $\delta \to 0$ and $n \to \infty$. Fix a small value of δ so that $\rho_\delta < \rho + \eta$. If we let $m(n, \delta)$ be the number of generations that are completed before the total number of births reaches δn then $m(n, \delta) \to \infty$. Since the probability the branching process dies out in the first k generations can be computed by iterating the generating function, it follows that the probability that the Poisson$((1-\delta)\lambda)$ survives until generation k and then dies out is $< \eta$ if k is large. □

An important application of breadth-first search is the qualitative property "Erdős-Rényi clusters are locally tree-like." Let $T_\alpha = \min\{t : U_t \le n - n^\alpha\}$, and note that at time m when we make connections from i_m to its neighbors we cannot make a connection to a vertex in R_m since we have identified all of their neighbors, but we can make a connection to another member of the set of active vertices A_m. If this occurs, an event we call a **collision**, then we create a loop in the graph and it is no longer a tree.

Theorem 1.2.3 *The expected number of collisions at times $t < T_\alpha$ is $\lambda n^{2\alpha-1}$. Taking $\alpha = 1/2 - \epsilon$ this result implies that the cluster will with high probability be a tree until it contains $n^{1/2-\epsilon}$ vertices.*

Proof At times $t < T_\alpha$, $|A_t| \le n^\alpha$. If $T_\alpha < \infty$ then $T_\alpha \le n^\alpha$ since at each time we add a vertex to the removed set. These two observations imply that the number of pairs of vertices that can be connected is $n^{2\alpha}$ while the probability a connection is made is $\le \lambda/n$. □

Remark 1.2.4 A better upper bound on the probability results if we note that the number of collisions is dominated by a binomial$(n^{2\alpha}, \lambda/n)$ random variable. In some cases we will need a conclusion that holds for every cluster in the graph. We can do that for Erdős-Rényi graphs by choosing α to be smaller and using the domination by a binomial.

1.2.1 Distances on the Giant Component

The branching process results derived earlier in this section can be used to study the typical distance between two points on the giant component.

Theorem 1.2.5 *Suppose $\lambda > 1$ and pick two points x and y at random from the giant component. Then $d(x, y)/\log n \to 1/\log\lambda$ in probability.*

The answer in Theorem 1.2.5 is intuitive. Consider the cluster growing from 1. The size of the branching process at time t is $\sim \lambda^t W$. When $t = (1-\epsilon)(\log n)/(\log \lambda)$ only $n^{1-\epsilon}$ can be reached, but when $\epsilon = 0$ the number is $\Theta(n)$.

Proof The lower bound on the distance comes from the last observation and the fact that the branching process is an upper bound on the growth of the cluster. To prove the upper bound requires more work since the branching process approximation is not valid out to time $t = (\log n)/\log \lambda$. To get around this difficulty we will grow clusters starting from x and y for time $t(\epsilon, n) = (1/2+\epsilon)(\log n)/\log \lambda$. If there were no collisions then the number of offspring is of order $n^{1/2+\epsilon}$. Lemma 1.2.3 implies that the expected number of collisions up to time $t(\epsilon, n)$ is $\lambda n^{2\epsilon}$. When we have a collision we lose not only that vertex but also all of its children. Theorem 1.2.3 shows that with high probability there is no collision before time $(1/2-\epsilon)(\log n)/\log \lambda$. If we are conservative and suppose that the clusters of the vertices lost to collisions all have time $2\epsilon(\log n)/\log \lambda$ to grow, then the number of lost particles is $n^{4\epsilon} \ll n^{1/2+\epsilon}$ if ϵ is small.

At time $t(\epsilon, n)$ there are two cases to consider: (i) the clusters grown from x and y have already intersected (in this case the distance from x to y is $\leq 2t(\epsilon)$) or (ii) they have not intersected. If the clusters grown from x and y have not died out and δ is small then with high probability there will be $\geq \delta n^{1/2+\epsilon}$ active vertices in the growing cluster at time $t(\epsilon, n)$, so the probability they will not intersect on the next step is

$$\leq \left(1 - \frac{\lambda}{n}\right)^{\delta^2 n^{1+2\epsilon}} \to 0,$$

which completes the proof. □

Dangling ends and the diameter

Our next goal is to show that the diameter $D = \max d(x, y)$ of the giant component in an Erdős-Rényi random graph is larger by a constant factor than the distance between two randomly chosen points on it. To do this we define a **dangling end** of length k to be a self-avoiding path $v_0, v_1, \ldots, v_k$ with the degree $d(v_0) = 1$ and $d(v_i) = 2$ for $1 \leq i < k$, and $d(v_k) \geq 3$.

These requirements are imposed so that two dangling ends of length k cannot intersect, except possibly at the final vertex v_k. To explain, note that since $d(v_0) = 1$ and $d(v_i) = 2$ for $1 \leq i < k$ only one dangling end can start at a given vertex. The vertices v_i with $1 \leq i < k$ cannot be in two dangling ends since if we look at the first vertex to be in two then it would have to have degree 3. The limited ability of dangling ends to overlap will lead to the conclusion that the limiting number of dangling ends with a fixed length k will be Poisson. To prove this we use the following.

Lemma 1.2.6 *Let $x^{(k)} = x(x-1)\cdots(x-k+1)$ and let Z_n be a sequence of non-negative integer-valued random variables with $EZ_n^{(k)} \to \lambda^k$ for all positive integers k. Then Z_n converges in distribution to $Z_\infty =$ Poisson(λ).*

Proof It is easy to check that if $Z = \text{Poisson}(\lambda)$ then $EZ_\infty^{(k)} = \lambda^k$. Since there are constants $c_{k,j}$ so that

$$x^k = \sum_{j=1}^{k} c_{k,j} x^{(j)},$$

we have $EZ_n^k \to EZ_\infty^k$. The Poisson distribution is determined by its moments so the conclusion follows, see e.g., Sect. 3.3.5 in PTE5. □

Let $\lfloor x \rfloor$ be the largest integer $\leq x$.

Theorem 1.2.7 *Let $p_1 = e^{-\lambda}\lambda$ and choose $k(n) = \lfloor \alpha \cdot \log n / \log(1/p_1) \rfloor$ so that $c_n = np_1^{k(n)}$ stays bounded away from 0 and ∞. The number of dangling ends of length $k(n)$, $M_n \approx \text{Poisson}(c_n)$. If α is small then $M_n > 0$ with high probability.*

Proof To use Lemma 1.2.6 we need to compute the moments of $M = M_n$. Let $A_{(x,y)}$ be the event that there is dangling end $v_0 = x, v_1, \ldots, v_{k+1} = y$. Let $P_{n,k+1} = n \cdot (n-1) \cdots (n-k)$ (where P is for permutations) be the number of ways to pick the vertices $v_0, \ldots, v_k$, which are all distinct

$$EM = \sum_{(x,y)} P(A_{(x,y)}) \sim P_{n,k+1}(\lambda/n)^k (1-\lambda/n)^{nk} \cdot q_2 \sim n(\lambda e^{-1})^k,$$

where $q_2 = P(\text{Poisson}(\lambda) \geq 2)$.

Here $(\lambda/n)^k$ accounts for the k connections that exist between the v_i, $0 \leq i \leq k$, and the factor $(1-\lambda/n)^{nk}$ gives the probability that no other connections exist. To explain the second formula, note that 0 is not connected to $n-1$ vertices and the v_i with $1 \leq i < k$ are not connected to $n-2$ vertices. However in this list of nonexisting connections $\binom{k}{2}$ of them appear twice in the list, so the exact exponent is $(n-1) + (n-2)(k-1) - \binom{k}{2}$. This differs from nk by $\leq k^2$, so if $k = O(\log n)$ the difference is insignificant.

Our next step is to investigate the second factorial moment $E[M(M-1)]$. To simplify the calculation we will first consider the case in which the two dangling ends do not end at the same vertex. We will then check that the probability for two ending at the same place is smaller. Thus our goal is to compute

$$\sum_{x \neq w, y \neq z} P(A_{(x,y)} \cap A_{(w,z)}).$$

Let $v_0 = x, v_1, \ldots, v_k = y$ be the path that produces the first one and let $u_0 = w, u_1, \ldots, u_k = z$ be the path that produces the second one. Note that the degrees of the vertices in the dangling end imply that $x \neq z$ and $y \neq w$.

- Since the paths are disjoint and self-avoiding, there are $P_{m,2(k+1)}$ ways of picking the two sequences (including the choice of x and w). Note that if $k = O(\log n)$,

$$P_{m,2(k+1)}/P^2_{m,(k+1)} \to 1.$$

- As in the formula for the mean, a factor of $(\lambda/n)^{2k}$ accounts for the $2k$ connections that exist between the vertices in the dangling ends and a factor $(1-\lambda/n)^{2kn}$ gives the probability (up to a factor $1+o(1)$) that no other connections exist.
- Given the existence of a path satisfying the first two conditions, we have $d(v_k) \geq 3$ with probability $\to q_2$.
- Finally note that the number of ways of picking vertices for two dangling ends that intersect in the final vertex is $P_{m,2k+1}$, so the probability of this event is of smaller order than that of two disjoint dangling ends.

It is straightforward to extend the last argument to show asymptotic independence of m disjoint dangling ends for any m.

If α is small then EM is large, so with high probability there is at least one dangling end. To show that it is connected to the giant component with positive probability run the cluster growth process starting from v_k and notice that the only effect of the dangling end is to reduce the number of vertices. If one wants the construction to succeed with high probability replace 3 by a much larger number. □

The result we have just proved is not very precise. Riordan and Wormald (2010) have found the exact order. To state the result we will need a fact from Sect. 1.7, see (1.7.4). If ρ is the extinction probability of a Poisson(λ) branching process then

$$\lambda\rho e^{-\lambda\rho} = \lambda e^{-\lambda}.$$

Theorem 1.2.8 *Let $\lambda > 1$ and $\lambda_* = \lambda\rho$.*

$$\text{diameter}(G(n,\lambda/n)) = \frac{\log n}{\log\lambda} + \frac{2\log n}{\log(1/\lambda_*)} + O_p(1),$$

where $O_p(1)$ denotes a term X_n with $P(X_n > K_n) \to 0$ for any $K_n \to \infty$.

The first term on the right is the typical distance between two points while the second gives the contribution of two dangling ends.

There are a large number of results about what happens when $\lambda \to \infty$ at different rates. See Chung and Lu (2001) for details. For example if $\lambda/(\log n) \to \infty$ the diameter is concentrated on at most 2 values while if $\lambda/\log n = c$ then it is concentrated on at most 2 if $c > 8$, 3 if $c \in (2, 8]$, and 4 if $c \in (1, 2]$. In addition there are four other results.

1.3 Cluster Growth as a Random Walk

Although the connection with branching processes is intuitive, it is more convenient technically to grow the cluster by adding the neighbors of one site at a time in order to obtain something that can be approximated by a random walk. In this section we will introduce that approach, use it to prove the existence of a giant component when $p = \lambda/n$ with $\lambda > 1$ and show that as $n \to \infty$ it has $\sim (1 - \rho(\lambda))n$ vertices where $\rho(\lambda)$ is the extinction probability for the Poisson(λ) branching process.

As in the previous section, we let R_t be the set of removed sites, U_t the unexplored sites and A_t the active sites. However, here we do not need a close connection with a branching process, so we do not need to impose an order on A_t. The initial conditions are $R_0 = \emptyset$, $U_0 = \{2, 3, \ldots, n\}$, and $A_0 = \{1\}$. At time $\tau = \inf\{t : A_t = \emptyset\}$ the process stops, and R_τ is the cluster containing 1. Since one vertex is added to the removed set at each time, there are τ points in the cluster. If $A_t \neq \emptyset$, pick i_t from A_t according to some rule that is measurable with respect to $\mathcal{A}_t = \sigma(A_0, \ldots A_t)$ and let

$$\begin{aligned} R_{t+1} &= R_t \cup \{i_t\}, \\ A_{t+1} &= A_t - \{i_t\} \cup \{y \in U_t : \eta_{i_t,y} = 1\}, \\ U_{t+1} &= U_t - \{y \in U_t : \eta_{i_t,y} = 1\}. \end{aligned} \tag{1.3.1}$$

We call this the **exploration process**.

Upper bound for $\lambda < 1$. To define a comparison random walk, we introduce a new set of variables ζ_y^t, $t \geq 1$, $y \leq n$, that are independent, $= 1$ with probability λ/n, and 0 otherwise. Let $S_0 = 1$ and for $t \geq 0$, let $U_t^c = \{1, 2, \ldots, n\} - U_t$

$$S_{t+1} = S_t - 1 + \begin{cases} \sum_{y \in U_t} \eta_{i_t,y} + \sum_{y \in U_t^c} \zeta_y^t & \text{if } A_t \neq \emptyset, \\ \sum_{y=1}^n \zeta_y^t & \text{if } A_t = \emptyset. \end{cases}$$

S_t is a random walk with $S_t \geq |A_t|$ if $t \leq \tau$, so if $T = \inf\{t : S_t = 0\}$ then $\tau \leq T$. We continue the definition after $A_t = \emptyset$ because it is convenient to have the random walk defined for all time.

The increments $X_t = S_t - S_{t-1}$, $t \geq 1$ of the random walk are -1 + binomial $(n, \lambda/n)$. If $\lambda < 1$ stopping the martingale $S_t - (\lambda - 1)t$ at the bounded stopping time $T \wedge t$ (where $x \wedge y = \min\{x, y\}$) gives

$$ES_{T\wedge t} - (\lambda - 1)E(T \wedge t) = ES_0 = 1.$$

Since $ES_{T\wedge t} \geq 0$, it follows that $E(T \wedge t) \leq 1/(1 - \lambda)$. Letting $t \to \infty$ we have $ET \leq 1/(1 - \lambda)$. Having verified that $ET < \infty$ we can now use Wald's equation, see Theorem 4.8.6 in PTE5 to conclude $E(S_T - S_0) = (\lambda - 1)ET$ and hence

$$ET = 1/(1 - \lambda).$$

We can get a much better result by using the exponential martingale:

Theorem 1.3.1 *Suppose $\lambda < 1$ and let $\alpha = \lambda - 1 - \log(\lambda) > 0$. If $a > 1/\alpha$ then*

$$P\left(\max_{1\le x\le n} |\mathcal{C}_x| \ge a\log n\right) \to 0.$$

Remark. This bound is very accurate. Corollary 5.11 of Bollobás (2001) shows that the size of the largest component is asymptotically

$$\frac{1}{\alpha}\left(\log n - \frac{5}{2}\log\log n\right) + O(1).$$

Proof We begin by computing the moment generating function of $X_i = -1 +$ binomial$(n, \lambda/n)$

$$\begin{aligned} E\exp(\theta X_i)) &= e^{-\theta}(1-\lambda/n+(\lambda/n)e^{\theta})^n \\ &\le \exp(-\theta+\lambda(e^{\theta}-1)) = \psi(\theta) \end{aligned} \tag{1.3.2}$$

since $1 + x \le e^x$. Note that the right-hand side is the moment generating function of -1 + Poisson(λ). The mean $\psi'(0) = EX_i = \lambda - 1$, so if $\lambda < 1$ then $\psi(\theta) < 1$ when $\theta > 0$ is small. To optimize we set the derivative

$$\frac{d}{d\theta}(-\theta+\lambda(e^{\theta}-1)) = -1+\lambda e^{\theta} = 0$$

which occurs for $\theta_1 = -\log\lambda$. At this point

$$\psi(\theta_1) = \exp(\log(\lambda)+1-\lambda) \equiv e^{-\alpha} < 1.$$

$\exp(\theta_1 S_t)/\phi(\theta_1)^t$ is a nonnegative martingale, so using the optional stopping theorem for the nonnegative supermartingale $M_t = \exp(\theta_1 S_t)/\psi(\theta_1)^t$, see e.g., Theorem 4.8.4 in PTE5.

$$1/\lambda = e^{\theta_1} \ge E(\psi(\theta_1)^{-T}) = E(e^{\alpha T}).$$

Using Chebyshev's inequality now

$$e^{k\alpha}P(T\ge k) \le 1/\lambda.$$

Letting $\mathcal{C}_x$ denote the cluster containing x, noting that $T \ge |\mathcal{C}_x|$ in distribution, and taking $k = (1+\epsilon)(\log n)/\alpha$

$$P(|\mathcal{C}_x| \ge (1+\epsilon)(\log n)/\alpha) \le n^{-(1+\epsilon)}/\lambda$$

from which the desired result follows. □

Lower Bound for $\lambda > 1$. To get a lower bound on the growth of the cluster let $\hat{U}_t^{\delta}$ consist of the $(1-\delta)n$ vertices in $\hat{U}_t$ with the smallest indices. As long as $\hat{A}_t \ne \emptyset$ and $\hat{U}^t \ge (1-\delta)n$, which corresponds to $|\hat{A}_t| + t \le n\delta$, we can define

$$\begin{aligned}\hat{R}_{t+1} &= \hat{R}_t \cup \{j_t\},\\ \hat{A}_{t+1} &= \hat{A}_t - \{j_t\} \cup \{y \in \hat{U}_t^\delta : \eta_{j_t,y} = 1\},\\ \hat{U}_{t+1} &= \hat{U}_t - \{y \in \hat{U}_t^\delta : \eta_{j_t,y} = 1\},\end{aligned}$$

where in this construction $i_t = \min \hat{A}_t$. Since we have reduced the size of U_t, we have $|A_t| \geq |\hat{A}_t|$. To define a comparison random walk, we let $W_0 = 1$, let

$$T_W = \inf\{s : W_s = 0, \text{ or } W_s + s \geq n\delta\}$$

and define

$$W_{t+1} = W_t - 1 + \begin{cases} \sum_{y \in \hat{U}_t^\delta} \eta_{i_t,y} & \text{if } t < T_W, \\ \sum_{y=1}^{n(1-\delta)} \zeta_y^t & \text{if } t \geq T_W. \end{cases}$$

It is easy to see that for $t \leq T_W$, $|\hat{A}_t| = W_t$, so $\tau \geq T_W$.

We will use the new lower bound W_t and the earlier upper bound S_t to prove

Theorem 1.3.2 *Suppose* $\lambda > 1$. *There is a constant* β *so that with high probability, there is only one component of the random graph with more than* $\beta \log n$ *vertices. The size of this component* $\sim (1 - \rho(\lambda))n$ *where* $\rho(\lambda)$ *is the extinction probability for the Poisson*(λ) *branching process.*

Here *with high probability* means that the probability tends to 1 as $n \to \infty$ (Fig. 1.4).

Proof There are four steps.

Step 1. There is a constant γ so that if $W_s \geq \gamma \log n$ then the probability W_t hits 0 at some time $t \geq s$ is $\leq n^{-2}$.

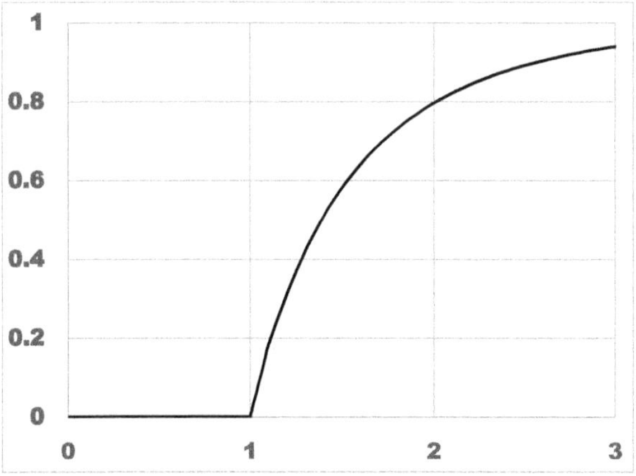

Fig. 1.4 Graph of the survival probability $1 - \rho(\lambda)$

Step 2. There is a constant β so that $P(0 < |A(\beta \log n)| < \gamma \log n) = o(n^{-1})$.

Step 3. If $A(\beta \log n) > \gamma \log n$ then with probability $\geq 1 - \exp(-\eta n^{2/3})$ we have $\epsilon n^{2/3} \leq |A(n^{2/3})| \leq 2\lambda n^{2/3}$. Combined with the first two steps, this shows that with probability $\to 1$, all clusters reaching size $\beta \log n$ will grow to size $O(n^{2/3})$ and will intersect producing a giant component.

Step 4. We show that the number of sites x with clusters $\mathcal{C}_x$ of size $\geq \beta \log n$ is asymptotically $(1 - \rho)n$. Here the fact that we only have to run the cluster exploration process to time $\beta \log n$ reduces the dependence between the events $\{|\mathcal{C}_x| \geq \beta \log n\}$ so that we can prove the result by computing second moments and using Chebyshev's inequality.

Turning to the details of the proofs.

Step 1. The increments of W have the distribution $-1 + \text{binomial}((1 - \delta)n, \lambda/n)$.

By (1.3.2) the moment generating function of an increment

$$\phi_\delta(\theta) \leq \exp(-\theta + \lambda(1 - \delta)(e^\theta - 1)). \tag{1.3.3}$$

Choose $\delta > 0$ so that $\lambda(1 - \delta) > 1$. $\phi'_\delta(0) = -1 + \lambda(1 - \delta) > 0$ so $\psi_\delta(-\theta) < 1$ when $\theta > 0$ is small. Since $\psi_\delta(\theta)$ is convex and tends to ∞ as $\theta \to \infty$ there is a unique positive solution of $\phi_\delta(-\theta_\delta) = 1$. $M_t = \exp(-\theta_\delta W_t)$ is a nonnegative (super)martingale. We put super in parentheses since it is a martingale but we will only use that it is a supermartingale. Suppose $W_0 = k$ and let $T_0 = \inf\{t : W_t = 0\}$. Using Theorem 4.8.4 in PTE5 (the stopping theorem for nonnegative supermartingales) and noting that $W_t \to \infty$ on $\{T_0 = \infty\}$ we have

$$e^{-k\theta_\delta} \geq P_k(T_0 < \infty). \tag{1.3.4}$$

If we take $k_\delta = (2/\theta_\delta) \log n$ we make the left-hand side $= n^{-2}$.

Step 2. To control the behavior of W_t and S_t with good error bounds we use the following:

Lemma 1.3.3 *Let $Z = X_1 + \cdots + X_t$ where the X_i are i.i.d. and have $E \exp(\theta X_i) < \infty$ for $|\theta| < \theta_0$. Let $\mu = EX_s$. If $b < \mu < a$ then there are constants $\eta(b), \eta(a) > 0$ so that*

$$P(Z \geq ta) \leq \exp(-t\eta(a)),$$
$$P(Z \leq tb) \leq \exp(-t\eta(b)).$$

Proof These are standard results on large deviations, see Lemma A.1.8. □

Let $X_1, \ldots, X_t$ be i.i.d. $-1 + \text{binomial}(n, \lambda)$, write S_t instead of Z. Let $a = 2\lambda - 1 > \mu = \lambda - 1$. Using Lemma 1.3.3 there is an $\eta > 0$ so that

$$P(S_t \geq (2\lambda - 1)t) \leq \exp(-\eta t).$$

Let $r = \beta \log n$ where $\beta(\lambda - 1)/2 > 2\gamma$ and $\exp(-\eta\beta \log n) \le n^{-2}$. When $|A_r| > 0$, S_t has not hit 0 at times $t \le r$ so it provides an upper bound on A_t. When $S_r + r \le 2\lambda r$ we have $|U_s| \ge n - 2\lambda r$ for all $t \le r$. The number of times that the upper bound process adds a vertex not added to A_t is at most $B(r, \lambda)$ which is the sum of $r \times 2\lambda r$ Bernoulli random variables with success probability λ/n. From this it follows that

$$P(|A_r| > 0, S_r - |A_r| \ge 2) \le \binom{2\lambda r^2}{2}(\lambda/n)^2 \le C\frac{(\log n)^2}{n}.$$

Combining our results

$$P(0 < |A(\beta \log n)| \le \gamma \log n) \le C\frac{(\log n)^2}{n^2}.$$

Step 3. Let $\epsilon_\delta = (\lambda(1-\delta) - 1)/2$. Using Lemma 1.3.3 twice with

$$X_i = \text{binomial}((1-\delta)n, \lambda/n),$$

we have

$$P(W(n^{2/3}) - W(0) \le \epsilon_\delta n^{2/3}) \le \exp(-\eta_2 n^{2/3}), \tag{1.3.5}$$

$$P(W(n^{2/3}) - W(0) + n^{2/3} \ge 2\lambda n^{2/3}) \le \exp(-\eta_3 n^{2/3}). \tag{1.3.6}$$

We take $W(0) = |A_r| \le 2\lambda\beta \log n$ in (1.3.6). Since $W_t + t$ is nondecreasing this shows that with probability $1 - O(n^{-2})$, $W_s + s \le \delta n$ for all $s \le n^{2/3}$, and the coupling between W_s and $|A(s + r)|$ remains valid for $0 \le s \le n^{2/3}$.

The bound in (1.3.5) implies that if a cluster reaches size $r = \beta \log n$ then the set of active sites at time $r + n^{2/3}$ is $\ge \epsilon_\delta n^{2/3}$ with high probability. Thus if we have two such clusters of size $\ge \beta \log n$ then either (a) they will intersect by time $r + n^{2/3}$ or (b) at time $r + n^{2/3}$ they have disjoint sets V_1 and V_2 of active sites of size $\ge \epsilon_\delta n^{2/3}$. The probability of no edge connecting V_1 and V_2 is

$$= \left(1 - \frac{\lambda}{n}\right)^{\epsilon_\delta^2 n^{4/3}} \le \exp(-\eta_4 n^{1/3}). \tag{1.3.7}$$

This proves the first assertion in Theorem 1.3.2.

Step 4. To prove the second assertion it suffices to show that

$$|\{x : |\mathcal{C}_x| \le \beta \log n\}|/n \to \rho(\lambda). \tag{1.3.8}$$

The first step is to note that the branching process approach to cluster growth implies

$$P(|\mathcal{C}_x| \le \beta \log n) \to \rho(\lambda).$$

To complete the proof we bound the correlation of the random variables $Y_x = 1$ if $|\mathcal{C}_x| \le \beta \log n$ and 0 otherwise and use Chebyshev's inequality. We isolate the reasoning as

Lemma 1.3.4 *Let F be an event that involves exposing J vertices starting at* 1, *and let G be an event that involves exposing K vertices starting at* 2. *Then*

$$|P(F \cap G) - P(F)P(G)| \le JK \cdot \frac{2\lambda}{n}.$$

Proof Let R_t, U_t and A_t be the process of exposing the cluster of 1. Introduce independent copies of the basic indicator random variables $\eta'_{x,y}$. Let $R'_0 = \emptyset$, $A'_0 = \{2\}$ and $U'_0 = \{1, 2, \ldots, n\} - \{2\}$. If $A'_t \neq \emptyset$, pick $i'_t = \min A'_t$. If $i'_t \notin R_{\beta \log n}$ let

$$\begin{aligned} R'_{t+1} &= R'_t \cup \{i'_t\}, \\ A'_{t+1} &= A'_t - \{i'_t\} \cup \{y \in U'_t : \eta'_{i'_t,y} = 1\}, \\ U'_{t+1} &= U'_t - \{y \in U'_t : \eta'_{i'_t,y} = 1\}. \end{aligned} \tag{1.3.9}$$

However if $i'_t \in R_J$, an event we call a collision, we use $\eta_{i'_t,y}$ instead of $\eta'_{i'_t,y}$. In words, if while growing cluster 2 we choose a site that was used in the growth of cluster 1, we use the original random variables $\eta_{x,y}$. Otherwise we use independent random variables. It should be clear from the construction that

$$|P(F \cap G) - P(F)P(G)| \le P(R_J \cap R'_K \neq \emptyset) \le JK \cdot \frac{\lambda}{n},$$

which proves the desired result. □

Using Lemma 1.3.4 with $J = K = \beta \log n$, the probability of a collision is at most $\lambda(\beta \log n)^2/n$. Using our bound on the covariance with the formula for the variance of a sum of random variables

$$\text{var}\left(\sum_{x=1}^{n} Y_x\right) \le n + n(n-1)\frac{\lambda(\beta \log n)^2}{n} \le Cn \log n$$

so it follows from Chebyshev's inequality that

$$P\left(\sum_{x=1}^{n}(Y_x - EY_x) \ge n^{2/3}\right) \le \frac{Cn \log n}{n^{4/3}} \to 0.$$

This proves (1.3.8) and completes the proof of Theorem 1.3.2.

1.4 Threshold for Being Connected

In this section we will answer the question: How large does λ have to be so that the probability Erdős-Rényi$(n, \lambda/n)$ is connected (i.e., ALL vertices in ONE component) tends to 1. I learned the following slick proof from the book by Frieze and Karonski, who attribute it to Erdős and Rényi in 1959.

Theorem 1.4.1 *Let $p = (c_n + \log n)/n$ and let $H(n, p)$ be the event that Erdős-Rényi(n, p) is connected.*

$$P(H(n,p)) \to \begin{cases} 0 & \text{if } c_n \to -\infty, \\ \exp(-e^{-c}) & \text{if } c_n \to c \in (-\infty, \infty), \\ 1 & \text{if } c_n \to \infty. \end{cases}$$

The proof will be completed by proving the next two lemmas. By monotonicity of the event $H(n, p)$ in p it suffices to prove the result when $c_n \to c \in (-\infty, \infty)$.

Lemma 1.4.2 *Consider $G =$ Erdős-Rényi$(n, \lambda/n)$ with $\lambda = \log n + c_n + o(1)$ and $c_n \to c$. Then the number of isolated vertices I_n converges to a Poisson distribution with mean e^{-c}.*

Proof The probability x is isolated is

$$\left(1 - \frac{\lambda}{n}\right)^n \sim \exp(-\log n - c_n) \sim e^{-c}/n,$$

so $EI_n \to e^{-c}$. The expected number of ordered k-tuples of isolated vertices is

$$EI_n^{(k)} = (n \cdot (n-1) \cdots (n-k+1)) \left(1 - \frac{\lambda}{n}\right)^{n+(n-1)+\cdots+(n-k+1)} \to e^{-ck}$$

so the Poisson convergence follows from Lemma 1.2.6. □

The next result completes the proof. For simplicity we suppose $c_n \equiv c$.

Lemma 1.4.3 *Let X_k be the number of clusters with k vertices. Let $p = (c + \log n)/n$ and let $H(n, p)$ be the event that Erdős-Rényi(n, p) is connected. As $n \to \infty$*

$$P(H(n,p)) = P(X_1 = 0) + o(1).$$

Proof If the graph is not connected then there is at least one component of size $\le n/2$, so $P(H(n, p)^c) = P(\cup_{k=1}^{n/2}\{X_k > 0\})$ and we have

$$P(X_1 > 0) \le P(H(n,p)^c) \le P(X_1 > 0) + \sum_{k=2}^{n/2} P(X_k > 0).$$

Computing as in Sect. 1.7 (see (1.7.1))

$$\sum_{k=2}^{n/2} P(X_k > 0) \le \sum_{k=2}^{n/2} \cdot \binom{n}{k} k^{k-2} p^{k-1} (1-p)^{k(n-k)} \equiv \sum_{k=2}^{n/2} u_k.$$

Now for $2 \le k \le 10$ using $\binom{n}{k} \le n^k/k!$ and $e^k \ge k^k/k!$

$$\begin{aligned} u_k &\le e^k n^k \left(\frac{c+\log n}{n}\right)^{k-1} e^{-k(n-10)(c+\log n)/n} \\ &\le e^k n^k \left(\frac{c+\log n}{n}\right)^{k-1} e^{-k(c+\log n)} e^{10k(c+\log n)/n} \\ &\le e^{k(1-c)} \left(\frac{c+\log n}{n}\right)^{k-1} \cdot (1+o(1)) \end{aligned}$$

while for $10 < k \le n/2$

$$\begin{aligned} u_k &\le (ne/k)^k k^{k-2} \left(\frac{c+\log n}{n}\right)^{k-1} e^{-k(\log n + c)/2} \\ &\le n \left(\frac{e^{1-c/2}(c+\log n)}{n^{1/2}}\right)^k \end{aligned}$$

so we have $\sum_{k=2}^{n/2} u_k \to 0$, which completes the proof. □

Our final result, which we state without proof, gives a formula for the diameter.

Theorem 1.4.4 *If* $\liminf np/(\log n) > c > 1$ *and* $(\log p)/(\log n) \to 0$ *then the diameter of Erdős-Rényi*(n, p), $D(n, p) \sim (\log n)/(\log np)$.

Remark 1.4.5 The first condition guarantees that the probability that the graph is connected tends to 1? as $n \to \infty$. The formula for the diameter is the same as in the sparse case when $np = \lambda$. To explain the second suppose $p = n^{-15/17}$. In this case $np = n^{2/17}$. It is not hard to show that the diameter will be 9 with probability approaching 1, but the formula gives 17/2. A more delicate situation occurs when

$$p = n^{(1/d)-1} (\log(n^2/c))^{1/d}.$$

In this case in the limit the diameter is d with probability $e^{-c/2}$ and $d+1$ with probability $1 - e^{-c/2}$. See Theorem 10.10 in Bollobás (2001). We mentioned other work by Chung and Lu (2001) and Riordan and Wormald (2010) at the end of Sect. 1.2. We leave it to the interested reader to explore this topic further.

1.5 Long Paths

At this point we know that the giant component exists for $\lambda > 1$ and is unique, but we do not know much about what it looks like. Ajtai, Komlós, and Szemerédi (1981) proved that there is a function $\alpha : (1, \infty) \to (0, 1)$ so that the Erdős-Rényi random graph with $\lambda = c/n$ contains a path of length at least $\alpha(c)n$. When we study the contact process on graphs, this result can be used to give a very easy proof that when λ is large enough the contact survives for time $\exp(\theta n)$ on an Erdős-Rényi graph with n vertices.

1.5.1 Krivelevich and Sudakov (2013)

have given a simple proof based on **depth-first search**. Given a graph $G = (V, E)$, which we will assume is undirected, the algorithm uses an ordering σ on the vertices of V and divides them into three disjoint sets that change over time. Here we have changed the notation in their paper to match that in the next section.

- R_t is the set of **retired vertices** for which the exploration is complete.
- S_t is the set of **sleeping vertices** that have not yet been touched.
- $A_t = V - (R_t \cup S_t)$ are the **active vertices** kept in a last-in, first-out data structure.

It follows from the construction given below and induction, that at any time t

(B1) There are no edges between R_t and S_t.
(B2) The vertices in A_t are a path.

On each step of the algorithm a vertex might move from S to A or from A to R, but on most steps nothing happens. To be precise, a step consists of ONE query - when we inspect the graph to see if there is an edge from i to j. The edge is present if $X_{ij} = 1$, where $X_{ij} = X_{ji}$ are independent Bernoulli(p) random variables.

- If $A \neq \emptyset$, we let v be the vertex at the top of the stack. If v does not have a neighbor in S (which we know because we have queried all the possible connections to S) then v is removed from A and added to R. Once vertices enter R they never leave.
- If we have not queried all the possible connections from v to S then we query the next vertex u in S in the ordering. If our query finds that u is a neighbor of v then u is deleted from S and added to A and replaces v as the top vertex in the stack.
- If $A = \emptyset$, then we have just completed computing one of the components of the graph. The algorithm chooses the first vertex in S according to the ordering σ. When $A \cup S = \emptyset$ all the connected components have been determined.

The next estimate is an important part of the proof.

Lemma 1.5.1 *Let X_{ij} with $1 \le i < j \le n$ be i.i.d. Bernoulli($(1+\epsilon)/n$) indexed by the edges of the graph. Let F be a set of edges. If $|F| = cn^2$ then whp*

$$\left|\sum_{ij\in F} X_{ij} - c(1+\epsilon)n\right| \le n^{2/3}.$$

Proof The sum has mean $c(1+\epsilon)n$ and variance $\le c(1+\epsilon)n$ so

$$P\left(\left|\sum_{ij\in F} X_{ij} - c(1+\epsilon)n\right| \le n^{2/3}\right) \le \frac{c(1+\epsilon)n}{n^{4/3}}$$

by Chebyshev's inequality, proving the desired result. □

Theorem 1.5.2 *If $b < 1/4$ and $\eta > 0$ is small enough then when $\epsilon < \epsilon_0(b,\eta)$ whp Erdős-Rényi $(n,(1+\epsilon)/n)$ contains a path of length $b\epsilon^2 n$.*

Improved lower bounds. The results in the next two sections will give a lower bound as $\epsilon^2 n/10$ and $\sim C\epsilon^2 n$. Ding et al. (2014) and Robinson and Wormald (1992) have proved $(4/3)\epsilon^2$ as a lower bound.

Upper bounds. In Sect. 1.8 we will see that the size of the giant component when $\lambda = 1+\epsilon$ is $\sim C\epsilon$ as $\epsilon \to 0$. To improve this upper bound we note that Pittel (1990) has shown that the fraction of vertices in the **2-core**, which is the maximal subgraph in which all vertices have degree ≥ 2, is $\le C\epsilon^2$. A long path may not have any subgraph in which all vertices have degree ≥ 2. We leave it to the reader to show that if there is path of length cn when $\lambda = 1+\epsilon$ then with high probability there is a circle of length $cn - 2n^{0.6}/\sqrt{\epsilon}$ when $\lambda = 1+2\epsilon$ by finding an edge that is open with probability ϵ/n connecting near the beginning to near the end.

Proof We will lower bound the size of A_N at time $N = \epsilon n^2/2$ and then the result follows from (B2). We first show that if η is small and $\eta > 2\epsilon$ then $|R_N| \le \eta n$. Suppose not, and let t be the first time $|R_t| = \eta n$. At that moment

$$|A_t| \le 1 + \sum_{s-1}^{t} X_s \le \epsilon(1+\epsilon)n/2 + n^{2/3} < \eta n,$$

by Lemma 1.5.1 with $c = \epsilon/2$ and the choice of η. This implies

$$|S_t| = n - |R_t| - |A_t| \ge n(1-2\eta) \ge 2n/3,$$

if η is small enough. At this point the algorithm must have queried all of the

$$|R_t| \cdot |S_t| \ge \frac{2n}{3}\eta nn \ge^2 /3 > N$$

edges connecting R_t to S_t and found them to be missing, but this is impossible, since there has not been enough time to do that.

We now return to time $N = cn^2/2$. If

$$|R_N| \leq \eta n \quad \text{and} \quad |A_N| \leq b\epsilon^2 n$$

then we have $S_N \neq \emptyset$. This means that we are still revealing connected components of G. Each positive answer we get resulted in moving a vertex from S to A (some of which may have been later moved to R). By Lemma 1.5.1 with $c = \epsilon/2$, the number of positive answers to queries at time $N = \epsilon n^2/2$ must be at least $\epsilon(1+\epsilon)n/2 - n^{2/3}$ and hence we have

$$|R_N \cup A_N| \geq \frac{\epsilon n}{2} + \frac{\epsilon^2 n}{2} - n^{2/3}. \tag{1.5.1}$$

The next step is where we use that $\lambda = 1 + \epsilon > 1$. If $|A_N| \leq b\epsilon^2 n$ then using (1.5.1) and the fact that all vertices are in A_N, R_N or S_N we have

$$|R_N| \geq \frac{\epsilon n}{2} + (0.5 - b)\epsilon^2 n - n^{2/3}, \tag{1.5.2}$$

$$|S_N| = n - |R_N| - |A_N| \geq n - |R_N| - b\epsilon^2 n. \tag{1.5.3}$$

Using $R_N \leq \eta n$, $|A_N| \leq b\epsilon^2 n$, and $|S_N| = n - |R_N| - |A_N|$ gives

$$\begin{aligned} |R_N| \cdot |S_N| &= |R_N|(n - |R_N| - |A_N|) \\ &\geq |R_N|(n - |R_N|) - \eta n \cdot b\epsilon^2 n. \end{aligned}$$

Using the fact that $m \to m(n-m)$ is increasing on $[0, n/2]$ together with (1.5.2) we have that the above is

$$\begin{aligned} &\geq \left(\frac{\epsilon n}{2} + (0.5 - b)\epsilon^2 n - n^{2/3}\right) \cdot \left(n - \frac{\epsilon n}{2} - (0.5 - b)\epsilon^2 n + n^{2/3}\right) - \eta b\epsilon^2 n^2 \\ &\geq \frac{\epsilon n^2}{2} + (0.5 - b)\epsilon^2 n^2 - \frac{\epsilon^2 n^2}{4} - \eta b\epsilon^2 n^2 - O(\epsilon^3 n^2) - O(n^{5/3}). \end{aligned}$$

To check the second inequality call the terms in the first parentheses 1, 2, and 3, and those in the second w, x, y, and z. The first term in the second equation is $1w$, the second $2w$ and the third $1x$. The terms $2x$, $1y$, $2y$ are $O(\epsilon^3 n^2)$ while terms involving 3 or z are $O(n^{5/3})$. The sum of the first four terms in the last line of the display above is

$$N + [(0.5 - b) - 0.25 - b\eta] \cdot \epsilon^2 n.$$

If $b < 1/4$ and η is small, then the second term is positive, so the right-hand side is again $> N$ and we have a contradiction. □

1.5.2 A Constructive Lower Bound Proof

Here we borrow ideas from Enriquez et al. (2017), or EFM for short, to give a simple proof of the existence of long paths that is not based on an argument by contradiction. In the next section we will describe their proof of a scaling limit that gives sharp results for depth-first search on the Erdős-Rényi graph.

Theorem 1.5.3 *If $\epsilon > 0$ is small enough then for large n an Erdős-Rényi$(n, (1+\epsilon)/n)$ contains a path of length $\epsilon^2 n/10$.*

The definition of depth-first search in this section, which is also used in the next, is almost the same as the one used in the previous section once one skips the queries that do nothing. At each step, we have the following objects.

- A_m is a set of **active vertices** in a last-in first-out stack.
- a_m is the last element in A_m
- S_m is a set of **sleeping vertices**.
- $R_m = \{1, 2 \dots n\} - (A_m \cup S_m)$ is a set of **retired vertices**.

As before (i) the vertices in R_m have no neighbors in S_m, and (ii) at any time A_m is a path.

We order the vertices by numbering them $1, \dots, n$. Initially $A_0 = \{1\}$, $S_0 = \{2, 3, \dots, n\}$, $R_0 = \emptyset$.

- If a_m has a neighbor in S_m, we set $a_{m+1} = \min\{b \in S_m : b \sim a_m\}$, where $x \sim y$ means x is a neighbor of y. We set $A_{m+1} = A_m \cup \{b\}$ i.e., b is added as the last element of A_{m+1}, set $S_{m+1} = S_m - \{a_{m+1}\}$, and $R_{m+1} = R_m$.
- If a_m has no neighbor in S_m, we set $A_{m+1} = A_m - \{a_m\}$, and $R_{m+1} = R_m \cup \{a_m\}$.

Let $X_m = |A_m| - 1$ and $\alpha_m = |A_m \cup R_m|/n$ be the fraction of the vertices explored up to time m. These quantities are related by

$$\alpha_m = (X_m + m + 2)/2n. \tag{1.5.4}$$

Proof To prove the formula by induction, we note that $|A_0| = 1$, $|R_0| = 0$, $X_0 = 0$ so

$$\alpha_0 = 1/n = (0 + 0 + 2)/2n = (X_m + m + 2)/2n.$$

If a_m has a neighbor in S_m, then $|R_m|$ does not change while X_m and m increase by 1, and α_m increases by $1/n$. If a_m does not have a neighbor in S_m, a_m moves from A_m to R_m, so α_m does not change while m increases by 1 and X_m decreases by 1. □

When A_m shrinks, the evolution of X_m is quite complicated, since the walk has returned to a vertex whose neighbors have been partially explored. In contrast, when the process visits a vertex a_m for the first time, things are very simple. At such a time the algorithm has never examined the connections between a_m and the vertices of S_m, so the number of neighbors of a_m in S_m,

$$\mathcal{N}(a_m) = \text{binomial}((1-\alpha_m)n, c/n) \approx \text{Poisson}((1-\alpha_m)c).$$

Note that as the algorithm progresses the neighborhood of a_m in S_m looks like an Erdős-Rényi graph with mean degree $(1-\alpha_m)c$.

If $(1-\alpha_m)c > 1$ then the graph restricted to S_m has a giant component. To avoid getting too close to the critical value we will only continue the construction while $(1-\alpha_m)c > 1+\eta$ where $\eta > 0$ and fixed. If a_m is part of a giant component $\mathcal{G}_m$ in S_m then we will never erase a_m in our construction. To find the next point in the path, we explore the clusters containing the neighbors of a_m one at a time until we find one that is part of $\mathcal{G}_m$. There must be one since $a_m \in \mathcal{G}_m$.

Let $\tau_0 = 0$, and

$$\tau_{i+1} = \inf\{m > \tau_i : X_m = i+1, \inf\{k > 0 : X_{m+k} = i\} > n^{1/10}\}.$$

Intuitively, τ_{i+1} is the $i+1$ time we have added a vertex that is part of the giant component. Here $n^{1/10}$ is chosen to be a size larger than all of the nongiant components. Recall that Theorem 1.3.2 has shown that the second largest component has size $O(\log n)$.

Lemma 1.5.4 *Let $\lambda_i = (1-\alpha(\tau_i))c$, and $\sigma_i = \sigma(\lambda_i)$ be the size of the giant component in an Erdős-Rényi graph with mean degree λ_i.*

$$E(\tau_{i+1} - \tau_i|\tau_i) = \frac{2}{\sigma_i} - 1.$$

Here we are using the notation of EFM, where $|\tau_i)$ is short for conditioning on $\mathcal{F}_{\tau_i}$, the information known at time τ_i.

Proof If we let F be the number of failures that occur before we find a neighbor in the giant component and let W_m be the size of the mth small cluster, then

$$\tau_{i+1} - \tau_i = 1 + \sum_{m=1}^{F} 2W_m. \tag{1.5.5}$$

The 2 comes from the fact that in exploring the small clusters we have to visit each vertex twice, once when we add it to the path and a second time to delete it. Finally the $+1$ is for adding the new vertex to $A(\tau_{i+1})$. Taking the expected value we have

$$E(\tau_{i+1} - \tau_i|\tau_i) = 1 + \hat{E}(W) \cdot E(F|\Omega_\infty), \tag{1.5.6}$$

where $\hat{E}$ is expectation with respect to the process conditioned to die out.

Theorem 1.1.9 shows that if $\lambda_i > 1$ and we condition a branching process with a Poisson(λ_i) offspring distribution to die out, then the result is a branching process with a Poisson $(\lambda_i(1-\sigma_i))$ offspring distribution. Summing we have

$$\hat{E} W_m = \sum_{k=0}^{\infty} \lambda_i^k (1-\sigma_i)^k = \frac{1}{1-\lambda_i(1-\sigma_i)}. \tag{1.5.7}$$

Formula (1.1.7) implies

$$E(F|\Omega_\infty) = (1-\sigma_i) E g(Z_1 - 1) = \frac{1-\sigma_i}{\sigma_i}(1-\lambda_i \sigma_i).$$

Combining the last three results, we have

$$\begin{aligned} E(\tau_{i+1} - \tau_i | \tau_i) &= 1 + 2\frac{1-\sigma_i}{\sigma_i}(1-\lambda_i(1-\sigma_i)) \cdot \frac{1}{1-\lambda_i(1-\sigma_i)} \\ &= \frac{2(1-\sigma_i)}{\sigma_i} + 2 - 1 = \frac{2}{\sigma_i} - 1, \end{aligned}$$

which proves the desired result. □

Remark 1.5.5 The cancellation makes the final result look mysterious, but it is not. Let N_i be the number of distinct sites we have to examine to find the one that starts branching process that lives forever. If we let I_x be the event that the branching process started by x does not die out, then I_x in $(\tau_i, \tau_{i+1}]$ are not independent but those in different intervals are, so if we have a large number of intervals $i_0 \le i \le i_1$ in which the fraction of explored sites is $\approx \alpha$ then

$$\frac{1}{i_1 - i_0} \sum_{i=i_0}^{i_1} N_i \approx E N_i \frac{1}{\sigma_{c(1-\alpha)}}$$

because the frequency of successes would be

$$\frac{i_1 - i_0}{\sum_{i=i_0}^{i_1} N_i} \approx \sigma_{c(1-\alpha)}.$$

See Lemma 1.5.12 for more on this point.

There is no formula for the survival probability $\sigma(\lambda)$, so we will consider the situation when $\lambda = 1 + \epsilon$ and ϵ is small. Computation of the critical exponent β, see (1.9.7), shows that $\sigma(1+\epsilon) \sim b\epsilon$ as $\epsilon \to 0$ and hence that if $\epsilon < \epsilon_0$ we have $\sigma(1+\epsilon) \ge b\epsilon/2$. The constant in the Erdős-Rényi case is 2, since the μ_2 in (1.9.7) is equal to 1 for the Poisson(1) distribution. The value is not important for the proof, so we leave it unspecified.

Suppose $\lambda = 1 + \epsilon$ and let

$$T_\epsilon = \max\{m : (1-\alpha_m)(1+\epsilon) \ge (1+\epsilon/2)\}.$$

If $\tau_i \le T_\epsilon$ then $\sigma_i \ge b\epsilon/4$ and hence by Lemma 1.5.4

$$E(\tau_{i+1} - \tau_i | \tau_i) \le \frac{8}{b\epsilon}. \tag{1.5.8}$$

Our next step is to show that

Lemma 1.5.6 *With high probability,* $T_\epsilon \ge \tau(b\epsilon^2 n/20)$ *.*

Since the path gets longer by 1 from time τ_i to τ_{i+1}, this completes the proof.

Proof To prove the lower bound on $T - \epsilon$ we note that

$$M_k = \tau_k - \sum_{j=0}^{k-1} E(\tau_{j+1} - \tau_j | \tau_j)$$

is a martingale. Since no small cluster has size $\ge n^{1/10}$ and no vertex has degree $\ge \log n$ the jumps in this martingale are bounded by $n^{1/10} \log n$.

Since the increments of our martingale are bounded we can use the **Azuma-Hoeffding inequality** given in Theorem A.2.2.

Theorem 1.5.7 *If* M_ℓ *is a martingale and* $|M_\ell - M_{\ell-1}| \le c_\ell$ *for* $1 \le \ell \le m$ *then*

$$P(M_m - M_0 \ge y) \le \exp\left(-y^2 \Big/ 2 \sum_{\ell=1}^{m} c_\ell^2\right).$$

Taking $c_\ell = n^{1/10} \log n$ for all ℓ and using $k \le n$ gives

$$P(M_k > n^{3/4}) \le \exp\left(-\frac{(n^{3/4})^2}{2n(n^{1/10}\log n)^2}\right) \le C\exp(-n^{1/4}).$$

Therefore by a union bound, for any $B < \infty$

$$P\left(\sup_{k \le B\epsilon^2 n} M_k > n^{3/4}\right) \le B\epsilon^2 n \cdot C\exp(-n^{1/4}) \to 0 \quad \text{as } n \to \infty. \tag{1.5.9}$$

To complete the proof, we will use induction to show that $T_\epsilon \ge \tau_i$ for all $0 \le i \le b\epsilon^2 n/20$. This trivially holds if $i = 0$. If $T_\epsilon \ge \tau_i$ then $E(\tau_{j+1} - \tau_j | \tau_j) \le 8/b\epsilon$ for all $j \le i$ and hence

$$\sum_{j=0}^{i} E(\tau_{j+1} - \tau_j | \tau_j) \le \frac{b\epsilon^2 n}{20} \cdot \frac{8}{b\epsilon} \le 0.4\epsilon n.$$

When the good event in (1.5.9) holds $M_{i+1} \le 0.4\epsilon n + n^{3/4}$ so we have $S(\tau_{i+1}) \ge (1 + \epsilon/2)n$, i.e., the statement holds for $i + 1$. □

1.5.3 Enriquez, Faraud, and Ménard (2017)

Their main results show that the profile of the tree constructed by the depth-first search algorithm described in the previous section on the giant component of an Erdős-Rényi graph with mean degree c converges to an explicitly computable deterministic shape. This gives a good but not sharp lower bound on the length of the longest path.

Following our usual style the number of the results in the paper are in parentheses. Earlier in this chapter we have used ρ for the extinction probability and $\sigma = 1 - \rho$ for the survival probability so we have changed their ρs to σs. When we want to indicate that the mean degree is c we will write σ_c.

Theorem 1.5.8 *(Theorem 1) Consider an Erdős-Rényi$(n, c/n)$ graph with $c > 1$ and let H_n be the length of the longest simple path.*

$$\liminf_{n\to\infty} H_n/n \geq \sigma_c - Li_2(\sigma_c)/c.$$

Here $Li_2(x)$ is the **dilogarithm function** which can be defined as $\sum_{k=1}^{\infty} z^k/k^2$ or

$$Li_2(z) = -\int_0^z \frac{\log(1-t)}{t}\,dt. \tag{1.5.10}$$

To explain the remark after Theorem 1.5.2 in Sect. 1.5.1 recall that $\log(1-t) = -t + t^2/2 - t^3/3 + \cdots$, so

$$Li(z) \approx \int_0^z 1 - t/2 + O(t^2)\,dt = z - z^2/4 + O(z^3),$$

and $z - Li_2(z) \sim z^2/4$.

Since we are primarily interested in the geometry of the giant component, we study the behavior of X_m conditioned on **S**, the event that 1 belongs to the giant component.

Theorem 1.5.9 *(Theorem 2). Conditional on* **S**

$$\lim_{n\to\infty} X_{[tN]}/N = h(t),$$

where h is continuous and defined on $[0, 2\sigma_c]$.

To define h we introduce functions $f(\sigma)$ and $g(\sigma)$ defined on $[0, \sigma_c]$ by

$$f(\sigma) = \frac{1}{c}\left[Li_2(\sigma_c) - Li_2(\sigma) + \log\frac{1-\sigma_c}{1-\sigma} - 2\left(\frac{\log(1-\sigma_c)}{\sigma_c} - \frac{\log(1-\sigma)}{\sigma}\right)\right], \tag{1.5.11}$$

$$g(\sigma) = \frac{1}{c}\left[Li_2(\sigma) - Li_2(\sigma_c) + \log\frac{1-\sigma}{1-\sigma_c}\right]. \tag{1.5.12}$$

The graph of h has an increasing part and a decreasing part which are defined by

$$(t, h(t))_{0\le t\le f(0)} = (f(\sigma), g(\sigma))_{0\le\sigma\le\sigma_c}, \tag{1.5.13}$$

$$(t, h(t))_{f(0)\le t\le 2\sigma_c} = (f(\sigma) + 2\sigma[1 - (f(\sigma) + g(\sigma))/2], g(\sigma))_{0\le\sigma\le\sigma_c}. \tag{1.5.14}$$

Theorem 1 is obtained from this by computing the maximum height of the curve given in Theorem 2, which is

$$g(0) = \frac{1}{c}(-\log(1-\sigma_c) - Li_2(\sigma_c)) = \sigma_c - \frac{Li_2(\sigma_c)}{c},$$

where in the second equality we have used the equation satisfied by the survival probability

$$1 - \sigma_c = \exp(-c\sigma_c) \quad \text{which implies} \quad \frac{\log(1-\sigma_c)}{\sigma_c} = -c. \tag{1.5.15}$$

To understand the formulas for f and g it is useful to have a concrete example. We choose $c = 1.5$ since there is a graph in the paper for this value of c. A little computation shows that

$$\sigma_c = 0.58282, \quad Li_2(\sigma_c) = 0.701584, \quad \text{and} \quad \log(1-\sigma_c) = -0.874227.$$

As a check on our computations we note that $\log(1-\sigma_c)/\sigma_c = -1.5$

$$f(0) = \frac{1}{c}[Li_2(\sigma_c) + \log(1-\sigma_c) - 2(-c+1)] = 0.551565$$

$$g(0) = \frac{1}{c}[-Li_2(\sigma_c) - \log(1-\sigma_c)] = 0.115102$$

while $f(\sigma_c) = g(\sigma_c) = 0$.

Main ideas that lead to the proof

To keep the discussion self-contained we repeat a small part of Sect. 1.5.2. When A_m shrinks, the evolution of X_m is quite complicated, since the walk has returned to a vertex whose neighbors have been partially explored. In contrast when the process visits a vertex a_m for the first time, things are very simple. At such a time the algorithm

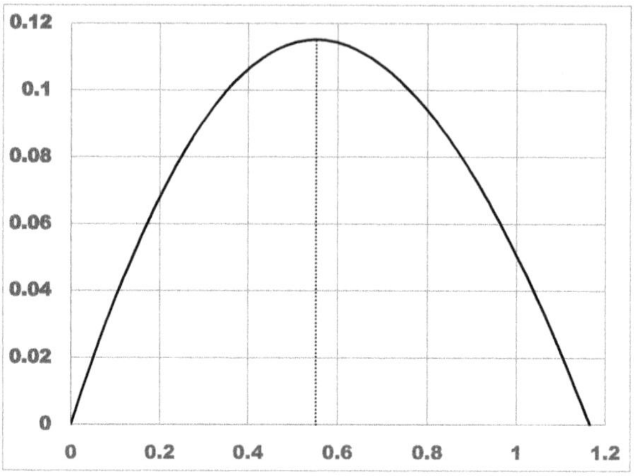

Fig. 1.5 Graph of $(t, h(t)$ for $0 \le f(0) \le 2\sigma_c$. The graph looks symmetric but $f(0) = 0.551565$ while $2\sigma_c = 2(0.58282)$

has never examined the connections between a_m and the vertices of S_m, so the number of neighbors of a_m in S_m,

$$\mathcal{N}(a_m) = \text{binomial}((1-\alpha_m)n, c/n) \approx \text{Poisson}((1-\alpha_m)c).$$

This a very important observation: as the algorithm progresses the neighborhood of a_m in S_m looks like an Erdős-Rényi graph with mean degree $(1-\alpha_m)c$. This motivates the division of the process into two phases, which can be seen in Fig. 1.5.

- If $(1-\alpha_m)c > 1$ then the graph restricted to S_m has a giant component, so the path is getting longer. EFM say that we are "**on the way up**." i.e., h is increasing.
- If $(1-\alpha_m) < c$ then the graph restricted to S_m is subcritical. There are only small components which are dead ends, so we explore them and then delete their vertices. EFM say that we are "**on the way down**," i.e., h is decreasing.

Renewal times: three key lemmas

The evolution on the way up is simplified by the fact that if a_m is part of the giant component then the path before a_m will not be erased until we are in the second part of the construction. Let $\tau_0 = 0$, and

$$\tau_{i+1} = \inf\{m > \tau_i : X_m = i+1, \inf\{k : X_{m+k} = i\} > n^{1/10}\} \wedge 2n.$$

The next result explains our interest in the elapsed time being $> n^{1/10}$. This result follows from Theorem 1.3.2 if we notice that the graph with vertex set S_m is Erdős-Rényi$((1-\alpha_m)n, c/Nn)$.

Lemma 1.5.10 *(Lemma 3). Let* **G** *be the event that at every time with*

$$\alpha_n < 1 - (1 + \eta)/c$$

the graph S_m *has no components between* $n^{1/10}$ *and* $n^{9/10}$ *and maximum degree vertex is at most* $\log n$. $\lim_{N\to\infty} P(\mathbf{G}) = 1$.

Denote by $\mathbf{F}_n$ the event that this is the first visit to a_n. Conditional on $\mathbf{F}_n$ the number of neighbors $\mathcal{N}(a_m)$ and clusters of the neighbors $x_1 < \cdots < x_{\mathcal{N}(a_m)}$ are independent of the excursions outside S_m. When we move from a_m to x_i then we have to explore the entire component of x_i before we return to a_m. If the component containing x_j is small ($< n^{1/10}$ this takes time $< 2n^{1/10}$), while if x_k is part of the giant component in S_m we will add it to the path and not erase x_k until we begin to go back down.

The key to achieving a much better result is to divide the time line into segments according to the value of α_m.

$$h_k = \inf\{i : \alpha_{\tau_i} > k\epsilon\}. \tag{1.5.16}$$

The parameter α varies only slightly between two successive h_k, so $\tau_{i+1} - \tau_i$, $h_k \le i \le h_{k+1}$ are almost i.i.d. In order to have $(1 - \alpha_m)c > 1 + \eta$ we only look at

$$k \le K = [1 - (1 + \eta)/c]/\epsilon.$$

The next result can be proved using the argument for Lemma 1.5.4 to get a result for each i and then taking into account that α_m varies by an amount of order ϵ across the interval.

Lemma 1.5.11 *(Lemma 4). There is a constant* C *so that if* n *is large enough then for all* $i \in [h_k, h_{k+1}]$ *with* $k \le K$

$$\frac{2}{\sigma_{(1-k\epsilon)c}} - 1 - C\epsilon \le E(\tau_{i+1} - \tau_i|\tau_i) \le \frac{2}{\sigma_{(1-k\epsilon)c}} - 1 + C\epsilon.$$

Using the Azuma-Hoeffding inequality as in the proof of Lemma 1.5.6 and the fact that α is almost constant on the interval, leads easily to:

Lemma 1.5.12 *(Lemma 5). There is a constant* C *depending only on* η *so that for every* $k \le K$ *with high probability*

$$\epsilon\sigma_{(1-k\epsilon)c} - C\epsilon^2 \le \liminf_{n\to\infty} \frac{h_{k+1} - h_k}{n} \le \limsup_{n\to\infty} \frac{h_{k+1} - h_k}{n} \le \epsilon\sigma_{(1-k\epsilon)c} + C\epsilon^2.$$

Proof of Theorem 1.5.9. Introducing a superscript ϵ to allow us to keep track of the dependence on this parameter, and applying Lemma 1.5.11.

$$\sum_{i=0}^{k-1} \epsilon\sigma_{(1-k\epsilon)c} - C\epsilon^2 \le \liminf_{n\to\infty} \frac{h_k^\epsilon}{n} \le \limsup_{n\to\infty} \frac{h_k^\epsilon}{n} \le \sum_{i=0}^{k-1} \epsilon\sigma_{(1-k\epsilon)c} + C\epsilon^2.$$

Taking $k = \lceil u/\epsilon \rceil$

$$\int_0^u \sigma_{(1-x)c}\, dx - C\epsilon^2 \le \liminf_{n\to\infty} \frac{h_k^\epsilon}{n} \le \limsup_{n\to\infty} \frac{h_k^\epsilon}{n} \le \int_0^u \sigma_{(1-x)c}\, dx + C\epsilon^2.$$

The k points of the **normalized profile** can be written as

$$\left(\frac{\tau_{h_k^\epsilon}}{n}, \frac{X(\tau_{h_k^\epsilon})}{n}\right) = \left(\frac{\tau_{h_k^\epsilon}}{n}, \frac{h_k^\epsilon}{n}\right) = \left(2k\epsilon + O(n^{-4/5}) - \frac{h_k^\epsilon}{n}, \frac{h_k^\epsilon}{n}\right),$$

the last equality coming from the fact that on the event **G** each increment $\tau_{i+1} - \tau_i$ is bounded by $n^{1/10} \log n$. Combining the last two equations we see that as $n \to \infty$ these K points are uniformly within $C\epsilon$ of the following parametrized curve

$$t(u) = 2u - \int_0^u \sigma_{(1-x)\epsilon}\, dx, \tag{1.5.17}$$

$$x(u) = \int_0^u \sigma_{(1-x)\epsilon}\, dx.$$

These are called x and y in the paper but here we would like the emphasize that the first coordinate is time and the second is the height. Finally, our definitions imply

$$\left(\frac{\tau_{h_{k+1}^\epsilon}}{n} - \frac{\tau_{h_{k+1}^\epsilon}}{n}\right) + \left(\frac{X(\tau_{h_{k+1}^\epsilon})}{n} - \frac{X(\tau_{h_k^\epsilon})}{n}\right) = 2\epsilon.$$

This shows that the slope of the renormalized profile is smaller than 1 in absolute value. We can conclude that the whole renormalized profile stays within $C\epsilon$ of the curve defined in (1.5.17).

Denouement

At this point the interesting part of the proof is done and it only remains to relate the curve defined in (1.5.17) to the one given in (1.5.13), which is an exercise in calculus. To do this, we parametrize the curve by $\sigma_{(1-u)c}$ instead of by u and note that since $\sigma_{(1-u)c}$ is the extinction probability of a branching process with a Poisson($(1-u)c$) distribution then recalling the formula for the generating function

$$1 - \sigma_{(1-u)c} = \exp(-(1-u)c\sigma_{(1-u)c}).$$

Taking logs we have

$$\log(1 - \sigma_{(1-u)c}) = -(1-u)c\sigma_{(1-u)c}, \tag{1.5.18}$$

which rearranges to

$$u = 1 + \frac{\log(1-\sigma_{(1-u)c})}{c\sigma_{(1-u)c}} = \frac{\log(1-\sigma_{(1-u)c})}{c\sigma_{(1-u)c}} - \frac{\log(1-\sigma_c)}{c\sigma_c},$$

where in the second step we have used (1.5.18) with $u = 0$ to conclude $\log(1-\sigma_c) = -c\sigma_c$. The last calculation shows that $2u$ is the last term in (1.5.11).

To obtain the formula in terms of $f(\sigma)$ and $g(\sigma)$, we need to prove

Lemma 1.5.13

$$\int_{1-u}^{1} \sigma_{x\epsilon}\, dx = \frac{1}{c}\left[Li_2(\sigma_{(1-u)\sigma_c}) - Li_2(\sigma_c) + \log(1-\sigma_{(1-u)c}) - \log(1-\sigma_c)\right].$$

Proof To do this we note that replacing $(1-u)$ by x in (1.5.18) gives

$$\frac{c\sigma'_{xc}}{1-\sigma_{xc}} = c\sigma_{xc} + c^2 x\sigma_{xc}.$$

Rearranging the last equation and dividing by c gives

$$\begin{aligned}\int_{1-u}^{1} \sigma_{x\epsilon}\, dx &= \int_{1-u}^{1} \frac{\sigma'_{xc}}{1-\sigma_{xc}} - \int_{1-u}^{1} cx\sigma_{xc} \\ &= \frac{1}{c}\left(\ln(1-\sigma_{(1-\mu)c}) - \ln(1-\sigma_c)\right) - \int_{1-\mu}^{1} \frac{\log(1-\sigma_{xc})}{\sigma_{xc}}\sigma'_{xc}\, dx,\end{aligned}$$

where we have used (1.5.18) in the second term. Integrating using the definition of Li_2 given in (1.5.10) shows that the last term is

$$-\frac{1}{c}\left(Li_2(\sigma_c) - Li_2(\sigma_{(1-\mu)c}\right).$$

We would like to thank Laurent Ménard for taking time to write out the proof and email it to us. □

Remark 1.5.14 Enriquez et al. (2022) have proved the existence of long paths in configuration model graphs. The proof is more complicated because the distribution of the lengths of unused edges changes in addition to the number. However the additional complexity is not troublesome if you just want to use the results.

1.6 CLT for the Size of the Giant Component

Up to this point we have been content to study the growth of clusters while they are $o(n)$. In this section we will use an idea of Martin-Löf (1986) to follow the random walk approach all of the way to the end of the formation of the giant component and prove a central limit theorem for the size of the giant component.

To avoid the problem caused by the process dying out, it is convenient to modify the rules so that if $A_t = \emptyset$ we pick $i_t \in U_t$, and rewrite the recursion as

$$\begin{aligned} R_{t+1} &= R_t \cup \{i_t\}, \\ A_{t+1} &= A_t - \{i_t\} \cup \{y \in U_t : \eta_{i_t,y} = 1\}, \\ U_{t+1} &= U_t - (\{i_t\} \cup \{y \in U_t : \eta_{i_t,y} = 1\}). \end{aligned}$$

In words, when one cluster is finished we pick a new vertex and start exposing its cluster.

When $A_t = \emptyset$ we subtract 1 + binomial$(|U_t|, \lambda/n)$ points from U_t versus binomial$(|U_t|, \lambda/n)$ points when $A_t \neq \emptyset$. However, we will experience only $O(1)$ visits to $\emptyset$ before finding the giant component, so this difference can be ignored. Let $\mathcal{F}_t$ be the σ-field generated by the process up to time t. Let $u_t^n = |U_t|$.

Lemma 1.6.1 *$u_{[ns]}^n/n$ converges in distribution to u_s, the solution of*

$$\frac{du_s}{ds} = -\lambda u_s \qquad u_0 = 1$$

and hence $u_s = \exp(-\lambda s)$.

Proof Let $\Delta u_t^n = u_{t+1}^n - u_t^n$. If $A_t \neq \emptyset$ then

$$\begin{aligned} E(\Delta u_t^n | \mathcal{F}_t) &= -u_t^n \frac{\lambda}{n}, \\ \text{var}\,(\Delta u_t^n | \mathcal{F}_t) &= u_t^n \frac{\lambda}{n}\left(1 - \frac{\lambda}{n}\right). \end{aligned}$$

If we let $t = [ns]$ for $0 \le s \le 1$ and divide by n then

$$\begin{aligned} E\left(\frac{\Delta u_{[ns]}^n}{n} \middle| \mathcal{F}_{[ns]}\right) &= -\frac{u_{[ns]}^n}{n} \cdot \lambda \cdot \frac{1}{n}, \\ \text{var}\left(\frac{\Delta u_{[ns]}^n}{n} \middle| \mathcal{F}_{[ns]}\right) &= \frac{u_{[ns]}^n}{n} \cdot \lambda \left(1 - \frac{\lambda}{n}\right) \cdot \frac{1}{n^2}. \end{aligned} \tag{1.6.1}$$

Dividing each right-hand side by $1/n$, the time increment in the rescaled process, we see that $\Delta u_{[ns]}^n$ has

$$\text{infinitesimal mean} = -\frac{u^n_{[ns]}}{n}\lambda,$$

$$\text{infinitesimal variance} = \frac{u^n_{[ns]}}{n}\lambda\left(1-\frac{\lambda}{n}\right)\cdot\frac{1}{n}.$$

Letting $n \to \infty$, the infinitesimal variance $\to$ 0, so the result follows from the degenerate case of Theorem 4.1 in Sect. 7.4 of Ethier and Kurtz (1986), in which the infinitesimal variance is 0. □

The last proof is simple and intuitive, but may be too sophisticated for some reader's tastes, so we now give

Alternative Proof The calculations above show that

$$M^n_t = \left(1-\frac{\lambda}{n}\right)^{-t} u^n_t/n$$

is a martingale with

$$\begin{aligned} E(M^n_t - M^n_0)^2 &= \sum_{s=0}^{t-1} E(M^n_{s+1} - M^n_s)^2 \\ &\le \sum_{s=0}^{t-1}\left(1-\frac{\lambda}{n}\right)^{-s+1}\lambda/n^2 \to 0, \end{aligned}$$

so by Kolmogorov's maximal inequality

$$E\left(\max_{0\le s\le n}(M^n_s - M^n_0)^2\right) \to 0.$$

Since $M^n_0 = 1$, this says that when n is large $M^n_s \approx 1$ uniformly in s, so $u^n_{[ns]}/n \approx (1-\lambda/n)^{[ns]} \to e^{-\lambda s}$. □

To determine the size of the giant component, we note that when $u^n_t + r^n_t = n$, $A_t = \emptyset$. This may occur several times for small t while we are searching for the giant component, but the solution we are looking for is the first occurrence after an excursion of $O(n)$. To locate roughly the time at which this occurs, we note that scaling $r^n_t = |R_t| \equiv t$ as we did u^n_t, $r^n_{[ns]}/n \to s$. (Here and for the rest of this section we will use t for the original integer time scale and $s \in [0, 1]$ for rescaled time.]) After scaling

$$u^n_t + r^n_t = n \quad \Rightarrow \quad e^{-\lambda s} + s = 1.$$

Solving we have $1 - s = \exp(\lambda((1-s) - 1))$, which is the fixed point equation for the extinction probability, $1 - s$. As the graph in Fig. 1.6 shows $e^{-\lambda s} + s > 1$ for $s > 1 - \rho$, so we are interested only in $u^n_{[ns]}/n$ for $0 \le s \le 1 - \rho + \epsilon$. In this part

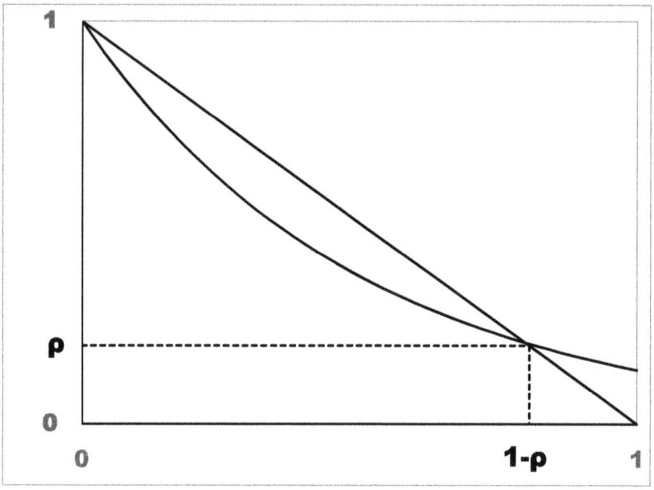

Fig. 1.6 Example of Lemma 1.6.1 with $\lambda = 2$

of the process we first generate a geometrically distributed number of small clusters and then find the giant component.

Consider now $y^n_{[ns]} = (u^n_{[ns]} - n\exp(-\lambda s))/\sqrt{n}$ for $0 \le s \le 1 - \rho$.

Theorem 1.6.2 *As $n \to \infty$, $y^n_{[ns]}$ converges in distribution to a normal with mean 0 and variance $e^{-\lambda s} - e^{-2\lambda s}$.*

Proof If $A_{[ns]} \neq \emptyset$ then using the formulas in (1.6.1)

$$E(\Delta y^n_{[ns]}|\mathcal{F}_{[ns]}) = -\frac{1}{\sqrt{n}}\left(-u^n_{[ns]} \cdot \frac{\lambda}{n} - n\exp(-\lambda s)(\exp(-\lambda/n) - 1)\right)$$
$$\sim -\frac{\lambda}{n}\left(\frac{u^n_{[ns]} - n\exp(-\lambda s)}{\sqrt{n}}\right) = -\lambda y^n_{[ns]} \cdot \frac{1}{n},$$
$$\text{var}\,(\Delta y^n_{[ns]}|\mathcal{F}_{[ns]}) = \text{var}\left(\frac{\Delta u^n_{[ns]}}{\sqrt{n}}\middle|\mathcal{F}_{[ns]}\right)$$
$$= \frac{1}{n} \cdot u^n_{[ns]} \cdot \frac{\lambda}{n}\left(1 - \frac{\lambda}{n}\right) \sim \lambda e^{-\lambda s} \cdot \frac{1}{n}.$$

Using Theorem 4.1 in Sect. 7.4 of Ethier and Kurtz (1986), we see that $y^n_{[ns]}$ converges in distribution to the solution of the following stochastic differential equation.

$$dy_s = -\lambda y_s\,ds + \sqrt{\lambda e^{-\lambda s}}\,dB_s \quad y_0 = 0.$$

The solution to this equation is

$$y_s = \int_0^s e^{-\lambda(s-r)} \sqrt{\lambda e^{-\lambda r}} \, dB_r. \tag{1.6.2}$$

To check this, note that if one continuously invests an amount g_s in an exponentially decaying stock market then your net wealth x_s satisfies

$$\frac{dx_s}{ds} = -\lambda x_s + g_s.$$

Since computation of interest is linear, each amount decays exponentially from its date of investment, and this differential equation has solution

$$x_s = \int_0^s e^{-\lambda(s-r)} g_r \, dr.$$

Readers who want a more rigorous proof can use stochastic calculus to check this.

Since the integrand in the stochastic integral (1.6.2) is deterministic, y_s has a normal distribution with mean 0 and variance

$$\int_0^t \exp(-2\lambda(s-r))\lambda e^{-\lambda r} \, dr = e^{-2\lambda s} \int_0^s \lambda e^{\lambda r} \, dr = e^{-\lambda s} - e^{-2\lambda s},$$

which proves the result. To prove the fact about the stochastic integral note that any approximating sum has a normal distribution and then pass to the limit. □

Remark. Again if one wants to avoid stochastic calculus, the theorem can be proved by applying the martingale central limit theorem to

$$M_t^n - M_0^n = \left(1 - \frac{\lambda}{n}\right)^{-t} u_t^n / n - 1.$$

The key observation is that

$$\sum_{r=0}^{[ns]-1} E((M_{r+1}^n - M_r^n)^2 | \mathcal{F}_r) \to \int_0^s \lambda e^{\lambda u} \, du = e^{\lambda s} - 1,$$

i.e., the quadratic variation has a deterministic limit. Multiplying the martingale by $e^{-\lambda s}$ multiplies the variance by $e^{-2\lambda s}$ and we arrive at the same limit as before.

We have analyzed the fluctuations of $u_{[ns]}^n$. To determine the fluctuations of the point where $u_t^n + t = n$, we can now prove the result as we do the central limit theorem of renewal theory. To briefly recall that approach, let $\xi_1, \xi_2, \dots$ be i.i.d. positive random variables with $E\xi_i = \mu$ and $\text{var}(\xi_i) = \sigma^2 \in (0, \infty)$. Let $T_n = \xi_1 + \cdots + \xi_n$ and $N(t) = \inf\{n : T_n > t\}$. The central limit theorem implies

$$T_n \approx n\mu + \sigma\sqrt{n}\chi,$$

where χ is a standard normal. Setting $n = t/\mu$

$$T_{t/\mu} \approx t + \sigma\sqrt{\frac{t}{\mu}}\chi.$$

If $\chi > 0$ then $N(t) < t/\mu$. The law of large numbers implies $T_n - T_m \approx (n - m)\mu$ when $n - m$ is large so we will have

$$\frac{t}{\mu} - N(t) \approx \frac{\sigma}{\mu}\sqrt{\frac{t}{\mu}}\chi.$$

The same reasoning applies in the current situation. Taking $s = 1 - \rho$ in Lemma 1.6.2 and letting $\mathcal{Z}$ denote a normal with variance $e^{-\lambda(1-\rho)} - e^{-2\lambda(1-\rho)}$ we have

$$u^n_{[n(1-\rho)]} \approx n\exp(-\lambda(1 - \rho)) + \sqrt{n}\mathcal{Z}. \tag{1.6.3}$$

$A_{[ns]} = \emptyset$ when $u^n_{[ns]} = n - [ns]$. To find out when this occurs, we suppose equality holds at $s_0 = (1 - \rho) + \mathcal{Y}/\sqrt{n}$. Using (1.6.3) and noting $s_0 \to (1 - \rho)$ as $n \to \infty$

$$n\exp(-\lambda\{(1 - \rho) + \mathcal{Y}/\sqrt{n}\}) + \sqrt{n}\mathcal{Z} = u^n_{[ns_0]} = n - [ns_0],$$

or rearranging

$$\exp(-\lambda s_0) - 1 + s_0 = -\sqrt{\mathcal{Z}}/\sqrt{n}.$$

Let $h(t) = e^{-\lambda t} - 1 + t$, which is $= 0$ at $t = 1 - \rho$. $h'(t) = -\lambda e^{-\lambda t} + 1$, so we can write the above as

$$h'(1 - \rho)\mathcal{Y}/\sqrt{n} \approx -\sqrt{\mathcal{Z}}/\sqrt{n},$$

or $\mathcal{Y} \approx \mathcal{Z}/h'(1 - \rho)$. $h'(1 - \rho) = 1 - \lambda\rho$. Putting the pieces together.

Theorem 1.6.3 *Suppose $\lambda > 1$. The size of the largest component $\mathcal{C}^{(1)}$ satisfies*

$$\frac{|\mathcal{C}^{(1)}| - n(1 - \rho)}{\sqrt{n}} \Rightarrow \chi$$

where $\Rightarrow$ means convergence in distribution and χ has a normal distribution with mean 0 and variance $(\rho - \rho^2)/(1 - \lambda\rho)^2$.

For other approaches to this result see Pittel (1990) and Barraez, Boucheron, and Fernandez de la Vega (2000). To compare variances note that Pittel's $c = \lambda$ and $T = \rho/\lambda$.

1.7 Combinatorial Approach

Combinatorial methods give more refined results about the Erdős-Rényi model. However, we first need some graph-theoretic preliminaries about the nature of components. It is easy to see by using induction that a tree with k vertices has $k - 1$ edges. We call a graph with k vertices and k edges a **unicyclic graph**, since it will have exactly one cycle, i.e., a path of adjacent vertices $x_0, x_1, \ldots, x_k = x_0$, and $x_j \neq x_0$ for $1 \leq j < k$. We call a graph with k vertices and $k + \ell$ edges with $\ell \geq 1$ a **complex component** with complexity ℓ. Section 5.4 of Bollobás (Bollobás, 2001) gives combinatorial results for unicyclic and complex components. In Sect. 1.8 we will use Aldous' (1997) result to give some much less explicit results about complex components in the critical regime. See Theorem 1.8.5. Repeating the proof of Theorem 1.2.3

Lemma 1.7.1 *Let $A < \infty$ be a constant, and consider only "small" components with $\leq A \log n$ vertices. The probability that a small component is unicyclic is $\leq (A \log n)^2/n$. The probability that a small component has complexity $\ell \geq 1$ is $\leq (A \log n)^3/n^2$.*

Proof We run the exploration process defined in (1.3.1) for $\leq A \log n$ steps and have a probability of collision of $\leq (A \log n)/n$ on each step. The probability of at least one collision is $\leq (A \log n)^2/n$, while the probability of at least two collisions is bounded above by the expected number of pairs of steps on which collisions occur, which is

$$\leq \binom{A \log n}{2} (A \log n/n)^2,$$

proving the desired result. □

Theorem 1.7.2 *Complex components are rare unless λ is close to 1.*

Proof Theorem 1.3.1 implies that if $\lambda < 1$ there is a C_λ so that with high probability the largest component is $\leq C_\lambda \log n$. Theorem 1.3.2 implies that if $\lambda > 1$ there is a β_λ so that with high probability the second largest component is $\leq \beta_\lambda \log n$. Lemma 1.7.1 implies that when the largest component is $\leq A \log n$ then with high probability there are no complex components. The expected number of unicyclic components is $\leq (A \log n)^2$. In this section we are interested in the fraction of clusters that have size k, so we can also ignore unicyclic components. □

Theorem 1.7.2 implies that for fixed $\lambda \neq 1$, we can restrict our attention to tree components. In 1889 Cayley showed that there are k^{k-2} trees with k labeled vertices. When $p = \lambda/n$ the expected number of trees of size k present is

$$\binom{n}{k} k^{k-2} \left(\frac{\lambda}{n}\right)^{k-1} \left(1 - \frac{\lambda}{n}\right)^{k(n-k)+\binom{k}{2}-(k-1)} \tag{1.7.1}$$

since each of the $k-1$ edges in the tree needs to be present and there can be no edges connecting its k vertices to its complement or any other edges connecting the k vertices. For fixed k, we can drop $-k^2 + \binom{k}{2} - k + 1$ from the exponent of the last term and the above is asymptotic to nq_k where

$$q_k = \frac{k^{k-2}}{k!} \lambda^{k-1} e^{-\lambda k}. \tag{1.7.2}$$

Recalling that in the subcritical regime cluster sizes have the same distribution as the total progeny in a Poisson(λ) branching process, we get the following corollary, which is "well-known," but not easy to prove directly from the definition of the branching process:

Corollary 1.7.3 *The probability distribution of the total progeny τ of a* Poisson(λ) *branching process with $\lambda < 1$ is given by*

$$P(\tau = k) = kq_k = \frac{1}{\lambda} \frac{k^{k-1}}{k!} (\lambda e^{-\lambda})^k. \tag{1.7.3}$$

There is an extra factor of k due to the fact that a tree of size k is $\mathcal{C}_x$ for k values of x.

This distribution was first discovered by Borel in 1942. It is called the Borel-Tanner distribution, since Tanner (1961) showed that when $\lambda < 1$ it gave the distribution of the total number of customers served in the first busy period of a queue with Poisson rate λ arrivals and service times always equal to 1. Of course, this becomes a branching process if we think of the customers that arrive during a person's service time as their children.

1.7.1 Super/subcritical Duality

Suppose $\lambda > 1$ and let ρ be the extinction probability for the branching process with a Poisson(λ) offspring distribution. The title of this subsection refers to the fact that there is a close relationship between Erdős-Rényi random graphs with mean degrees $\lambda > 1$ and $\lambda\rho < 1$. Using the fixed point equation $\rho = e^{\lambda(\rho-1)}$ for the extinction probability ρ gives

$$\lambda\rho e^{-\lambda\rho} = \lambda e^{\lambda(\rho-1)} e^{-\lambda\rho} = \lambda e^{-\lambda}. \tag{1.7.4}$$

Let $m = n\rho$ and consider Erdős-Rényi($m, \lambda\rho/m$), an Erdős-Rényi graph with number of vertices equal to the number of vertices in non-giant components of Erdős-Rényi($n, \lambda/n$). Changing variables in (1.7.2) we see that

$$\frac{m}{\lambda\rho}\frac{k^{k-2}}{k!}(\lambda\rho e^{-\lambda\rho})^k = \frac{n}{\lambda}\frac{k^{k-2}}{k!}(\lambda e^{-\lambda})^k.$$

In words, the expected number of trees of size k is the same in Erdős-Rényi$(m, \lambda\rho/m)$ and Erdős-Rényi$(n, \lambda/n)$. Changing variables in the same way in (1.7.3)

$$\frac{1}{\lambda\rho}\frac{k^{k-1}}{k!}(\lambda\rho e^{-\lambda\rho})^k = \frac{1}{\lambda}\frac{k^{k-1}}{k!}(\lambda e^{-\lambda})^k \cdot \frac{1}{\rho}.$$

In words, the total progeny of a Poisson(λ) branching process conditioned on extinction is the same as that of a Poisson$(\lambda\rho)$ branching process, which is Theorem 1.1.9.

(1.7.2) is a result about the expected number of trees. The next result is a law of large numbers, which says that the actual number is $1 + o(1)$ times the expected value.

Theorem 1.7.4 *Let T_k^n be the number of tree components of size k in the Erdős-Rényi graph with n vertices. As $n \to \infty$, $T_k^n/n \to q_k$ in probability, where q_k is defined in (1.7.2)*

Proof This proof comes from Bollobás (2001), 106–107. The expected number of ordered pairs of tree components of size k (with the second tree different from the first) is

$$\binom{n}{k}k^{k-2}\left(\frac{\lambda}{n}\right)^{k-1}\left(1-\frac{\lambda}{n}\right)^{k(n-k)+\binom{k}{2}-k+1}$$
$$\cdot\binom{n-k}{k}k^{k-2}\left(\frac{\lambda}{n}\right)^{k-1}\left(1-\frac{\lambda}{n}\right)^{k(n-2k)+\binom{k}{2}-k+1}.$$

The second formula differs from the first only in two places: first we have only $n-k$ vertices to choose from, and the first term already takes into account the fact that there are no connections from the first tree to the second. Since $\binom{n-k}{k} \le \binom{n}{k}$ the above is

$$\le (ET_k)^2\left(1-\frac{\lambda}{n}\right)^{-k^2} \le (ET_k)^2 e^{\lambda k^2/n}.$$

From this we get

$$\text{var}\,(T_k) = E(T_k(T_k-1)) + ET_k - (ET_k)^2 \le ET_k + (ET_k)^2(e^{\lambda k^2/n} - 1).$$

Using Chebyshev's inequality

$$P(|T_k - ET_k| \ge n^{2/3}) \le \frac{ET_k + (ET_k)^2(e^{\lambda k^2/n}-1)}{n^{4/3}} \to 0,$$

since $ET_k \sim nq_k$ and $e^{\lambda k^2/n} - 1 \sim \lambda k^2/n$. This gives the desired result. Note that we could replace $n^{2/3}$ in the last display by $\omega(n)n^{1/2}$ where $\omega(n) \to \infty$ as $n \to \infty$. □

The results above allow us to verify the remark we made about the largest non-giant component for $\lambda > 1$.

Theorem 1.7.5 *Suppose $\lambda > 1$ and let $\mathcal{C}^{(2)}$ be the second largest component. If $\alpha = \lambda - 1 - \log\lambda$ and $a > 1/\alpha$ then as $n \to \infty$*

$$P\left(|\mathcal{C}^{(2)}| \geq a \log n\right) \to 0.$$

Proof For simplicity we will do our calculations for the limit (1.7.2) rather than for the exact formula (1.7.1). Stirling's formula tells us that

$$k! \sim k^{k+1/2} e^{-k} \sqrt{2\pi} \quad \text{as } k \to \infty, \tag{1.7.5}$$

so we have (Lemma 1.7.6 will show this is valid for $k = o(n^{1/2})$)

$$q_k = \frac{1}{\lambda} \cdot \frac{k^{k-2}}{k!} (\lambda e^{-\lambda})^k \sim \frac{1}{\lambda\sqrt{2\pi}} k^{-5/2} (\lambda e^{1-\lambda})^k.$$

Now $g(\lambda) \equiv \lambda e^{1-\lambda} = 1$ when $\lambda = 1$ and $g'(\lambda) = (1-\lambda)e^{1-\lambda}$. Thus $g(\lambda)$ is increasing for $\lambda < 1$, decreasing for $\lambda > 1$, and has $g(\lambda) < 1$ when $\lambda \neq 1$. Summing and using the fact that $k^{-5/2}$ is decreasing and $\lambda e^{1-\lambda} < 1$

$$Q_K = \sum_{k=K}^{\infty} q_k \sim \frac{1}{\lambda\sqrt{2\pi}} K^{-5/2} \frac{(\lambda e^{1-\lambda})^K}{1 - \lambda e^{1-\lambda}}.$$

Taking $K = a \log n$

$$(\lambda e^{1-\lambda})^{a \log n} = \exp((\log\lambda - \lambda + 1) a \log n) = n^{-(1+\epsilon)}$$

when $a = (1+\epsilon)/\alpha$, which proves the desired result. □

1.7.2 Cluster Sizes Near Criticality

In the next section we will use weak convergence to look at component sizes near $\lambda = 1$. Here we will use combinatorics to derive some results. We begin with a calculation that is simple and gives the right answer, but is not completely rigorous. (1.7.2) tell us that the expected number of trees of size k is, for large k,

$$n \frac{k^{k-2}}{k!} \lambda^{k-1} e^{-\lambda k} \sim \frac{n}{\lambda\sqrt{2\pi}} k^{-5/2} (\lambda e^{1-\lambda})^k. \tag{1.7.6}$$

By **Stirling's formula**

$$k! \sim k^k e^{-k}\sqrt{2\pi k} \tag{1.7.7}$$

When $\lambda = 1$, $\lambda e^{1-\lambda} = 1$ so summing from $k = K$ to ∞, the expected number of trees of size $\geq K$ is

$$\sum_{k=K}^{\infty} \frac{n}{\sqrt{2\pi}} k^{-5/2} \sim \frac{2}{3\sqrt{2\pi}} n K^{-3/2}.$$

This is small when $K >> n^{2/3}$, suggesting that the largest tree components are of order $n^{2/3}$.

Having figured out what to guess, we will now go back and do the calculation carefully. For the moment λ is a general parameter value, which you should think of as close to 1.

Theorem 1.7.6 *Let $\alpha(\lambda) = \lambda - 1 - \log(\lambda)$. If $k \to \infty$ and $k = o(n^{3/4})$ then the expected number of tree components of size k in an Erdős-Rényi graph with n vertices*

$$\gamma_{n,k}(\lambda) \sim n \cdot \frac{k^{-5/2}}{\lambda\sqrt{2\pi}} \exp\left(-\alpha(\lambda)k + (\lambda - 1)\frac{k^2}{2n} - \frac{k^3}{6n^2}\right). \tag{1.7.8}$$

Proof (1.7.1) tells us that

$$\gamma_{n,k}(\lambda) = \binom{n}{k} k^{k-2} \left(\frac{\lambda}{n}\right)^{k-1} \left(1 - \frac{\lambda}{n}\right)^{k(n-k)+\binom{k}{2}-k}.$$

Using Stirling's formula (1.7.7) and $k = o(n)$ in the last exponent gives

$$\gamma_{n,k}(\lambda) \sim n \left[\prod_{j=1}^{k-1}\left(1 - \frac{j}{n}\right)\right] \cdot \frac{k^{-5/2}}{e^{-k}\sqrt{2\pi}} \cdot \lambda^{k-1} \left(1 - \frac{\lambda}{n}\right)^{kn-k^2/2}.$$

Using the expansion $\log(1-x) = -x - x^2/2 - x^3/3 - \cdots$ we see that if $k = o(n)$ then

$$\left(1 - \frac{\lambda}{n}\right)^{kn-k^2/2} \sim \exp(-\lambda k + \lambda k^2/2n),$$

while if $k = o(n^{3/4})$ we have

$$\prod_{j=1}^{k-1}\left(1 - \frac{j}{n}\right) = \exp\left(-\frac{1}{n}\sum_{j=1}^{k-1} j - \frac{1}{2n^2}\sum_{j=1}^{k-1} j^2 + o\left(\frac{k^4}{n^3}\right)\right) \sim \exp\left(-\frac{k^2}{2n} - \frac{k^3}{6n^2}\right).$$

Combining our calculations gives the desired formula. □

Taking $\lambda = 1$ in Lemma 1.7.6 we have

$$\gamma_{n,k}(\lambda) \sim \frac{nk^{-5/2}}{\sqrt{2\pi}} e^{-k^3/6n^2}. \tag{1.7.9}$$

Note that the right-hand side behaves like a power law when $k = o(n^{2/3})$ but decays to 0 exponentially fast on $[n^{2/3}, \infty)$.

1.8 Critical Regime

In contrast to the previous section, probabilistic methods, specifically martingales and weak convergence will take center stage. We begin with a result for the critical case $\lambda = 1$ that is simple to prove but then move on to a result that describes the emergence of the giant component in the critical regime $\lambda = 1 + \theta n^{-1/3}$ where $-\infty < \theta < \infty$.

1.8.1 An Inequality of Nachmias and Peres

We will prove what they call the easy upper bound.

Theorem 1.8.1 *Suppose* $\lambda = 1$. *Let* $\mathcal{C}_1$ *be the largest component. If* $x > 1$

$$P(|\mathcal{C}_1| > xn^{2/3}) \leq \frac{6}{x^{3/2}}.$$

By working harder and using exponential martingales one can show that

Theorem 1.8.2 *For any* $n > 1000$ *and* $x > 8$

$$P(|\mathcal{C}_1| > xn^{2/3}) \leq \frac{4}{x} \exp(-x^2(x-4)/32).$$

As the authors say in their paper (2010), upper bounds were proved by Bollobás (1984) and Luczak (1990) but their methods are remarkable for their simplicity, especially in comparison to the machinery used in Sect. 1.8.2.

Proof of Theorem 1.8.1. To begin, we recall the recursion defined in (1.3.1) in Sect. 1.3 for the set of active sites when we add the neighbors of one site at each step:

$$|A_{t+1}| = |A_t| - 1 + \xi_{t+1}, \tag{1.8.1}$$

where ξ_{t+1} is the number of vertices added on step $t + 1$, and we start with $A_0 = 1$. Since the number of vertices to which a connection can be made decreases with time, $\xi_i \leq_d X_i = \text{binomial}(n, 1/n)$, where $\leq_d$ us a stochastic ordering between the distributions. Before we forget we would like to record the facts

$$EX = n \cdot (1/n) = 1 \quad \text{and} \quad \text{var}\,(X) = n(1/n)(1-1/n) = 1 - 1/n. \tag{1.8.2}$$

The size of the cluster is $\tau = \min\{i \geq 1 : A_i = 0\}$. If we let $S_t = 1 + \sum_{k=1}^{t}(X_k - 1)$ then $S_t \geq_d A_t$ and

$$\sigma = \min\{i \geq 1 : S_i = 0\}$$

has $\sigma \geq_d \tau$. If we arrange things so that $\xi_i \leq X_i$ almost surely for $i \leq \tau$ then we have $\tau \leq \sigma$ almost surely.

We are interested in how large the cluster can get so we let

$$\gamma = \min\{i \geq 1 : S_i = 0 \text{ or } S_i \geq H\}.$$

Since $E(X_i - 1) = 0$, $S_{i \wedge \sigma}$ is a nonnegative martingale. Using the optional stopping theorem for nonnegative supermartingles (Theorem 4.8.4 in PTE5) we conclude that

$$1 = ES_0 \geq ES_\gamma \geq HP(S_\gamma \geq H),$$

so the probability S_i reaches $\geq H$ before hitting 0 is

$$P(S_\gamma \geq H) \leq 1/H. \tag{1.8.3}$$

Lemma 1.8.3 *Let $p \in (0, 1)$, let X_i be i.i.d. random variables with binomial(n, p) distributions, and let $S_t = 1 + \sum_{i=1}^{t} X_i$. Fix an integer H and define*

$$\gamma = \min\{t \geq 1 : S_t = 0 \text{ or } S_t \geq H\}.$$

Given $S_\gamma \geq H$ the overshoot $S_\gamma - H$ is dominated by the binomial(n, p) distributions.

Proof Suppose we condition instead on $\{S_\gamma \geq H, \gamma = k\}$. Let $r = H - S_{k-1} + 1$. Let $I_1, \ldots I_n$ be i.i.d. Bernoulli(p) so that $I_1 + \cdots + I_n = X_k$. If $S_k \geq H$ let J be the smallest value of m so that $I_1 + \cdots + I_m = r$. The overshoot $S_\gamma - H = I_{J+1} + \cdots + I_n$, which is clearly smaller in distribution than binomial(n, p). Since the result holds if we condition on the time $\gamma = k$ then it also holds if we only condition on $S_\gamma \geq H$. □

Using Lemma 1.8.3 with $f(x) = 2Hx + x^2$ we see that if $X =$ binomial$(n, 1/n)$ then

$$\begin{aligned} E(2H(S_\gamma - H) + (S_\gamma - H)^2 | S_\gamma \geq H) &\leq E(2HX + X^2) \\ &\leq 2H + 1 + (1 - 1/n) \leq 2H + 2, \end{aligned}$$

where we have used $EX = 1$ and $EX^2 = (EX)^2 + \text{var}\,(X)$ and (1.8.2). Writing $S_\gamma^2 = H^2 + 2H(S_\gamma - H) + (S_\gamma - H)^2$ we have for $H \geq 2$

$$E(S_\gamma^2 | S_\gamma \geq H) \leq H^2 + 2H + 2 \leq H^2 + 3H. \tag{1.8.4}$$

We now introduce a second martingale. Consulting Sect. 4.8.1 in PTE5 we see that

$$S_t^2 - (1 - 1/n)t$$

is the **quadratic martingale** associated with the random walk S_t. Since Lemma 1.8.3 controls how far the martingale jumps over H we can use the optional stopping theorem to conclude that

$$E\left(S_\gamma^2 - (1 - 1/n)\gamma\right) = 1.$$

To prove this, we note that since $\gamma \wedge m$ is a bounded stopping time, so

$$E\ S_{\gamma \wedge m}^2 = 1 + (1 - 1/n)E(\gamma \wedge m).$$

Using the dominated convergence theorem on the left and the monotone convergence theorem on the right we get

$$1 + (1 - 1/n)E(\gamma) = ES_\gamma^2 = P(S_\gamma \geq H)E((S_\gamma)^2 | S_\gamma \geq H) \leq H + 3$$

by (1.8.3) and (1.8.4). Rearranging gives

$$E\gamma \leq \frac{H+2}{1 - 1/n} \leq H + 3 \qquad \text{if } H \leq n - 3.$$

We conclude that for $2 \leq H \leq n - 3$

$$P(\gamma > H^2) \leq \frac{H+3}{H^2} \leq \frac{2}{H}.$$

Define $\bar{\gamma} = \gamma \wedge H^2$. If $\gamma < H^2$ and $S_\gamma < H$ then $S_{\bar{\gamma}} = S_\gamma = 0$, so we have

$$P(S_{\bar{\gamma}} > 0) \leq P(S_\gamma \geq H) + P(\gamma \geq H^2) \leq 3/H. \tag{1.8.5}$$

Let $K = H^2$ and note that if $\mathcal{C}(v) > H^2$ we must have $S_{\bar{\gamma}} > 0$, because if not then the cluster size would be $\leq \bar{\gamma} \leq H^2$. From this we deduce that if $9 \leq K \leq (n-3)^2$ then

$$P(|\mathcal{C}(v)| > K) \leq \frac{3}{\sqrt{K}}.$$

Let N_K be the total number of vertices contained in components of size larger than K. Then three trivial inequalities give

$$P(|\mathcal{C}_1| > K) \leq P(N_K > K) \leq \frac{EN_K}{K} \leq \frac{nP(|\mathcal{C}(v)| > K)}{K}.$$

Putting $K = (\lfloor \sqrt{xn^{2/3}} \rfloor)^2$ where $x > 1$ yields

$$P(C_1 > xn^{2/3}) \leq P(C_1 > K) \leq \frac{3n}{(\lfloor \sqrt{xn^{2/3}} \rfloor)^3} \leq \frac{6}{x^{3/2}},$$

which completes the proof of Theorem 1.8.1.

1.8.2 Aldous' Theorem for Critical Regime Cluster Sizes

In this section we continue to use the recursion in (1.8.1) but we combine it with some less elementary probability. In our first use of this recursion here, we will stop at $\tau = \inf\{t : A_t = \emptyset\}$, but to make sure we find a large component we will start with $|A_0|$ large. The ideas here are from Martin-Löf (1998), but we carry out the details somewhat differently.

As in our study of the giant component in Sect. 1.5.1, we will speed up time and rescale the sizes of our sets to get a limit. To see what to guess, note that the combinatorial calculations suggest that the largest components are of order $n^{2/3}$. Since $R_t = t$ and R_τ is the size of the clusters containing A_0, we will scale time by $n^{2/3}$. When $\lambda = 1 + \theta n^{-1/3}$, $|A_t|$ will be almost a mean zero random walk. In this case $|A_t| - |A_0|$ will be $O(t^{1/2})$ so we will scale the number by $n^{1/3}$.

Having decided on the scaling, we compute the infinitesimal mean and variance. Let $a_t = |A_t|$, $\Delta a_t = |A_{t+1}| - |A_t|$, and note that $u_t = |U_t| = n - t - a_t$. Since the number of new vertices is binomial$(u_t, \lambda/n)$ with $\lambda = 1 + \theta n^{-1/3}$ we have

$$\begin{aligned} E(\Delta a_t | \mathcal{F}_t) &= -1 + (n - t - a_t)(1 + \theta n^{-1/3})/n \\ &= -\frac{t + a_t}{n} + (\theta n^{-1/3})(1 - (t + a_t)/n), \\ \text{var}\,(\Delta a_t | \mathcal{F}_t) &= (n - t - a_t)\frac{1 + \theta n^{-1/3}}{n}\left(1 - \frac{1 + \theta n^{-1/3}}{n}\right). \end{aligned}$$

Speeding up time by $n^{2/3}$, dividing by $n^{1/3}$, and using $a_{[sn^{2/3}]} = O(n^{1/3})$,

$$\begin{aligned} E\left(\frac{\Delta a_{[sn^{2/3}]}}{n^{1/3}} \middle| \mathcal{F}_{[sn^{2/3}]}\right) &= \frac{-[sn^{2/3}] - a_{[sn^{2/3}]}}{n \cdot n^{1/3}} + \frac{\theta n^{-1/3}(1 + \theta n^{-1/3})}{n^{1/3}} \\ &= -sn^{-2/3} + O(n^{-1}) + \theta n^{-2/3} + O(n^{-1}). \end{aligned}$$

The variance is much easier

$$\text{var}\left(\frac{\Delta a_{[sn^{2/3}]}}{n^{1/3}} \middle| \mathcal{F}_{[sn^{2/3}]}\right) \sim \frac{1}{n^{2/3}}.$$

Letting $n \to \infty$ we see that $a_{[sn^{2/3}]}/n^{1/3}$ converges in distribution to the solution of

$$dB_s^\theta = (-s + \theta)\,ds + dB_s,$$

which is simply $B_s^\theta = B_s + \theta s - s^2/2$ (run until the first time it hits zero).

The calculation above leads to a remarkable result of Aldous (1997), stated here as two theorems, that gives the joint distribution of the sizes of the large clusters divided by $n^{2/3}$. Consider now the version of the exploration process in which we choose $i_t \in U_t$ when $A_t = \emptyset$. This adds one to $|A_t|$ each time it hits zero, which has the effect of creating a reflecting barrier at 0:

$$W_t^\theta = B_t^\theta - \min_{0 \le s \le t} B_s^\theta .$$

We say that (u, v) is an excursion interval of W_t^θ if $W_u^\theta = W_v^\theta = 0$ but $W_t^\theta \neq 0$ for $t \in (u, v)$. During excursions $|A_t|$ does not hit 0. Since we expose one vertex at a time, the lengths of the excursion intervals represent cluster sizes in the random graph.

Theorem 1.8.4 *Let $K_1^n \ge K_2^n \ge \cdots$ be the ordered component sizes of Erdős-Rényi$(n, (1 + \theta n^{-1/3})/n)$. Then as $n \to \infty$, $\{n^{-2/3} K_j^n : j \ge 1\}$ converges in distribution to $\{L_j : j \ge 1\}$ where $L_1 > L_2 > L_3 > \cdots$ are the ordered lengths of excursion intervals in $\{W_s^\theta : s \ge 0\}$.*

By working harder in this framework it is possible to prove a result about the "topology" of the clusters. A cluster of size k must have at least $k - 1$ edges. Changing from the notation used in the previous section, we define the number of edges in excess of this number to be the **surplus** of the cluster, since that is the number of "collisions" needed to create such a component. Let $N^\theta(t)$ be a counting process with intensity W_t^θ. That is N^θ has jumps of size 1 and

$$N^\theta - \int_0^t W_u^\theta \, du \quad \text{is a martingale.}$$

The arrivals in this point process are called marks. Intuitively connections between vertices in the cluster are produced at a rate proportional to the size of the active set.

Theorem 1.8.5 *Let $\sigma_1^n \ge \sigma_2^n \ge \cdots$ be the surpluses of the ordered component sizes defined in Theorem 1.8.4. Then as $n \to \infty$,*

$$\{(n^{-2/3} K_j^n, \sigma_j^n) : j \ge 1\}$$

converges in distribution to $\{(|\gamma_j|, \mu(\gamma_j)) : j \ge 1\}$ where the γ_j are the excursion intervals listed in order of decreasing length and $\mu(\gamma_j)$ is the number of marks in γ_j.

Much is known about the number of complex components. See Janson (1993), Janson et al. (1993), Luczak et al. (1994), and the (Bollobás 2001) second edition of Bollobás' book on *Random Graphs.*

1.8.3 Multiplicative Coalescent

While Theorem 1.8.5 is a nice limit theorem, the truly remarkable part of the multiplicative contribution was to note that if we divide the sizes of clusters in Erdős-Rényi$(n, (1+tn^{-1/3})/n)$ by $n^{2/3}$ to create a process indexed by $-\infty < t < \infty$ then there is a simple and intuitive description of the limit. Each pair of clusters of sizes (x, y) merges at rate xy to a cluster of size $x+y$. To see why this is true consider two clusters of sizes $xn^{2/3}$ and $yn^{2/3}$. In a short interval of time $(t, t+h)$, an edge is added between two vertices with probability $hn^{-4/3}$, so the probability of making a connection between clusters of size x and y in time h is $\approx xyh$.

To have an honest Markov process, we need a state space. Aldous chose $\ell^2_\searrow$, the collection of decreasing sequences $x_1 \geq x_2 \geq x_3 \geq \cdots$ with $\sum_k x_k^2 < \infty$. The other thing that should be noticed is that the time interval is $-\infty < t < \infty$, so there is no initial distribution. In Aldous' original paper he solved this problem by showing that there was only one "standard multiplicative coalescent" that had the one-dimensional distributions consistent with Theorem 1.8.4. Aldous and Limic (1998) later characterized all processes on $-\infty < t < \infty$ in which each pair of clusters of sizes (x, y) merges at rate xy to a cluster of size $x+y$. (In terms of the theory of Markov processes one is finding all of the entrance laws.) In addition to the constant process $(x_1 = v > 0,\ x_i = 0,\ i \geq 2)$ there are some nonstandard ones, which are irrelevant for the following application.

Theorem 1.8.6 *Let $K_1^n(t) \geq K_2^n(t) \geq \cdots$ be the ordered component sizes of Erdős-Rényi$(n, (1+tn^{-1/3})/n)$. As $n \to \infty$, $\{K_j^n(t)/n^{2/3} : j \geq 1\}$, $-\infty < t < \infty$ converges in distribution to the standard multiplicative coalescent.*

The convergence of rescaled large components to the multiplicative coalescent provides a nice intuitive process of the growth of clusters in the critical regime.

1.9 Critical Exponents

Let $p(s, \mu)$ be the limit as $n \to \infty$ of the probability a randomly chosen vertex in an Erdős-Rényi graph with mean degree μ belongs to a cluster of size s. The critical exponents for the Erdős-Rényi random graph are defined much like they are for ordinary percolation on $\mathbb{Z}^d$. See e.g., page 236 in Grimmett (1999) book.

- The percolation probability $\theta(\mu) = 1 - \sum_{s=1}^{\infty} p(s, \mu)$ has

$$\theta \approx (\mu - \mu_c)^\beta \qquad \text{as } \mu \downarrow \mu_c. \tag{1.9.1}$$

In the physics literature the meaning of $\approx$ is not precisely defined. It could be something as weak as

$$\frac{\log \theta}{\log(\mu - \mu_c)} \to \beta \qquad \text{as } \mu \downarrow \mu_c.$$

In order to derive relations between exponents, we will suppose $\theta \sim C(\mu - \mu_c)^\beta$, where $a(t) \sim b(t)$ means $a(t)/b(t) \to 1$. This stronger result holds for Erdős-Rényi graphs, but even the weaker definition has not been shown for ordinary percolation in $d > 2$.

- Let $\mathcal{C}_x$ be the cluster containing x. To have the mean cluster size finite for $\mu \neq \mu_c$, we exclude the giant component

$$\chi(\mu) = E(|\mathcal{C}_x|; |\mathcal{C}_x| < \infty) \approx |\mu - \mu_c|^{-\gamma}. \tag{1.9.2}$$

It is not clear (to me at least) why the exponent should be the same when $\mu \uparrow \mu_c$ and $\mu \downarrow \mu_c$, but it is true and the constants are the same as well, see (1.9.8).

- We let $\chi_k = E(|\mathcal{C}_\rho|^k; |\mathcal{C}_x| < \infty)$ for integers $k \geq 2$,

$$\chi_k(\mu) \approx |\mu - \mu_c|^{-\Gamma(k)} \tag{1.9.3}$$

and let $\Delta_k = \Gamma(k) - \Gamma(k-1)$. As in the case of ordinary percolation it is expected that $\Delta_k = \Delta$ is independent of $k \geq 2$. These numbers will turn out to be different from $\Delta_1 = \gamma$.

- There is an exponent for the decay of the cluster size at criticality

$$p(s, \mu_c) \approx s^{1-\tau}. \tag{1.9.4}$$

In the first three cases the Erdős-Rényi Greek letters are the same as those for percolation. In this case the percolation exponent is

$$P_{p_c}(|\mathcal{C}| \geq n) \approx n^{-1/\delta}.$$

It would be sensible to use $p(s, \mu_c) \approx s^{-1-1/\delta_r}$ for the Erdős-Rényi case but to avoid having to rewrite Sect. 1.9.4, we will stick with definition given above and trust that the reader will not be confused by having two meanings for τ.

- Finally there is another exponent for the behavior near criticality

$$p(s, \mu) \approx s^{1-\tau} f(s\delta^{1/\sigma}) \quad \text{where} \quad \delta = \mu - \mu_c, \tag{1.9.5}$$

although it is probably more accurate to call this a **scaling relationship**, see Sect. 1.9.4.

***d*-dimensional percolation.** The most significant difference between the critical exponents for ordinary percolation and those for the Erdős-Rényi random graphs is that the Erdős-Rényi graph is not a spatial model, so we do not have a **correlation length** $\xi(p)$ that gives the length (and height and width) of a typical finite cluster. This quantity is often defined in terms of the exponential decay of probability 0 and x are in the same finite cluster

$$\tau^f(0,x) = P_p(0 \leftrightarrow x, |\mathcal{C}_0| < \infty).$$

For example, if e_1 is the first unit vector

$$\xi(p) = \lim_{n\to\infty} -\frac{1}{n} \log \tau^f(0, ne_1).$$

A simpler approach taken by Kesten (1987) is to define the correlation length by

$$\xi(p) = \left(\frac{1}{\chi_1(p)} \sum_y |y|^2 P(0 \to y; |\mathcal{C}_0| < \infty) \right)^{1/2}$$

and the critical exponent ν by $\xi(p) \approx |p - p_c|^{-\nu}$.

Finally on a d-dimensional graph we have another exponent η called the **anomalous dimension**

$$P_{cr}(0 \to x) \approx |x|^{2-d-\eta}.$$

Note that when $\eta = 0$ this has the behavior of the Green's function of Brownian motion. In this case summing give

$$\sum_{x:|x|\le R} P_{cr}(0 \to x) \approx \int_1^R r^{2-d} \cdot r^{d-1}\, dr \sim CR^2.$$

Mean-field theory makes the connection between the Erdős-Rényi critical exponents and those for percolation in high dimensions. In the case of the Ising model this term refers to a version in which each spin interacts equally with all the others. If one makes this modification to percolation then the Erdős-Rényi model results. Physicists tell us that above the upper critical dimension, which is $d_c = 6$ for ordinary percolation, the critical exponents take on their mean-field values:

$$\beta = 1, \qquad \gamma = 1, \qquad \Delta = 2.$$

To compute the others we need a spatial object which is usually taken to be percolation on trees. This leads to

$$\nu = 1/2 \qquad \eta = 0.$$

Thanks to work of Aizenman and Newman (1984), Barsky and Aizenman (1991), and Hara and Slade (1989, 1994), etc. there are a number of rigorous results confirming this picture. See Sect. 10.3 of Grimmett (1999) for more details. The key words are tree graph inequalities and the lace expansion.

1.9.1 Survival Probability, $\beta = 1$

Let G be Erdős-Rényi$(n, \mu/n)$ with $\mu > 1$. Let x be a randomly chosen $x \in G$ and let Z_m be the number of vertices at distance m from x. When the growing cluster containing x has size $\le n^{1/2-\epsilon}$ the cluster is whp a tree (see Theorem 1.2.3), so Z_m is a branching process in which each individual in generation m has a Poisson(μ) number of offspring. In Sect. 1.3 we showed that if $\mu > 1$ then there is a constant $C(\mu)$ so that a vertex is in the giant component if and only if it is in a cluster of size $\ge C(\mu)\log n$, so the probability of this event can be computed using the branching process.

Consider a branching process with offspring distribution r_k with mean $\mu > 1$ and finite second moment. Let $\phi(z) = \sum_{k=0}^{\infty} r_k z^z$ be the generating function. If r_K is Poisson(μ) then

$$\phi(z) = \sum_{k=0}^{\infty} e^{-\mu}\frac{\mu^k|}{k!}z^k = \exp(-\mu(1-z)). \tag{1.9.6}$$

Let ρ be the probability the system dies out. Breaking things down according to the number of children in the first generation

$$\rho = \sum_{k=0}^{\infty} r_k \rho^k = \phi(\rho)$$

$\phi(1) = 1$ is a trivial solution. In Theorem 1.1.4 we showed ρ is the unique solution of $\phi(\rho) = \rho$ in $[0, 1)$. If μ is close to 1 then ρ will be close to 1. To find an approximation to ρ in this case we note that

$$\phi(1) = 1, \qquad \phi'(1) = \sum_{k=0}^{\infty} k r_k = \mu, \qquad \phi''(1) = \sum_{k=0}^{\infty} k(k-1)r_k = \mu_2.$$

If x is close to 1 then expanding ϕ in power series around 1 and ignoring the error gives

$$\phi(1-x) = 1 - \mu x + \mu_2 x^2/2,$$

so for a fixed point at $(1-x)$ we want

$$(\mu - 1)x = \mu_2 x^2/2,$$

or $x = 2(\mu - 1)/\mu_2$. If we let $\theta(\mu) = P(|\mathcal{C}_x| = \infty)$, which is the same as the fraction of vertices in the giant component

$$\theta(\mu) \sim \frac{2}{\mu_2}(\mu - 1), \tag{1.9.7}$$

then the critical exponent $\beta = 1$. Note that this holds whenever the offspring distribution has finite variance. If the distribution is a power law with infinite variance then the power changes. See Sect. 2.7 for results for the configuration model with a power-law degree distribution.

1.9.2 Mean Cluster Size, $\gamma = 1$

Let $\mathcal{C}_x$ be the cluster containing x. Taking the branching process viewpoint, if $\mu < 1$

$$E|\mathcal{C}_x| = \sum_{n=0}^{\infty} EZ_m = \sum_{m=0}^{\infty} \mu^m = \frac{1}{1-\mu}, \tag{1.9.8}$$

so $\gamma = 1$. In the supercritical regime we consider

$$E(|\mathcal{C}_x|; |\mathcal{C}_x| < \infty).$$

The cluster size is the same as the total progeny in a supercritical branching process conditioned to die out. In Sect. 1.1, we proved the following result:

A supercritical branching process conditioned to become extinct is a subcritical branching process. If the original offspring distribution is Poisson(μ) with $\mu > 1$ then the conditioned one is Poisson($\mu\rho$), where ρ is the extinction probability.

Using the result for the subcritical case in (1.9.8)

$$E(|\mathcal{C}_x| \,|\, |\mathcal{C}_x| < \infty) = \frac{1}{1-\mu\rho}.$$

Since $\mu_2 = \mu^2$ for Poisson, (1.9.7) shows that if μ is close to 1

$$\rho \approx 1 - \frac{2(\mu-1)}{\mu^2},$$

so we have

$$1 - \mu\rho = 1 - \mu + \frac{2(\mu-1)}{\mu} = \frac{2(\mu-1) - \mu(\mu-1)}{\mu}.$$

From this we see that

$$E(|\mathcal{C}_x| \mid |\mathcal{C}_x| < \infty) = \frac{1}{1-\mu\rho} \sim \frac{1}{(1-\mu)} \tag{1.9.9}$$

thus the asymptotic behavior of the mean cluster size as $\mu \downarrow 1$ is the same as the behavior as $\mu \uparrow 1$, not just the exponent $\gamma = 1$ but also the constant $C = 1$.

1.9.3 Higher Moments, Δ = 2

To compute higher moments of the cluster size, it is convenient to convert the branching process into a random walk in which only one particle is allowed to give birth at each time. If S_m is the number of living particles after m births then as long as $S_m > 0$ and $m = o(n)$

$$S_{m+1} \approx S_m - 1 + \text{Poisson}(\mu).$$

When the number of living particles hits 0 it stays 0. To have the process defined for all time let $\xi_1, \xi_2, \ldots$ be i.i.d. $-1 + \text{Poisson}(\mu)$ and $S_{m+1} = S_m + \xi_{m+1}$.

We will use martingales to compute moments of the cluster size. To do this rigorously we need bounds on the tail of the distribution. To obtain the necessary bounds, we begin by computing the moment generating function of ξ_i

$$\psi(\theta) = e^{-\theta} \sum_{m=0}^{\infty} e^{-\mu} \frac{\mu^m}{m!} e^{\theta m} = e^{-\theta} \exp(\mu(e^{\theta} - 1)). \tag{1.9.10}$$

$\bar{M}_t = \exp(\theta S_t)/\psi(\theta)^t$ is a nonnegative martingale, so using the optional stopping theorem for the nonnegative supermartingales, Theorem 4.8.4 in PTE5,

$$\bar{M}_0 = e^{\theta} \geq E(\psi(\theta)^{-\tau}). \tag{1.9.11}$$

The mean $\psi'(0) = E\xi_i = 1 - \mu$ so if $\mu < 1$ then $\psi(\theta) < 1$ when $\theta > 0$ is small. To prepare to optimize the bound we will derive later, we note that the derivative of $\log \psi(\theta)$ (Fig. 1.7).

$$\frac{d}{d\theta}(-\theta + \mu(e^{\theta} - 1)) = -1 + \mu e^{\theta} = 0$$

when $\theta_1 = -\log \mu$. At this point $e^{\theta_1} = 1/\mu$ and

$$\psi(\theta_1) = \exp(\log(\mu) + 1 - \mu) \equiv e^{-\alpha} < 1.$$

Since $\psi(\theta_1^{-1}) = e^{\alpha} > 1$, using Chebyshev's inequality with (1.9.11)

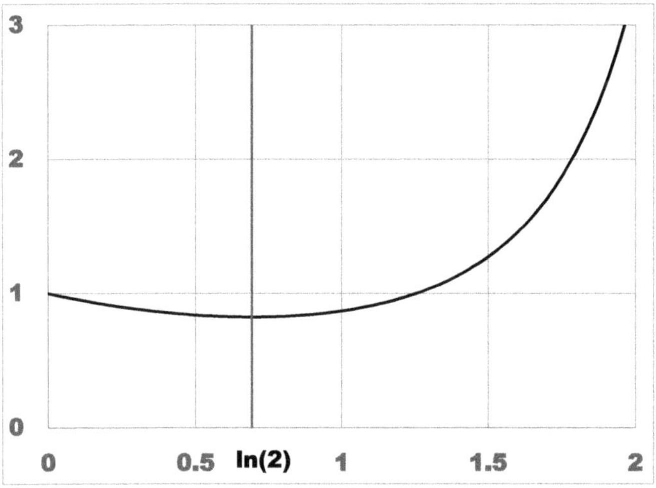

Fig. 1.7 Graph of $\psi(\theta)$ when $\mu = 1/2$, showing a minimum at $\theta = \ln(2)$

$$e^{\alpha m} P(\tau \geq m) = \psi(\theta_1)^{-m} P(\tau \geq m) \leq E(\psi(\theta_1)^{-\tau}) \leq e^{\theta_1}.$$

One particle dies on each time step so $1 + \xi_1 + \cdots + \xi_\tau = \tau$ and we have

$$P\left(\max_{0 \leq t \leq \tau} S_t \geq m\right) \leq e^{-m\alpha}/\mu. \tag{1.9.12}$$

Computing moments of the cluster size. Recall that the random walk of interest is

$$S_t = S_0 + \xi_1 + \cdots + \xi_t,$$

where ξ_i have distribution $-1 + \text{Poisson}(\mu)$, so we will let $\nu = E\xi_i = \mu - 1$. The simplest martingale is the **linear martingale**

$$M_n^{(1)} = S_n - n\nu.$$

Using the domination in (1.9.12) we can conclude that

$$1 = S_0 = E(S_\tau - (\mu - 1)\tau) = (1 - \mu)E\tau,$$

so the expected cluster size is $E\tau = 1/(1 - \mu)$, as already computed in (1.9.8)

If T_m is a random walk in which steps have mean 0 and variance σ^2 then $T_m - m\sigma^2$ is a martingale, called the **quadratic martingale**. Applying this result to $T_m = S_m - m\nu$ and recalling that $-1 + \text{Poisson}(\mu)$ has variance $\nu_2 = \mu$ we see that

$$M_m^{(2)} = (S_m - m\nu)^2 - \mu m$$

is a martingale. Using the domination in (1.9.12) we can stop at time τ and conclude

$$1 = E(S_\tau - \tau\nu)^2 - \mu E\tau = (\mu - 1)^2 E\tau^2 - \frac{\mu}{1-\mu}.$$

Rearranging we have

$$E\tau^2 = \frac{\mu}{(1-\mu)^3}. \tag{1.9.13}$$

Since $E\tau = 1/(1-\mu)$ the gap exponent $\Delta_2 = 2$.

To compute all the moments we require an infinite sequence of martingales, and hence need to take a more systematic approach to finding them. The **cumulant generating function** is

$$\kappa(\theta) = \log E \exp(\theta\xi_i) = -\theta + \mu(e^\theta - 1). \tag{1.9.14}$$

$\bar{M}_m = \exp(uS_m)/\phi(u)^m = \exp(uS_m - m\kappa(u))$ is a martingale, so differentiating we see that if

$$g_k(u; m, x) = \frac{\partial^k}{\partial u^k} \exp(ux - m\kappa(u))$$

and $f_k(m, x) = g_k(0; m, x)$ then $f_k(m, S_m)$ is a martingale. For more details see page 80 of Neveu (1975). To compute the f_k note that

$$g_k(u; m, x) = h_k(u; m, x) \exp(ux - m\kappa(u)),$$

where the h_k satisfy the recursion

$$h_k(u; m, x) = h_{k-1}(u; m, x) \cdot (x - m\kappa'(u)) + (\partial/\partial u) h_{k-1}(u; m, x). \tag{1.9.15}$$

From this we get

$$\begin{aligned}
h_1(u; m, x) &= x - m\kappa'(u), \\
h_2(u; m, x) &= (x - m\kappa'(u))^2 - m\kappa''(u), \\
h_3(u; m, x) &= (x - m\kappa'(u))^3 - 3m\kappa''(u)(x - m\kappa'(u)) - m\kappa'''(u).
\end{aligned}$$

The derivatives of the cumulant generating function $\kappa(u) = -u + \mu(e^u - 1)$ are $\kappa'(u) = -1 + \mu e^u$ and $\kappa'^{(k)}(u) = \mu e^u$ for $k \geq 2$. Writing $\nu = \mu - 1$ we have

$$\begin{aligned}
f_1(m, x) &= x - m\nu, \\
f_2(m, x) &= (x - m\nu)^2 - m\mu, \\
f_3(m, x) &= (x - m\nu)^3 - 3m\mu(x - m\nu) - m\mu.
\end{aligned}$$

We have already seen the martingales that result from the first two functions. The third one is given by

$$M_m^{(3)} = (S_m - m\nu)^3 - 3\,m\mu(S_m - m\nu) - m\mu.$$

Using the domination in (1.9.12) we can stop at time τ and conclude

$$\begin{aligned} 1 &= E(S_\tau - \tau\nu)^3 - 3\mu E[\tau(S_\tau - \tau\nu)] - \mu E\tau \\ &= -(\mu-1)^3 E\tau^3 + 3\mu(\mu-1)E\tau^2 - \mu E\tau. \end{aligned}$$

Since $E\tau = O(1/(1-\mu))$ while $E\tau^2 = O(1/(1-\mu)^3)$, the 1 and the $\mu E\tau$ are of much smaller order than the other terms in the equation, so we have

$$\frac{E\tau^3}{E\tau^2} \sim \frac{3\mu}{(\mu-1)^2} \qquad \text{and } \Delta_3 = 2.$$

From the last computation we see that we don't need all of the terms, just the largest ones. Using the recursion in (1.9.15) when $k = 4$ and letting $y = x - n\kappa'(u)$ to reduce the size of the formula

$$h_4(u; n, x) = y^4 - 3n\kappa''(y)y^2 + \cdots - 3n\kappa''(y)y^2 \cdots$$

so repeating the last argument $E\tau^3/E\tau^2 \sim C(\mu-1)^{-2}$ and $\Delta_4 = 2$. The pattern persists for higher moments, proving the result for Poisson branching processes. The method requires

($\star$) the moment generating function is finite for small θ,

but the simple formula for the cumulants is not important, so it will hold for all offspring distributions satisfying ($\star$).

1.9.4 Scaling Theory

We have introduced five critical exponents for the phase transition in Erdős-Rényi random graphs: $\beta, \gamma, \Delta, \tau$ and σ. The values of these exponents are not independent. In this section we will show that they satisfy three equalities, so if σ and τ are known then the other three can be computed. See (1.9.18), (1.9.19), and (1.9.20).

Formula (1.7.9) implies that when $\mu = 1$ the probability of a cluster of size k is

$$\sim \frac{nk^{-5/2}}{\sqrt{2\pi}} e^{-k^3/6n^2}.$$

This says that the largest components are of size $n^{2/3}$. The critical exponent τ defined in (1.9.4)

$$P(s, t_c) \sim f(0)s^{1-\tau}$$

has the value $\tau = 5/2$

To get a result for μ close to 1, we use (1.7.8)

$$P(s, \mu) \approx \frac{s^{1-5/2}}{\sqrt{2\pi}} \cdot \exp(-s(\mu - 1)^2/2). \tag{1.9.16}$$

Thus (1.9.5) holds with $\tau = 5/2, \sigma = 1/2$ and

$$f(x) = e^{-x/2}. \tag{1.9.17}$$

It has long been known for percolation (see Stauffer (1979)) that (1.9.5) implies

$$\beta = (\tau - 2)/\sigma \tag{1.9.18}$$

$$\gamma = (3 - \tau)/\sigma \tag{1.9.19}$$

$$\Delta = 1/\sigma \tag{1.9.20}$$

Since (1.9.5) gives the asymptotics for the cluster size distribution as $\mu \to \mu_c$ it should not be surprising that we can compute the asymptotic behavior of the moments and of the percolation probability. Kesten (1987) has proved a number of rigorous results concerning scaling relations concentrating on the two-dimensional case.

Lemma 1.9.1 *If $2 < \tau < 3$ and f is bounded, and is Lipschitz continuous at 0, then*

$$\beta = (\tau - 2)/\sigma.$$

Proof Using (1.9.5) and replacing sum by integration.

$$\theta(t) \approx \int_1^\infty s^{1-\tau}[f(0) - f(s\delta^{1/\sigma})]\,ds.$$

Changing variables $s = x\delta^{-1/\sigma}$, $ds = \delta^{-1/\sigma}dx$ the above

$$= \delta^{(\tau-2)/\sigma} \int_{\delta^{1/\sigma}}^\infty x^{1-\tau}[f(0) - f(x)]\,dx.$$

Since f is bounded and $\tau > 2$ the integral over $[1, \infty)$ is finite. Since f is Lipschitz continuous the integrand is $\leq Cx^{2-\tau}$ near 0. Since $\gamma < 3$, the integral over [0, 1] is finite, and it follows that $\theta(t) \sim C\delta^{(\tau-2)/\sigma}$. □

The next result establishes (1.9.19) and (1.9.20).

Lemma 1.9.2 *If* $2 < \gamma < 3$ *and* $\int_1^\infty x^m f(x)\,dx < \infty$ *for all m then for all* $r \geq 1$

$$\Gamma(r) = (r + 2 - \tau)/\sigma.$$

It follows that for all integers $k \geq 2$,

$$\Delta_k = \Gamma(k) - \Gamma(k-1) = 1/\sigma.$$

Proof

$$E|\mathcal{C}_x|^r = \int_1^\infty s^{r+1-\tau} f(s\delta^{1/\sigma})\,ds. \tag{1.9.21}$$

Changing variables $s = x\delta^{-1/\sigma}$, $ds = \delta^{-1/\sigma}dx$ the above

$$= \delta^{(\tau-2-r)/\sigma} \int_{\delta^{1/\sigma}}^\infty x^{\rho+1-\tau} f(x)\,dx.$$

The assumption $\int_1^\infty x^m f(x)\,dx < \infty$ for all m implies that the integral over $[1, \infty)$ is finite. Since $\tau < 3$ and $\rho \geq 1$ the integral over $[0, 1]$ is finite □

References

Aizenman M, Newman CM (1984) Tree graph inequalities and critical behavior in percolation models. J Stat Phys 36:107–143

Ajtai M, Komlós J, Szemerédi E (1981) The longest path in a random graph. Combinatorica 1:1–17

Aldous D (1997) Brownian excursions, critical random graphs and the multiplicative coalescent. Ann Probab 25:812–854

Aldous D, Limic V (1998) The entrance boundary of the multiplicative coalescent. Electro J Probab 3(paper 3):59

Barraez D, Boucheron S, Fernandez de la Vega W (2000) On the fluctuations of the giant component. Comb Probab Comput 9:287–304

Barsky D, Aizenman M (1991) Percolation critical exponents under the triangle condition. Ann Probab 19:1520–1536

Bollobás B (1984) The evolution of random graphs. Trans Amer Math Soc 286:257–274

Bollobás B (2001) Random graphs, 2nd edn. Press, Cambridge U

Borel E (1942) Sur l'emploi du théorème de Bernoulli pour faciliter le calcul d'une infinité de coefficients. application au probléme de l'attente á un guichet. C R Acad Sci 214:452–456

Chung F, Lu L (2001) The diameter of sparse random graphs. Adv Appl Math 26:257–279

Ding J, Lubetzky E, Peres Y (2014) Anatomy of a giant component: the strictly supercritical regime. European J Comb 35:155–168

Enriquez N, Faraud G, Ménard L (2017) Limiting shape of depth first search on an Erdős-Rényi graph. Rand Struct Alg 56:501–516

Enriquez N, Faraud G, Ménard L, Noiry N (2022) Depth first exploration of a configuration model. Electron J Probab 27(article 53):27
Ethier SN, Kurtz TG (1986) Markov processes: characterization and convergence. Wiley Series in Probability and Statistics
Grimmett G (1999) Percolation, 2nd edn. Springer
Hara T, Slade G (1989) Mean-field critical behavior for percolation in high dimensions. In: Proceedings of the IXth international congress of mathematics physics. Adam Hilger, Boston, pp 450–453
Hara T, Slade G (1994) Mean-field critical behavior and the lace expansion. In: Grimmett GR (ed) Probability and phase transition. Kluwer, Dordrecht
Janson S (1993) Multicyclic components in a random graph process. Rand Struct Alg 4:71–84
Janson S, Knuth DE, Luczak T, Pittel B (1993) The birth of the giant component. Rand Struct Alg 4:233–358
Kendall DG (1956) Deterministic and stochastic epidemics in closed populations. In: Proceedings of the third Berkeley symposium on mathematical statistics and probability, vol 4, pp 149–165
Kesten H (1987) Scaling relations for 2D-percolation Commun. Math Phys 109:109–156
Kesten H, Stigum B (1966) A limit theorem for multidimensional Galton-Watson processes. Ann Math Stat 37:1211–1223
Krivelevich M, Sudakov B (2013) The phase transition in random graphs: a simple proof. Rand Struct Alg 43:131–138
Luczak T (1990) Component behavior near the critical point of a random graph process. Rand Struct Alg 1:287–310
Luczak T, Pittel B, Weirman JC (1994) The structure of a random graph at the point of phase transition. Trans Amer Math Soc 341:721–748
Lyons R, Pemantle R, Peres Y (1995) Conceptual proofs of $L \log L$ criteria for mean behavior of branching processes. Ann Probab 23:1125–1138
Martin-Löf A (1986) Symmetric sampling procedures, general epidemic processes and their threshold limit theorems. J Appl Prob 23:265–282
Martin-Löf A (1998) The final size of a nearly critical epidemic, and the first passage time of a Wiener process to a parabolic barrier. J Appl Prob 35:671–682
Nachmias A, Peres Y (2010) The critical random graph with martingales. Israel J Math 176:29–41
Neveu J (1975) Discrete parameter martingales. North Holland
Pittel B (1990) On tree census and the giant component in sparse random graphs. Rand Struct Alg 1:311–342
Riordan O, Wormald N (2010) The diameter of sparse random graphs. Comb Probab Comput 19:835–926
Robinson RW, Wormald NC (1992) Almost all cubic graphs are Hamiltonian. Rand Struct Alg 3:117–125
Stauffer D (1979) Scaling theory for percolation clusters. Phys Rep 54:1–79
Tanner JC (1961) A derivation of the Borel distribution. Biometrika 48:222–224

Chapter 2
General Degree Distributions

The motivation for the study of graphs with a pre-specified degree distribution comes from the fact that in an Erdős-Rényi random graph, vertices have degrees that have asymptotically a Poisson distribution. However, in social and communication networks, the distribution of degrees is much different from the Poisson, and in many cases has a power law form, i.e., the fraction of vertices of degree k,

$$p_k \sim Ck^{-\gamma} \quad \text{as} \quad k \to \infty.$$

Values in the range $2 < \gamma < 3$ are common. These networks exhibit a "scale-free" property because their structure appears similar at different scales.

In Sect. 2.1 we will introduce the **configuration model**, which is a simple algorithm for creating a graph with a given degree distribution p_k. Generalizing our approach to Erdős-Rényi graphs, if we look at the number of vertices Z_n at distance n from x then we get a branching process, but this time the offspring distributions after the first one are different. Z_1 has distribution p_k, but the number of children in subsequent generations has the **size-biased distribution**

$$q_{k-1} = kp_k/\mu \quad \text{where} \mu = \sum_k kp_k.$$

From this observation, we see that there is a giant component if (and only if) the mean of q_k, which we call ν, is > 1. The fraction of vertices in the giant component can be computed almost as before but we need to use the generating function G_0 for p_k and G_1 for q_k. See Theorem 2.1.3

In Sect. 2.2 we introduce the **limiting degree distribution approach**. It is more flexible and avoids some of the drawbacks of the configuration model, e.g., if we randomly delete edges from a configuration model graph, then the results is not one. In subcritical Erdős-Rényi random graphs the largest component is $O(\log n)$, but this is no longer true for graphs with power law degree distributions, for the trivial reason

R. Durrett, *Dynamics on Graphs*, Probability Theory and Stochastic Modelling 110,
https://doi.org/10.1007/978-3-032-21908-4_2

that the largest degree of a vertex is the graph is $O(n^{1/(\gamma-1)})$. In Sect. 2.3 we prove a result of Janson (2008), which shows that when the size-biased degree distribution is bounded by $Ck^{1-\gamma}$ the size of the largest component in a graph with n vertices $|\mathcal{C}_1| \leq An^{1/(\gamma-1)}$.

In Sect. 2.4 we will look at the distance between two randomly chosen vertices. When the degree distribution has finite variance the situation is much like the Erdős-Rényi case. Distances are of order $\log_\nu n$ with the fluctuations of order 1. When the distribution is a power law with $2 < \gamma < 3$, the world is very small: distances are

$$\sim (2 \log \log n) \log(1/(\gamma - 2)). \qquad (\star)$$

Also in Sect. 2.4 we introduce the **Chung-Lu model** in which vertex i is assigned weight w_i and the probability of a connection from i to j is $(w_i w_j / \sum_k w_k) \wedge 1$. If the weights are chosen so that the degree distribution has a power law distribution, $p_k \sim Ck^{-\gamma}$, then the distance between two randomly chosen vertices is given by $(\star)$. The proof is much easier since edges are independent, and it gives some information about the structure of the graph: The set of vertices with degree $\geq n^{1/\sqrt{\log \log n}}$ are a connected set.

In Sect. 2.5 we tackle the subject of **first passage percolation** on random graphs, which could almost be a book all by itself. The topic occupies 71 pages in van der Hofstad's notes for the 2017 St. Flour Summer School, while the four main papers we discuss have a total of 270 pages. Despite the fact that the graphs we consider are locally tree-like, which limits the ability to optimize the passage time, there are some surprises. When the degree distribution is a power law with $2 < \gamma < 3$ and passage times are exponential (see Sect. 2.5.2) the shortest path between two vertices has length $O(\log \log n)$, but the shortest passage time which is $O(1)$ occurs on a path of length $\sim \alpha n$. Branching process fans will enjoy the fact that the analysis of passage times for general continuous distributions in Sect. 2.5.3 uses **Bellman-Harris processes** (also called age-dependent branching processes), Malthusian parameters and the stable age distribution. In some papers the very general Crump-Mode-Jagers processes are used.

Section 2.6 considers the analogue of Aldous' result for cluster sizes in the critical case discussed in Sect. 1.8. When the degree distribution has finite third moment the results are almost the same as in the Erdős-Rényi case. However, the analysis of the power-law case with $3 < \gamma < 4$ (which means that the size-biased distribution has $2 < \gamma < 3$) seems like an unfinished symphony.

Finally Sect. 2.7 considers percolation on graphs. This is perhaps the simplest process to consider on graphs since if we delete edges with probability $1 - p$ we have another random graph for which we want the answer to the question: Is there a giant component? There are four cases to consider: finite variance degrees, and power laws with $3 < \gamma < 4$, $\gamma = 3$, and $2 < \gamma < 3$. When $\gamma = 3$ the critical value $p_c = 0$ and the percolation probability goes to 0 like $\exp(-(1 + o(1))/cp)$. We will see this case again in Chap. 3 where it will be called the square root model.

2.1 Configuration Model

Molloy and Reed (1995, 1998) were the first to study the phase transition in the cluster sizes of graphs with specified degree distributions. We will describe their work in the next section. In this one we will describe the configuration model, which is an efficient algorithm for constructing such graphs.

2.1.1 *Examples*

At the turn of this century attention focused on a number of examples. As the reader will see, the case in which $2 < \gamma < 3$ turns out to be very interesting. These are called **scale-free random graphs**.

By the **world wide web**, we mean the collection of web pages and the oriented links between them. Barabási and Albert (1999) found that the in-degree and out-degrees of web pages followed power laws with $\gamma_{\text{in}} = 2.1$, $\gamma_{\text{out}} = 2.7$.

By the **Internet**, we mean the physically connected network of routers that move email and files around the Internet. In 2000 there were about 150,000 routers connected by 200,000 links and a degree distribution that could be fit by a power law with $\gamma = 2.3$.

The **movie actor network**, in which two actors are connected by an edge if they have appeared in a film together, has a power law degree distribution with $\gamma = 2.3$.

Liljeros et al. (2001) analyzed data gathered in a study of **sexual behavior of 4,781 Swedes**, and found that the number of partners per year had $\gamma_{\text{male}} = 3.3$ and $\gamma_{\text{female}} = 3.5$.

The **collaboration graph** in a subject is a graph with an edge connecting two people if they have written a paper together. Barabási et al. (2002) studied papers in mathematics and neuroscience published in 1991–1998. The two databases that they used contained 70,901 papers with 70,975 authors, and 210,750 papers with 209,293 authors, respectively. The fitted power laws had $\gamma_{\text{M}} = 2.4$ and $\gamma_{\text{NS}} = 2.1$.

Newman's (2003) survey paper provides a wealth of information about "Networks in the Real World." His discussion makes it clear that the interest in these networks is not new. Alfred Lotka stated the **Law of Scientific Productivity** in 1926, which states that number of papers written by an individual follows a power law. Derek de Solla Price (1965) wrote in a paper published in Science that both the in-degree and out-degree of the collaboration graph had power laws. Newman's paper has 89 pages and 429 references for readers who want to see networks in a broader context.

2.1.2 Definitions

The first appearance of the **configuration model** came in the construction of **random regular graphs**. See Bollobás (1980, 1985). Suppose that we want a graph with n vertices, each of which has degree r. If nr is even (otherwise the graph is not possible) one partitions $\{1, \dots n\} \times \{1, \dots r\}$ into $rn/2$ pairs to define the edges of the graph, which can be done in

$$\frac{(rn)!}{(rn/2)!2^{m/2}} = (rn-1)(rn-3)\cdots 3 \cdot 1 = (rn-1)!! \quad \text{ways.}$$

To do the construction in general let $d_1, \dots, d_n$ be independent and have $P(d_i = k) = p_k$. Since we want d_i to be the degree of vertices in a graph, we condition on $E_n = \{d_1 + \cdots + d_n \text{ is even}\}$. If $P(E_1) \in (0, 1)$ then $P(E_n) \to 1/2$ as $n \to \infty$ so the conditioning will have little effect on the finite-dimensional distributions. To build the graph we think of d_i half-edges attached to i and then pair the half-edges at random. The picture in Fig. 2.1gives an example with 8 vertices. If you are using this algorithm in a computer program to construct the graph then it is silly to give up and start again if the sum of the degrees is odd. In practice one either deletes an edge chosen at random, or saves work by deleting an edge from vertex n.

As we will see in Sect. 2.3, it is often advantageous to allow loops and multiple edges and work with r-regular multigraphs $G^*(n, r)$ which are constructed by using the procedure described below with $d_i = r$ for $1 \le i \le r$. If we condition on the graph being simple then we get a random choice from the set of random regular graphs $G(n, r)$. For people who want clean graphs, we note that if one assumes

(A) p_k has a finite second moment.

then Theorem 2.1.1 below implies that the probability that the graph is simple has a positive limit. Using the method of moments we can prove.

Theorem 2.1.1 *Let* $\mu = \sum_k kp_k$ *and* $\mu_2 = \sum_k k(k-1)p_k$. *As* $n \to \infty$, *the number of self-loops* χ_0 *and the number of parallel edges* χ_1 *are asymptotically independent Poisson*$(\mu_2/2\mu)$ *and Poisson*$((\mu_2/2\mu)^2)$.

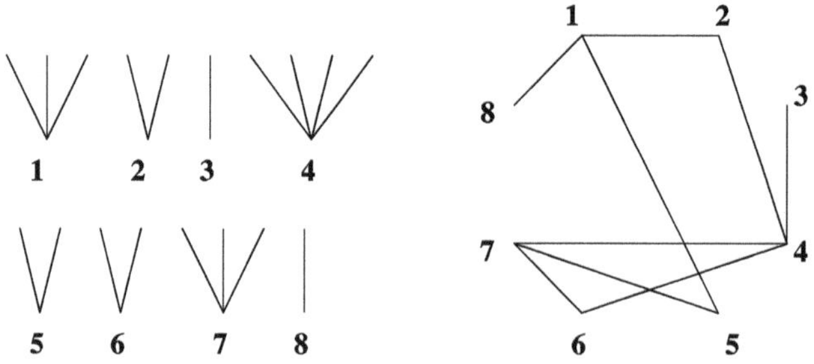

Fig. 2.1 Example of the construction

Proof If $D = d_1 + \cdots + d_n$ then the expected number of self-loops is

$$\sim \sum_i d_i(d_i - 1)/2D \to \mu_2/2\mu$$

and the expected number of pairs of parallel edges is

$$\sim \frac{1}{2}\sum_i \frac{d_i d_i - 1}{2} \sum_{j \neq i} \frac{d_j}{D}\frac{d_j - 1}{D} \to (\mu_2/2\mu)^2.$$

To check the counting in the second formula, suppose that we connect half-edges 3 and 7 from i to half-edges 5 and 2 from j. When we pick the two edges for i, order is not important, but when it comes to j it is. The 1/2 in front comes from the fact that (i, j) and (j, i) are both in the double sum.

This gives the asymptotics for the mean, but in almost the same way we can compute the expected number of ordered k-tuples $E(N(N-1)\cdots(N-k+1) \to \lambda^k$ to conclude that the self-loops and parallel edges have Poisson limits. After this, we can consider product moments to get convergence of the joint distribution to independent Poissons.

2.1.3 *Phase Transition*

As we have seen in our analysis of Erdős-Rényi graphs, the growth of clusters can be approximated in the early stages by a branching process. In this section we will follow Newman et al. (2001), who were playing by the rules of *Physical Review E* (i.e., they didn't have to prove things). If we start with a given vertex x then the number of neighbors (the first generation in the branching process) has distribution p_j. This is not true for the second generation. A first generation vertex with degree k is k times as likely to be chosen as one with degree 1, so the distribution of the number of children of a first generation vertex is for $k \geq 1$

$$q_{k-1} = \frac{kp_k}{\mu} \quad \text{where} \quad \mu = \sum_k kp_k.$$

This is called the **size-biased distribution** since vertices are chosen with probability proportional to their degrees. The $k-1$ on the left-hand side comes from the fact that we used up one edge connecting to the vertex. Note that since we have assumed p has finite second moment, q has finite mean $\nu = \sum_k k(k-1)p_k/\mu$.

The random r-regular graph is a trivial example. $p_r = 1$ and so $q_{r-1} = rp_r/\mu = 1$. For a little more excitement consider

Example 2.1.2 (Poisson distribution) $p_k = e^{-\lambda}\lambda^k/k!$. In this case $\mu = \lambda$, so

$$q_{k-1} = e^{-\lambda}\frac{k\lambda^k}{k!\lambda} = e^{-\lambda}\frac{\lambda^{k-1}}{(k-1)!},$$

and q is again Poisson. Conversely if $p = q$ we have $p_{k-1} = kp_k/\mu$. Iterating gives $p_k = p_0\mu^k/k!$, so p_k is Poisson with mean μ.

Having discovered that the growth of the cluster is a **two-phase branching process** in which the first generation has distribution p, and the second and subsequent generations have distribution q, it is now routine to compute the threshold for the existence of large components and the size of such components. Let $G_0(z) = \sum_k p_k z^k$ and $G_1(z) = \sum_k q_k z^k$ be the two generating functions. These are related:

$$\frac{G_0'(z)}{\mu} = \sum_{k=1}^{\infty}\frac{kp_k}{\mu}z^{k-1} = \sum_{k=1}^{\infty} q_{k-1}z^{k-1} = G_1(z).$$

Let Z_n be the number of vertices in the nth generation. For $n \geq 1$, $EZ_n = \mu\nu^{n-1}$ so when $\nu < 1$

$$E\left(\sum_{n=0}^{\infty} Z_n\right) = 1 + \sum_{n=1}^{\infty}\mu\nu^{n-1} = 1 + \frac{\mu}{1-\nu}, \tag{2.1}$$

and we have the first part of

Theorem 2.1.3 *The condition for the existence of a giant component is $\nu > 1$. In this case the fraction of vertices in the giant component is asymptotically $1 - G_0(\rho_1)$ where ρ_1 is the smallest fixed point of G_1 in $[0, 1]$.*

We will prove this result in a more general setting in Sect. 2.2.2. To explain the second claim: We know that ρ_1 is the extinction probability of the homogeneous branching process with offspring distribution q. Breaking things down according to the number of offspring in the first generation, the probability that the two phase branching process dies out is $\sum_{k=0}^{\infty} p_k\rho_1^k = G_0(\rho_1)$, since each of the k first generation families must die out and they are independent.

Example 2.1.4 (Miller (2011)) Figure 1 in his paper gives an example of the configuration model with 70 vertices and $p_1 = p_3 = 0.5$ There are five clusters of size 2, one of size 4, and the remaining $56/70 = 0.8$ vertices are in the giant component. To see if this is reasonable we note that $\mu = 2$, so the size-biased degree distribution has

$$q_0 = \frac{1p_1}{\mu} = 1/4, \qquad q_2 = \frac{3p_3}{\mu} = 3/4$$

$\nu = 2 \cdot 3/2$. $G_1(z) = 1/4 + (3/4)z^2 = (1-z)(1/4 - 3z/4)$ so $\rho_1 = 1/3$. and $G_0(\rho_1) = (\rho_1 + \rho_1^3)/2$ so $1 - G_0(\rho_1) = 0.814814$.

2.1.4 Cluster Sizes

Our next topic is using generating functions to study the cluster size distribution. This technology is a little "old school" but it allows us to compute the distribution of the size of subcritical and critical, or more generally the non-giant components. We begin by considering the homogeneous branching process with offspring distribution q. Let $H_1(z)$ be the generating function of the total progeny in the homogeneous branching process. If there are k offspring in the first generation then the total progeny $= 1 + Y_1 + \cdots + Y_k$ where the Y_i are independent and have the same distribution as the total progeny. From this it follows that for $z < 1$

$$H_1(z) = z \cdot \sum_{k=0}^{\infty} q_k (H_1(z))^k = zG_1(H_1(z)). \tag{2.2}$$

The condition $z < 1$ guarantees $z^\infty = 0$. Similarly, if $H_0(z)$ is the generating function of the total progeny in the two-phase branching process then

$$H_0(z) = z \cdot \sum_{k=0}^{\infty} p_k (H_1(z))^k = zG_0(H_1(z)). \tag{2.3}$$

We claim that

$$\lim_{z \uparrow 1} H_1(z) = \rho_1.$$

This should be obvious because the total progeny is finite if and only if the branching process dies out. To check this, note that when $z < 1$, $G_1(H_1(z))/H_1(z) = 1/z > 1$ so $H_1(z) < \rho_1$, but $\lim_{z \to 1} H_1(z)$ is a fixed point of G_1 so the limit must be ρ_1.

To find the mean size of finite clusters, we begin with the homogeneous branching process. Differentiating (2.2), setting $z = 1$ and recalling $H_1(1) = \rho_1$, $G_1(\rho_1) = \rho_1$,

$$\begin{aligned} H_1'(z) &= G_1(H_1(z)) + zG_1'(H_1(z))H_1'(z), \\ H_1'(1) &= \rho_1 + G_1'(\rho_1)H_1'(1). \end{aligned}$$

Rearranging we have

$$\frac{H_1'(1)}{H_1(1)} = \frac{1}{1 - G_1'(\rho_1)}. \tag{2.4}$$

When the branching process dies out, $\rho_1 = 1$, $G_1'(1) = \nu$ and $H_1(1) = 1$, and so this says that the mean cluster size is $1/(1 - \nu)$. To check the result for $\nu > 1$, recall that Theorem 1.1.8 implies that if the homogeneous branching process is supercritical and we condition it to die out then the result is a branching process with offspring distribution that has generating function $G_1(\rho_1 z)/\rho_1$. Differentiating and setting $z = 1$, we see that the mean of the offspring distribution is $G_1'(\rho_1)$, so the expected total progeny will be

$$\sum_{n=0}^{\infty} G_1'(\rho_1)^n = \frac{1}{1 - G_1'(\rho_1)}.$$

Turning to the two-phase branching process. Differentiating (2.3) and setting $z = 1$

$$H_0'(z) = G_0(H_1(z)) + zG_0'(H_1(z))H_1'(z),$$
$$H_0'(1) = H_0(1) + G_0'(\rho_1)H_1'(1),$$

since $H_1(1) = \rho_1$ and $G_0(\rho_1) = H_0(1)$ is the extinction probability for the two phase process. When the branching process dies out, $H_0(1) = 1, \rho_1 = 1, G_0'(1) = \mu$, $H_1(1) = 1$, and $H_1'(1) = 1/(1 - \nu)$ so we have

$$H_0'(1) = 1 + \frac{\mu}{1 - \nu},$$

as we have computed in (2.1). To get a result for $\nu > 1$, we divide each side by $H_0(1)$ and use (2.4) to get

$$\frac{H_0'(1)}{H_0(1)} = 1 + \frac{G_0'(\rho_1)H_1'(1)}{H_0(1)} = 1 + \frac{G_0'(\rho_1)H_1(1)}{H_0(1)(1 - G_1'(\rho_1))}. \tag{2.5}$$

To check this, note that in the two-phase branching process the probability $Z_1 = k$ given that the process dies out is $p_k \rho_1^k / G_0(\rho_1)$. (2.4) implies that the mean number of descendants of a first generation individual given the extinction of its family is $1/(1 - G_1'(\rho_1))$. Thus the expected total progeny is

$$1 + \sum_{k=1}^{\infty} \frac{p_k \rho_1^k}{G_0(\rho_1)} \frac{k}{1 - G_1'(\rho_1)} = 1 + \frac{\rho_1 G_0'(\rho_1)}{G_0(\rho_1)} \cdot \frac{1}{1 - G_1'(\rho_1)}.$$

Since $H_1(1) = \rho_1$ and $G_0(\rho_1) = H_0(1)$ this agrees with the answer in (2.5).

Cluster size distribution at the critical value
Taking $w = H_1(z)$ in (2.2) gives

$$z = H_1^{-1}(w) = w/G_1(w).$$

In the critical case $G_1(1) = 1$ and $G_1'(1) = 1$. Taking the derivative of the right-hand side and setting $w = 1$ gives

$$\frac{G_1(w) - G_1'(w)w}{G_1(w)^2} = 0.$$

Differentiating again and setting $w = 1$ gives

$$\frac{[G_1'(w) - G_1''(w)w - G_1'(w)]G_1(w)^2 - 2G_1(w)G_1'(w)[G_1(w) - G_1'(w)w]}{G_1(w)^4}$$
$$= -G_1''(1).$$

If q has a finite second moment then for some $x_\epsilon \in [1-\epsilon, 1]$

$$H_1^{-1}(1-\epsilon) = 1 - \frac{1}{2}G_1''(x_\epsilon)\epsilon^2.$$

$G_1''(x_\epsilon) \to G_1''(1)$ as $\epsilon \to 0$, so as $\delta \to 0$

$$1 - H_1(1-\delta) \sim \sqrt{2\delta/G_1''(1)}.$$

Combining this with (2.3) we have

$$\begin{aligned} 1 - H_0(1-\delta) &= 1 - (1-\delta) + (1-\delta)[G_0(1) - G_0(H_1(1-\delta))] \\ &\sim G_0'(1)\sqrt{2\delta/G_1''(1)}. \end{aligned} \tag{2.6}$$

Let h_k be the sequence with generating function $H(z)$. As we will now explain, the rate of convergence of $H_0(1-\delta)$ to 1 tells us about the rate of convergence of $\sum_{k=K}^{\infty} h_k$ to 0. Tauberian theorems (see e.g., Feller volume II) are the machinery for doing these results rigorously. Here we will be content to argue informally. Suppose that $h_k \sim Ck^{-\alpha}$, where $h_k \sim g_k$ means $h_k/g_k \to 1$. Use the less precise notion of $h_k \approx g_k$

$$1 - H_0(1-\delta) \approx \sum_k Ck^{-\alpha}\{1 - (1-\delta)^k\}.$$

Letting $x = k\delta$ the above is

$$\begin{aligned} \sum_{x:x/\delta\in\mathbb{Z}^+} C(x/\delta)^{-\alpha}\{1-(1-\delta)^{x/\delta}\} &\approx \delta^{\alpha-1}\cdot\delta \sum_{x:x/\delta\in\mathbb{Z}^+} Cx^{-\alpha}(1-e^{-x}) \\ &\approx \delta^{\alpha-1}\int_0^\infty x^{-\alpha}(1-e^{-x})\,dx. \end{aligned}$$

Comparing with (2.6) suggests that $\alpha = 3/2$.

While the calculations above have been for the branching process, they work remarkably well for the random graph. In Sect. 1.7 we showed that in Erdős-Rényi$(n, 1/n)$ the expected number of trees of size k is

$$\sim \frac{nk^{-5/2}}{\sqrt{2\pi}}e^{-k^3/3n^2}$$

so the probability of belonging to a component of size k is asymptotically $Ck^{-3/2}$ while $k \ll n^{2/3}$. When $k = n^{2/3}$ this probability is $O(1/n)$ and there is an exponential cutoff.

2.2 Limiting Degree Distribution Approach

2.2.1 *Molloy and Reed (1995, 1998)*

In their approach one specifies $v_i(n)$, the number vertices of degree i. They assumed

(i) $v_i(n) \geq 0$
(ii) $\sum_i v_i(n) = n$
(iii) the degree sequence is **feasible**, i.e., $\sum_i i v_i(n)$ is even
(iv) the degree sequence is **smooth**, i.e., $\lim_{n\to\infty} v_i(n)/n = p_i$
(v) the degree sequence is **sparse**, i.e., $\sum_{i\geq 0} i v_i(n) \to \sum_i i p_i$.

Advantages of this approach. It is more general since the configuration model defined in the previous section has these properties with probability one. The good news about the configuration model is that it is easy to implement. However, there are some problems:

(a) In the study of epidemics on graphs in Chap. 4 it is natural to pass to a subgraph by flipping a coin for each edge to determine if we keep it. It is not clear if the resulting graph is a configuration model. However it is obvious that the degree sequence of the thinned graph has the properties assumed above.

(b) One way to construct a random regular graph is to take r random perfect matchings to the edges. This is definitely not something generated by the configuration model, but its degree sequence has the desired asymptotic properties,

(c) Finally, it would be nice if the Erdős-Rényi model emerged when the degree distribution in the configuration model is chosen to be Poisson, but this is not the case. It does, however, satisfy the conditions listed above.

The initial result of Molloy and Reed imposed stronger conditions on the maximum degrees than needed, but that problem was later removed. To state their main result (Theorem 1 on pages 164–165) recall that our condition for criticality is

$$1 = \sum k(k-1)p_k/\mu.$$

Multiplying each side by μ, recalling that $\mu = \sum_k kp_k$, and then subtracting the left-hand side from the right converts the condition into

$$Q \equiv \sum_k k(k-2)p_k = 0.$$

The final detail before we can state their result is that they use **asymptotically almost surely (a.a.s)** for "with probability tending to 1 as $n \to \infty$," something we call with high probability (whp).

Theorem 2.2.1 *Suppose (i)–(v) hold, and that $v_i(n) = 0$ for $i > n^{1/4-\epsilon}$.*

(a) If $Q > 0$ there are constants $\zeta_1, \zeta_2 > 0$ so that a.a.s. G has a component with at least $\zeta_1 n$ vertices and $\zeta_2 n$ cycles.
(b) Suppose $Q < 0$ and $v_i(n) = 0$ for $i \geq w_n$ where $w_n \leq n^{1/8-\epsilon}$. Then there is a $\beta > 0$ so that a.a.s. there is no cluster with $\beta w_n^2 \log n$ vertices.

Molloy and Reed (1995) did not get the exact size of the giant component since they used their lower bound on the growth starting from one point until the cluster reached size δn. In their 1998 article they prove a precise result about the size and proved a "duality result" that relates small clusters in the supercritical regime to those in the subcritical regime, which we will return to later. To state their result for the size, we let $\mu = \sum_i i p_i$ and define

$$\chi(\alpha) = \mu - 2\alpha - \sum_{i \geq 1} i p_i \left(1 - \frac{2\alpha}{\mu}\right)^{i/2}.$$

Theorem 2.2.2 *Let α^* be the smallest positive solution of $\chi(\alpha) = 0$ in $[0, \mu/2]$. The size of the giant component is $\epsilon^* n$ where*

$$\epsilon^* = 1 - \sum_{i \geq 1} p_i \left(1 - \frac{2\alpha^*}{\mu}\right)^{i/2}. \tag{2.7}$$

At first glance these formulas may not look much like the ones in Sect. 2.1. However, rewriting $\chi(\alpha^*) = 0$, dividing by μ and then dividing by $(1 - 2\alpha^*/\mu)^{1/2}$ we have

$$\left(1 - \frac{2\alpha^*}{\mu}\right)^{1/2} = \sum_{i \geq 1} \frac{i p_i}{\mu} \left(1 - \frac{2\alpha^*}{\mu}\right)^{(i-1)/2},$$

which shows that $\xi = (1 - 2\alpha^*/\mu)^{1/2}$ is a fixed point of G_1 the generating function of $q_{k-1} = k p_k/\mu$. With this established it is easy to see that the expression for ϵ^* in (2.7) is just $1 - G_0(\xi)$, which is the result in Theorem 2.1.3.

In the supercritical regime the non-giant clusters have the same distribution as a random graph with $(1 - \epsilon^*)n$ vertices and degree distribution

$$\hat{p}_i = \frac{p_i}{1 - \epsilon^*} \left(1 - \frac{2\alpha^*}{\mu}\right)^{i/2}.$$

Since $\rho_1 = (1 - 2\alpha^*/\mu)^{1/2}$ this follows from our results for branching processes conditioned to die out. Note that $\hat{p}_k$ decays exponentially fast so the second largest cluster has size $O(\log n)$.

2.2.2 *Janson and Luczak (2009)*

For each n let $(d_i)_1^n$ be a sequence of nonnegative integers that give the degree sequence of the nth graph. Let $G(n, (d_i)_1^n)$ be a graph chosen at random from all of those with degree sequence $(d_i)_1^n$ (tacitly assuming that such a graph exists).

They also consider multigraphs (with self-loops and parallel edges) with the given degrees. Let $(d_i)_1^n$ be a degree sequence with $\sum_{i=1}^n d_i^n$ even and let $G^*(n, (d_i)_1^n)$ constructed by randomly pairing half-edges as in Sect. 2.1. $G^*(n, (d_i)_1^n)$ is not a random choice from the set of multigraphs with the given degree distribution but when we condition on the graph being simple the result has the same distribution as $G(n, (d_i)_1^n)$.

Following in the footsteps of Molloy and Reed they make the following assumptions, which we will call **condition A.**

(i) $n_k/n = |\{i : d_i = k\}|/n \to p_k$ as $n \to \infty$,
(ii) $\lambda = \sum_k kp_k \in (0, \infty)$,
(iii) $\sum_i d_i^2 = O(n)$,
(iv) $p_1 > 0$.

Let D_n have $P(D_n = k) = n_k/n$. This implies $ED_n = n^{-1}d_i = 2m/n$, where m is the number of edges in the graph. Let D have $P(D = k) = p_k$, and note that (ii) implies $\lambda = ED$. In the new notation (i) says that $D_n \Rightarrow D$ where $\Rightarrow$ denotes convergence in distribution. (iii) says $E(D_n^2) = O(1)$ so the dominated convergence theorem implies $ED_n \to ED$. Let

$$g(x) = \sum_{k=0}^{\infty} p_k x^k,$$

$$h(x) = xg'(x) = \sum_{k=1}^{\infty} kp_k x^k, \quad \text{and} \quad H(x) = \lambda x^2 - h(x). \tag{2.8}$$

Since $h(0) = 0$ and $h(1) = g'(1) = \lambda$ we have $H(0) - H(1) = 0$. Note also that

$$H'(1) = 2\lambda - \sum_k k^2 p_k = E(2D - D^2) = E[D(D-2)].$$

Theorem 2.2.3 *Suppose condition A holds and let C_1, C_2 be the two largest components in the resulting random graph $G(n, (d_i)_1^n)$, let $v(C_i)$ be the number of vertices in C_i, let $e(C_i)$ be the number of edges in C_i, and $v_k(C_i)$ be number of vertices of degree k in C_i.*

(i) if $E[D(D-2)] > 0$ then there is a unique $\xi \in (0, 1)$ so that $H(\xi) = 0$. The following convergences occur in probability

$$v(C_1)/n \to 1 - g(\xi) > 0, \qquad v_k(C_1)/n \to p_k(1 - \xi^k) > 0,$$
$$e(C_1)/n \to (\lambda/2)(1 - \xi^2), \qquad v(C_2)/n \to 0, \ e(C_2)/n \to 0.$$

(ii) If $E[D(D-2)] < 0$ *then* $v(\mathcal{C}_1)/n \to 0$ *and* $e(\mathcal{C}_1)/n \to 0$.

To compare with Theorem 2.1.3 we note that g here is the same as G_0 there so ξ must be the extinction probability ρ_1 of the branching process with the size-biased distribution $q_{k-1} = kp_k/\mu$ and $\mu = \sum_k kp_k$. For a long time I was mystified by the fact that the fixed point of G_1 occurs at the same location as the zero of H. The connection can found from (2.15) and (2.17) which together say that the number of active vertices $A(t)/n \approx H(e^{-t})$ so the exploration process terminates at time $\tau = -\log \xi$. The final piece of the puzzle comes from Lemma 2.2.6 and 2.2.7, which shows that the number of sleeping vertices $S(t)/n \approx g(e^{-t})$.

Example 2.2.4 (Binomial(3,p) distribution) $p_k = \binom{3}{k} p^k (1-p)^k$. The mean $\lambda = 3p$, so

$$
\begin{aligned}
g(x) &= (1-p+px)^3, \\
h(x) &= xg'(x) = x \cdot 3p(1-p+px)^2, \\
H(x) &= 3px^2 - h(x).
\end{aligned}
$$

$E[D(D-2)] = -p_1 + 3p_3 = -3p(1-p)^2 + 3p^3 = 3p(p^2 - (1-p)^2$ so $p_c = 1/2$. The size-biased distribution has

$$
\begin{aligned}
q_0 &= [1 \cdot 3p(1-p)^2]/3p = (1-p)^2, \\
q_1 &= [2 \cdot 3p^2(1-p)]/3p = 2p(1-p), \\
q_2 &= [3 \cdot p^2]/3p = p^2,
\end{aligned}
$$

which is binomial(2,p). This surprised me the first time I computed it but it really should not have. The size-biased mean $\nu = 2p$.

The extinction probability for the branching process with the size-biased distribution satisfies $(1-p+pz)^2 = z$, which expands to

$$
(1-p)^2 + [2p(1-p) - 1]z + p^2z^2 = 0,
$$

so using the quadratic formula

$$
\rho_1 = \frac{1 - 2p(1-p) \pm \sqrt{1 - 4p(1-p)}}{2p^2}.
$$

When $p = 1/2$, $z = 1$ is a double root. For this to be < 1 when $p > 1/2$ we want the minus root. For a graph of the numerical values see Fig. 2.2. To prepare for Fig. 2.3 which gives a graph of H(x) note that when $p = 0.6$ the extinction probability is 4/9.

Proof of Theorem 2.2.3. As in Sect. 1.3, we build up the cluster by exposing neighbors of vertices one at a time until we have found the cluster. However the details here are somewhat different than for Erdős-Rényi .

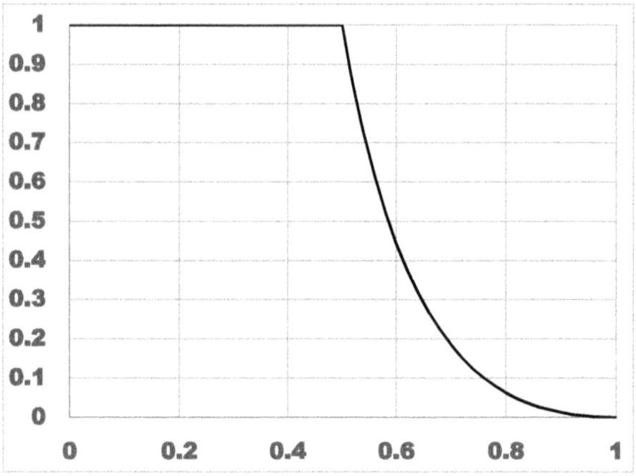

Fig. 2.2 Extinction probability for binomial(2,p) branching process

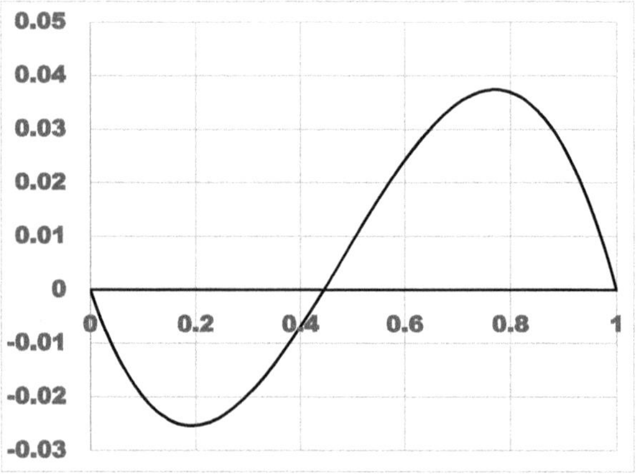

Fig. 2.3 $H(x)$, defined in (2.8), for a binomial(2,0.6) branching process. Note that $H(4/9) = 0$

Following Janson and Luczak we describe the procedure first informally and then give a more precise description. We regard each edge as two half-edges with each half-edge having one vertex. We label the vertices as sleeping or awake ($=$ used), and the half-edges as sleeping, active, or dead. We start with all vertices and half-edges sleeping.

- Pick a vertex, label it as awake, and label its half-edges active.
- Then take any half edge x and find its partner y in the graph, and label these two half-edges as dead.
- If the endpoint of y is sleeping, label it as awake and all other half-edges attached to y as active.

- Repeat until there are no more active half-edges. When this occurs we have found the first component. Pick a new sleeping vertex and start again.

To implement this for the random multigraph $G^*(n, (d_i)_1^n)$ we construct the graph at the same time as we build the cluster. Each time we need to find the partner of a half-edge it is uniformly distributed over all the living half-edges. To make these choices we assign each half-edge x a random variable τ_x that is exponential(1). The authors think of this as the time the half-edge dies. We will add "of natural causes" because some half-edges are killed in step 2.

1. If there is no active half-edge, select a sleeping vertex, declare it awake and all of its half-edges active. For definiteness we choose the vertex by choosing a sleeping half-edge at random. If there is no sleeping half-edge then the process stops and all the remaining sleeping vertices are isolated.
2. If there is at least one active half-edge, pick one at random and kill it. Note that this does not change the number of sleeping edges.
3. Wait until the next half-edge ties and then pair it with the edge just killed. If its vertex is sleeping we make it awake and make all of the other half-edges active. Then repeat step 1.

Let $S(t)$ and $A(t)$ be the number of sleeping and active half-edges at time t, and let

$$L(t) = S(t) + A(t) \quad \text{be the number of living edges.} \tag{2.9}$$

We start with all $2m$ half-edges sleeping but immediately we do steps 1 and 2 so $L(0) = 2m - 1$. At later stages once one half-edge dies we either perform step 2 or both steps 1 and 2. Since 1 does not change the number of living half-edges while 2 and 3 each decrease it by 1, $L(t)$ is decreased by 2 each time a living half-edge dies

Lemma 2.2.5 *As $n \to \infty$*

$$\sup_{\geq 0} \left| \frac{L(t)}{n} - \lambda e^{-2t} \right| \to 0.$$

Proof $\sum_i d_i^n = 2m$, so $L(0)/n = \sum_i d_i^n/n \to \lambda$. To see where the 2 comes from in e^{-2t} note that if half-edges x and y are eventually paired to make an edge then these half-edges are only alive if τ_x and τ_y are both larger than t. The convergence of

$$\sup_{t\geq 0} \left| \frac{1}{m} |\{(x,y) : \min\{\tau_x, \tau_y\} > t| - e^{-2t} \right| \to 0$$

follows from the proof of the Glivenko-Cantelli theorem (Theorem 2.4.9 in PTE5), i.e., if a sequence of decreasing functions converges to a bounded continuous limit then the convergence occurs uniformly.

Let $V_k(t)$ be the number of sleeping vertices of degree k, then

$$S(t) = \sum_{k=1}^{\infty} k V_k(t).$$

Step 2 does not affect the number of sleeping half-edges, while in step 3 each sleeping vertex of degree k is eliminated at rate k. To ignore the effect of step 1 we first let $\tilde{V}_k(t)$ be the number of vertices all of whose half-edges have $\tau_x > t$, and note that

$$V_k(t) \le \tilde{V}_k(t). \tag{2.10}$$

Lemma 2.2.6 *As $n \to \infty$*

$$\sup_{t\ge 0} \left| \frac{\tilde{V}_k(t)}{n} - p_k e^{-kt} \right| \to 0, \tag{2.11}$$

for every $k \ge 0$ and

$$\sup_{t\ge 0} \left| \sum_{k=0}^{\infty} \frac{\tilde{V}_k(t)}{n} - g(e^{-t}) \right| \to 0, \tag{2.12}$$

where $g(z) = \sum_k p_k z^k$. Hence if $\tilde{S}(t) = \sum_k k\tilde{V}_k(t)$ then

$$\sup_{t\ge 0} \left| \sum_{k=0}^{\infty} \frac{\tilde{S}(t)}{n} - h(e^{-t}) \right| \to 0, \tag{2.13}$$

where $h(z) = \sum_k p_k z^k$.

Proof The proof of the first conclusion is almost the same as the proof of Lemma 2.2.5. The next two results follow from the first by summing. To do this rigorously one must use the dominated convergence theorem, but its use is justified by the assumption that $\sum_k k^2 p_k$ is bounded.

Lemma 2.2.7 *If $d_{max} = \max_i d_i$ is the maximum degree of $G^*(n, (d_i)_1^n)$ then*

$$0 \le \tilde{S}(t) - S(t) \le \sup_{0\le s\le t} (\tilde{S}(t) - L(t)) + d_{max}.$$

Proof (2.10) implies $S(t) \le \tilde{S}(t)$. The difference $\tilde{S}(t) - S(t)$ can only increase as a result of step 1. If step 1 is performed at time t on a vertex of degree j then there are no active vertices. Step 2 applies immediately and we have $A(t) = j - 1 < d_{max}$ so using (2.9)

$$\tilde{S}(t) - S(t) = \tilde{S}(t) - L(t) + A(t) < \tilde{S}(t) - L(t) + d_{max}.$$

$\tilde{S}(t) - S(t)$ is never changed by step 2 and either unchanged or decreased by step 3 so $\tilde{S}(t) - S(t)$ does not increase until the next time step 1 is performed.

Let

$$\tilde{A}(t) = L(t) - \tilde{S}(t) = A(t) - (\tilde{S}(t) - S(t)). \tag{2.14}$$

Then by Lemmas 2.2.5 and 2.2.6

$$\sup_{t \geq 0} \left| \frac{\tilde{A}(t)}{n} - H(e^{-t}) \right| \to 0. \tag{2.15}$$

Lemma 2.2.7 can be written as

$$0 \leq (\tilde{S}(t) - S(t)) < -\inf_{s \leq t} \tilde{A}(s) + d_{max}. \tag{2.16}$$

From (2.14) and (2.16) we obtain the relation

$$\tilde{A}(t) \leq A(t) < \tilde{A}(t) - \inf_{s \leq t} \tilde{A}(s) + d_{max}. \tag{2.17}$$

Lemma 2.2.8 *Suppose condition A holds. Recall $H(x) = \lambda x^2 - h(x)$, where $h(x) = \sum_{k=1}^{\infty} k p_k x^k$.*

(i) If $E[D(D-2)] > 0$ then there is a unique $\xi \in (0, 1)$ so that $H(\xi) = 0$. Moreover $H(x) < 0$ for $x \in (0, \xi)$ and $H(x) > 0$ for $x \in (\xi, 1)$.
(ii) If $E[D(D-2)] \leq 0$ then $H(x) < 0$ for $x \in (0, 1)$.

Proof As remarked earlier $H(0) = H(1) = 0$ and $H'(1) = -E[D(D-2)]$. Let $\phi(x) = H(x)/x = \lambda x - \sum_{k=1}^{\infty} k p_k x^{k-1}$.

$$\phi'(x) = \lambda - \sum_{k=2}^{\infty} k(k-1) p_k x^{k-2},$$

$$\phi''(x) = -\sum_{k=3}^{\infty} k(k-1)(k-2) p_k x^{k-3},$$

so ϕ is strictly concave unless $p_k = 0$ for $k \geq 3$. In this case $H(x) = (\lambda - p_2)x^2 - p_1 x - p_0$ is a quadratic. $H'(1) = -E[D(D-2)] = p_1 > 0$ by assumption (iv) so we are in case (ii). Since $H(0) = 0$ there can be no roots in $(0, 1)$ and $H(x) > 0$ for $x \in (\xi, 1)$.

For the rest of the proof we can suppose ϕ is strictly concave. In case (ii) we have $\phi'(x) = H'(x)/x - H(x)/x^2$ so $\phi'(1) = H'(1) - H(1) \geq 0$. Since ϕ' is decreasing, $\phi'(y) > 0$ for $y < 1$ and $\phi(x) < \phi(1) = 0$.

In case (i) $H'(1) < 0$ and thus $H(x) > 0$ for x close to 1. $H'(0) = -h'(0) = -p_1 < 0$ and thus $H(x) < 0$ for x close to 0. From this we conclude that there is at least one $\xi \in (0, 1)$ with $H(\xi) = 0$. The strict concavity of $\phi(x) = H(x)/x$ implies there cannot be another solution in $(0, 1)$.

Proof of part (i) of Theorem 2.2.3. Let ξ be the zero of H given in Lemma 2.2.8 and let $\tau = -\log \xi$. Then by Lemma 2.2.8 $H(e^{-t}) > 0$ for all $0 < t < \tau$ and thus $\inf_{t \leq \tau} H(e^{-t}) = 0$. Consequently (2.15) implies

$$n^{-1} \inf_{t \leq \tau} \tilde{A}(t) = \inf_{t \leq \tau} n^{-1} \tilde{A}(t) - \inf_{t \leq \tau} H(e^{-t}) \to 0. \tag{2.18}$$

By Condition (iii) $d_{max} = O(n^{1/2})$ and thus $n^{-1} d_{max} \to 0$. Consequently (2.16) and (2.18) yield

$$\sup_{t \leq \tau} n^{-1} |A(t) - \tilde{A}(t)| = \sup_{t \leq \tau} n^{-1} |\tilde{S}(t) - S(t)| \to 0, \tag{2.19}$$

and thus by (2.15)

$$\sup_{t \leq \tau} |n^{-1} A(t) - H(e^{-t})| \to 0. \tag{2.20}$$

Let $0 < \epsilon < \tau/2$. Since $H(e^{-t}) > 0$ on the compact interval $[\epsilon, \tau - \epsilon]$, (2.20) implies that with high probability $A(t)$ remains positive on $[\epsilon, \tau - \epsilon]$ and thus no new component is started during this interval.

On the other hand by (i) of Lemma 2.2.8, $H(e^{-t-\epsilon}) < 0$ and (2.15) implies $\tilde{A}(t + \epsilon) \to H(e^{-t-\epsilon})$. Since $A(\tau + \epsilon) \geq 0$, if we let $\delta = |H(e^{-t-\epsilon})|/2$

$$\tilde{S}(\tau + \epsilon) - S(\tau + \epsilon) = A(\tau + \epsilon) - \tilde{A}(\tau + \epsilon) \geq -\tilde{A}(\tau + \epsilon) > n\delta, \tag{2.21}$$

while (2.19) gives $|\tilde{S}(\tau) - \tilde{S}(\tau)| < n\delta$ whp. Consequently whp

$$\tilde{S}(\tau + \epsilon) - S(\tau + \epsilon) > \tilde{S}(\tau) - S(\tau),$$

so step 1 is performed between τ and $\tau + \epsilon$. Let T_1 be the last time step 1 was performed before $\tau/2$ and let T_2 be the next time it is performed after $\tau/2$. We have shown that $T_1 \leq \epsilon$ and $\tau - \epsilon \leq T_2 \leq \tau + \epsilon$. In other words, $T_1 \to 0$ and $T_2 \to \tau$.

Lemma 2.2.9 *Let T_1^* and T_2^* be two (random) times when step 1 is performed with $T_1^* \leq T_2^*$ and assume that $T_1^* \to t_1$ and $T_2^* \to t_2$ where $0 \leq t_1 \leq t_2 \leq \tau$. If C^* is the union of all the components explored between T_1^* and T_2^* then*

$$v_k(C^*)/n \to p_k(e^{-kt_1} - e^{-kt_2}), \tag{2.22}$$

$$v(C^*)/n \to g(e^{-t_1}) - g(e^{-t_2}), \tag{2.23}$$

$$e(C^*)/n \to [h(e^{-t_1}) - h(e^{-t_2})]/2. \tag{2.24}$$

In particular if $t_1 = t_2$ then $v(C^*)/n \to 0$ and $e(C^*)/n \to 0$.

Proof C^* contains all of the vertices awakened in $[T_1^*, T_2^*)$ and no others, so for $k \geq 1$

$$v_k(C^*) = V_k(T_1^*-) - V_k(T_2^*-). \tag{2.25}$$

Since $T_2^* \to t_2$ and H is continuous

$$\inf_{t \leq T_2^*} H(t) \to \inf_{t \leq t_2} H(t) = 0,$$

and (2.15) and (2.16) imply, in analogy with (2.18) and (2.19), that $\inf_{t \leq T_2^*} n^{-1}\tilde{A}(t) \to 0$ and

$$\sup_{t \leq T_2^*} n^{-1}|\tilde{S}(t) - S(t)| \to 0. \tag{2.26}$$

Since $\tilde{V}_j(t) \geq V_j(t)$ for every j and $t \geq 0$ then for all $k \geq 1$

$$\tilde{V}_k(t) - V_k(t) \leq k^{-1} \sum_{j=1}^{\infty} j(\tilde{V}_j(t) - V_j(t)) \leq k^{-1}(\tilde{S}(t) - S(t)). \tag{2.27}$$

Hence (2.26) implies that for every $k \geq 1$

$$\sup_{t \leq T_2^*} n^{-1}|\tilde{V}_k(t) - V_k(t)| = o_p(n).$$

This is trivially true for $k = 0$. Consequently using Lemma 2.2.6 we have for $j = 1, 2$

$$\begin{aligned} V_k(T_j^*-) &= \tilde{V}_k(T_j^*-) \\ &= np_k \exp(-kT_j^*) + o_p(n) \\ &= np_k \exp(-kt_j) + o_p(n), \end{aligned} \tag{2.28}$$

and (2.22) follows from (2.25).

Similarly using $\sum_{k=0}^{\infty} \tilde{V}_k(t) - V_k(t)\tilde{S}(t) - S(t)$ we have

$$\begin{aligned} v(C^*) &= \sum_{k=1}^{\infty} V_k(T_1^*-) - V_k(T_2^*-) \\ &= \sum_{k=1}^{\infty} \tilde{V}_k(T_1^*-) - \tilde{V}_k(T_2^*-) + o_p(n) \\ &= ng(e^{-T_1^*}) - ng(e^{-T_1^*}) + o_p(n), \end{aligned}$$

and

$$
\begin{aligned}
2e(C^*) &= \sum_{k=1}^{\infty} k[V_k(T_1^*-) - V_k(T_2^*-)] \\
&= \sum_{k=1}^{\infty} k[\tilde{V}_k(T_1^*-) - \tilde{V}_k(T_2^*-)] + o_p(n) \\
&= nh(e^{-T_1^*}) - nh(e^{-T_1^*}) + o_p(n),
\end{aligned}
$$

(2.23) and (2.24) follow.

Let C' be the component created at T_1 and explored until T_2. By Lemma 2.2.9 with $t_1 = 0$ and $t_2 = \tau$,

$$v_k(C')/n \to p_k(1 - e^{-k\tau}), \tag{2.29}$$

$$v(C')/n \to g(1) - g(e^{-\tau}) = 1 - g(\xi), \tag{2.30}$$

$$e(C')/n \to [h(1) - h(\xi)]/2 = \frac{\lambda}{2}(1 - \xi^2). \tag{2.31}$$

At this point we have discovered the largest component and discovered its size. We leave it to the reader to show that all the other components have size $o(n)$, and to prove part (ii) of Theorem 2.2.3. The details can be found in Janson and Luczak.

2.3 Subcritical Cluster Sizes

In subcritical Erdős-Rényi random graphs the largest component is $O(\log n)$. This result is false for graphs with arbitrary degree distributions. Suppose for example that $p_k \sim ck^{-\gamma}$ where $\gamma > 3$ to make the second moment $\sum_k k^2 p_k < \infty$. The tail of the distribution $\sum_{k=K}^{\infty} p_k \sim cK^{1-\gamma}$ so the largest degree present in a graph with n vertices is $O(n^{1/(\gamma-1)})$. This is a trivial lower bound on the size of the largest component but it turns out to be accurate up to a constant factor.

In the previous version of this book, published in 2007, this statement was a conjecture. Now, thanks to Janson (2008) it is a theorem. He works in the set-up of the limiting degree distribution approach. Let

$$
\begin{aligned}
\mu_n &= ED_n = \frac{1}{n}\sum_{k=0}^{\infty} kn_k = \frac{1}{n}\sum_{i=1}^{n} d_i, \\
\nu_n &= \frac{1}{n\mu_n}\sum_{k=0}^{\infty} k(k-1)n_k = \frac{1}{n\mu_n}\sum_{i=1}^{n} d_i(d_i - 1).
\end{aligned}
$$

Theorem 2.3.1 *Suppose $\mu_n \to \mu_\infty > 0$, $\nu_n \to \nu_\infty < 1$, and*

$$P(D_n \ge k) \le C_1 k^{1-\gamma}, \tag{2.32}$$

for some $\gamma > 3$ and $C_1 < \infty$, then there is a constant A so that

$$|\mathcal{C}_1| \le A n^{1/(\gamma-1)}.$$

To explain this result note that if x is the vertex with largest degree and $\Delta_n = m$ then the law of large numbers implies that the number of vertices at distance 2 from x in $\mathcal{C}_x$ is $\sim m\nu_n$, or more generally the number of vertices at distance k from x is $\sim m^{k-1}\nu_n$. Summing the geometric series and ignoring the interchange of limit and sum we have the first term in the following sharper result, which is given in Janson (2008).

Theorem 2.3.2 *Let $\Delta_n = \max_{1\le i\le n} d_i$ be the maximum degree of $G(n, (d_i)_1^n)$. Then*

$$|\mathcal{C}_1| = \frac{\Delta_n}{1-\nu_n} + o_p(n^{1/(\gamma-1)}). \tag{2.33}$$

Proof of Theorem 2.3.1. As in the previous section Janson finds it convenient to work with multigraphs that allow multiple edges and loops. Let $(d_i)_1^n$ be a sequence of nonnegative integers such that $\sum_{i=1}^n d_i$ is even. Assign d_i half-edges to vertex i and then pair the half-edges at random to make a graph that we again call $G^*(n, (d_i)_1^n)$. Define $G(n, (d_i)_1^n)$ by conditioning $G^*(n, (d_i)_1^n)$ to be simple. The assumption (2.32) implies that the limiting probability G^* is simple is positive.

Fix a vertex v and explore the component $\mathcal{C}(v)$ by the standard exploration process. The process begins by declaring v *used* and all the half-edges at v to be *active*. We reveal the random configuration only as we need it, or equivalently we construct the random configuration as we need it. Then for $i = 1, 2, \ldots$, as long as there is an active half-edge, we pick one x_i. We choose its partner half-edge y_i uniformly among all the half-edges except x_i and the ones that have been paired and let v_i be its vertex. If the vertex is not already used we declare its other $d(v_i) - 1$ half-edges active. Finally declare x_i, y_i, v_i used. Repeat.

Let S_i be the number of active half-edges after i steps. $S_0 = d(v)$ while

$$S_i = S_{i-1} + \xi_i - 1,$$

where ξ_i is the number of new half-edges found at the ith step. $\xi_i = 0$ if v_i is used and $\xi_i = d(v_i) - 1$ if not. The process stops at $\tau = \inf\{t : S_t = 0\}$. The component $\mathcal{C}(v)$ has τ edges and at most $\tau + 1$ vertices. In particular, for any integer $M > 0$ if $|\mathcal{C}(v)| > M$ then $\tau \ge M$. This implies $S_M = d(v) + \sum_{i=1}^M (\xi_i - 1)$ and thus

$$d(v) + \sum_{i=1}^M \xi_i \ge M. \tag{2.34}$$

The random variables ξ_i are dependent but we can bound them by independent random variables. Suppose that $i \le \sqrt{n}$. ($\gamma > 3$ so $1/(\gamma-1) < 1/2$.) If $\xi_i = k \ge 1$ then $d(v_i) = k+1$ so there are at most $n_{k+1} = nP(D = k+1)$ possible choices for v_i and thus

$$(k+1)n_{k+1} = n(k+1)P(D = k+1)$$

choices for the half-edge y_i. When y_i is chosen we have revealed the pairing of $2(i-1)$ half-edges and chosen x_i but y_i is uniformly distributed over all

$$\sum_{j=1}^{n} d_i - 2(i-1) - 1 = n\mu - O(n^{1/2})$$

remaining half-edges. Hence given all the history of the process up to step i

$$\begin{aligned} P(\xi_i = k) &\le \frac{n(k+1)P(D = k+1)}{n\mu - O(n^{1/2})} \\ &\le \frac{(k+1)P(D = k+1)}{\mu}(1 + O(n^{-1/2})). \end{aligned} \tag{2.35}$$

Note that the correction term in the last equation is independent of k and $i \le \sqrt{n}$

Fix $\epsilon > 0$ with $4\epsilon < 1 - \nu_\infty$ and let $\nu' = \nu + \epsilon$. We suppose n is large enough so that $\nu = \nu_n < 1 - 3\epsilon$ so $\nu' < 1 - 2\epsilon$. Let X be an integer-valued random variable with

$$P(X \ge k) = \max\left(1, \frac{\nu'}{\nu} \sum_{k \ge x} \frac{(k+1)P(D = k+1)}{\mu}\right). \tag{2.36}$$

If n is large enough (not depending on k) the factor $(1 + O(n^{-1/2}))$ in (2.35) is smaller than ν'/ν so

$$P(\xi_x \ge x) \le P(X \ge x),$$

even if we condition on $\xi_1, \ldots, \xi_i$. Thus the ξ_i can be coupled with independent random variables X_i so that

$$\xi_i \le X_i \quad \text{for all } 1 \le i \le \sqrt{n}. \tag{2.37}$$

In addition we trivially have

$$\xi_i \le \Delta. \tag{2.38}$$

For the remainder of the proof we will only use the last two properties of ξ_i

By (2.36) and the assumptions of the theorem

$$\begin{aligned}EX &= \sum_{m=1}^{\infty} P(X \geq m) \leq \frac{\nu'}{\nu} \sum_{m=1}^{\infty} \sum_{k=m}^{\infty} \frac{(k+1)P(D=k+1)}{\mu} \\ &\leq \frac{\nu'}{\nu} \sum_{k=1}^{\infty} \frac{k(k+1)P(D=k+1)}{\mu} \\ &= \frac{\nu'}{\nu} \cdot \nu = \nu', \end{aligned} \tag{2.39}$$

where to go from the first line to the second we have interchanged the order of summation. If we let $m = \lceil x \rceil$ then using various definitions along with some simple arithmetic we conclude that for all $x \geq 0$

$$\begin{aligned}P(X \geq x) &\leq \frac{\nu'}{\nu} \sum_{k=m}^{\infty} \frac{(k+1)P(D=k+1)}{\mu} \\ &\leq C_3 \sum_{k=m+1}^{\infty} kP(D=k) \\ &= C_3 \left(mP(D > m) + \sum_{j=m+1}^{\infty} P(D \geq j) \right) \\ &\leq C_4 m^{2-\gamma} + C_4 \sum_{j=m+1}^{\infty} j^{1-\gamma} \leq C_5 m^{2-\gamma} \leq C_5 x^{2-\gamma}. \end{aligned} \tag{2.40}$$

For some A large and $\delta < 1/(\gamma - 1)$ small to be chosen later let

$$M = An^{1/(\gamma-1)} \quad \text{and} \quad M_1 = n^{1/(\gamma-1)-\delta}. \tag{2.41}$$

Note that M is the desired upper bound on $\mathcal{C}_1$. If we define the truncated random variables $Y_i = X_i 1_{(X_i \leq M_1)}$ then by (2.39), $EY_i \leq EX_i \leq \nu'$. If $|\mathcal{C}(v)| > M$ then (2.34) holds and so using $\xi_i \leq \Delta$

$$\begin{aligned}M &\leq d(v) + \sum_{i=1}^{M} d_i \leq \Delta + \sum_{i=1}^{M} \xi_i 1_{(X_i \leq M_1)} + \sum_{i=1}^{n} \xi_i 1_{(X_i > M_1)} \\ &\leq \sum_{i=1}^{M} Y_i + \Delta \left(1 + \sum_{i=1}^{M} 1_{(X_i > M_1)} \right) \\ &\leq \nu' M + \sum_{i=1}^{M} (Y_i - EY_i) + \Delta \left(1 + \sum_{i=1}^{n} 1_{(X_i > M_1)} \right), \end{aligned} \tag{2.42}$$

and thus recalling $\nu' < 1 - 2\epsilon$

$$P((|\mathcal{C}(v)| \geq M) \leq P\left(\sum_{i=1}^{M}(Y_i - EY_i) > \epsilon M\right)$$
$$+ P\left(\sum_{i=1}^{M} 1_{(X_i > M_1)} > \epsilon \frac{M}{\Delta} - 1\right). \tag{2.43}$$

For the first sum in (2.43) fix a number $r \geq \gamma$ so that $r\delta > 2$. **Rosenthal's inequality** for nonnegative random variables states (this is Theorem 3.9.1 of Gut (2005) but if you don't have that book you can consult (2) in Ibragimov and Sharakhmetov (2001))

$$E\left(\sum_{k=1}^{n} Z_i\right)^r \leq C_r \max\left(\left(\sum_{k=1}^{n} EZ_i^2\right)^{r/2}, \sum_{k=1}^{n} EZ_i^r\right). \tag{2.44}$$

Letting $Z_i = |Y_i - EY_i|$ we have

$$E\left(\sum_{k=1}^{n} |Y_i - EY_i|\right)^r \leq C_r\left((nE(Y_i - EY_i)^2)^{r/2} + nE|Y_i - EY_i|^r\right).$$

Taking $n = M$ and numbering the constants as Janson has done in his proof, we have

$$E\left|\sum_{i=1}^{M} |Y_i - EY_i|\right|^r \leq C_6 M^{r/2}(E|Y_1 - EY_1|^2)^{r/2} + C_7 ME|Y_1 - EY_1|^r.$$
$$\leq C_6 M^{r/2}(EY_1^2)^{r/2} + C_8 MEY_1^r. \tag{2.45}$$

To go from the first line to the second we note that $E(X - c)^2$ is minimized when $c = EX$. For the second term we note that using $E(Y; A)$ for the integral of Y over A

$$E|X - EX|^r = E(|X - EX|^r; X \geq EX) + E(|X - EX|^r; X \leq EX)$$
$$\leq E(|X|^r; X \geq EX) + |EX|^r P(X \leq EX) \leq 2E|X|^r,$$

where on the second line we have used the fact that $X \geq 0$ and then Jensen's inequality to put the rth power inside.

We estimate the second moment of Y_1 using (2.40)

$$EY_1^2 = \int_0^\infty 2xP(Y_1 > x)\,dx = \int_0^{M_1} 2xP(X > x)\,dx$$
$$\leq 1 + C_5 \int_1^{M_1} 2x^{3-\gamma}\,dx \leq C_9 M_1, \tag{2.46}$$

since $\gamma > 3$. Trivially $EY_1^r \le M_1^r$. Using Markov's inequality on (2.45), the second moment estimate in (2.46), $Y_1 \le M_1$ and $M_1/M = A^{-1}n^{-\delta}$

$$P\left(\sum_{i=1}^{M} |Y_i - EY_i| > \epsilon M\right) \le (\epsilon M)^{-r} E\left|\sum_{i=1}^{M} |Y_i - EY_i|\right|^r$$
$$\le C_{10}(EY_1^2/M)^{r/2} + C_{11}M^{1-r}EY_1^r \tag{2.47}$$
$$\le C_{12}(M_1/M)^{r/2} + C_{13}M(M_1/M)^r$$
$$\le C_{14}n^{-r\delta/2} + C_{15}n^{1-r\delta} = o(n^{-1}),$$

since $r\delta > 2$.

For the second sum in (2.43) write $I_i = 1_{(X_i > M)}$ and note that

$$P\left(\sum_{i=1}^{M} I_i \ge L\right) \le \binom{M}{L} P(I_k = 1,\ 1 \le k \le L)$$
$$= \binom{M}{L} P(I_1 = 1)^L = \binom{M}{L} P(X > M_1)^L. \le (MP(X > M_1))^L. \tag{2.48}$$

Now by (2.40) and the choice of M

$$MP(X_1 > M_1) \le C_5 A n^{1/(\gamma-1)+(2-\gamma)[1/(\gamma-1)-\delta]} = CAn^{(\gamma-2)\delta-(\gamma-3)/(\gamma-1)}.$$

Choose $\delta > 0$ so that $\delta_1 = (\gamma - 3)/(\gamma - 1) - (\gamma - 2)\delta > 0$. Then $MP(X_1 > M_1) = O(n^{-\delta_1})$ and

$$P\left(\sum_{i=1}^{M} I_i \ge L\right) = O(n^{-L\delta_1}). \tag{2.49}$$

Now choose $L > 1/\delta_1$ and then $A \ge C_2(L+1)/\epsilon$. Since $\Delta \le C_2 n^{1/(\gamma-1)}$

$$\frac{M}{\Delta} \ge \frac{A}{C_2} \ge \frac{L+1}{\epsilon}. \tag{2.50}$$

It follows from (2.49) that

$$P\left(\sum_{i=1}^{M} I_i \ge \frac{\epsilon M}{\Delta} - 1\right) \le P\left(\sum_{i=1}^{M} I_i \ge L\right) = O(n^{-L\delta_1}) = o(n^{-1}).$$

Consequently (2.43) shows that $P(|\mathcal{C}(v)| > M) = o(n^{-1})$. Since this holds for every vertex it follows that with high probability $|\mathcal{C}(v)| \le M = An^{1/(\gamma-1)}$ and the proof of Theorem 2.3.1 is complete. □

2.4 Distance Between Two Randomly Chosen Vertices

In this section we will consider the distance between two vertices chosen randomly from the giant component of a random graph generated by the configuration model.

2.4.1 *Finite Variance*

To be precise, we assume that the random degree D has $P(D \geq x) \leq x^{-\gamma+1}$ with $\gamma > 3$. The results here follow van der Hofstad, Hooghiemstra, and van Mieghem (2005). Let $\mu = ED$ and let $\nu = E(D(D-1))/ED$ be the mean of the size-biased distribution. Let Z_m be the two phase branching process and let

$$W = \lim_{m \to \infty} \frac{Z_m}{\mu \nu^{m-1}}. \tag{2.51}$$

Theorem 2.4.1 *Suppose $\nu > 1$. Let H_n be the distance between 1 and 2 in the random graph on n vertices. For $n \geq 1$, let $a(n) = \lfloor \log_\nu n \rfloor - \log_\nu n \in (-1, 0]$. As $n \to \infty$*

$$P(H_n - \lfloor \log_\nu n \rfloor = k | H_n < \infty) = P(R_{a(n)} = k) + o(1).$$

If $\kappa = \mu(\nu - 1)^{-1}$ then for $a \in (-1, 0]$

$$P(R_a > k) = E(\exp(-\kappa \nu^{a+k} W_1 W_2) | W_1 W_2 > 0),$$

where W_1 and W_2 are independent copies of W defined in (2.51).

Note that $x = \log_\nu k$ has $\nu^x = k$, i.e., $x = \log n / \log \nu$, so the answer has the same intuitive explanation as the one in Theorem 1.2.5 for Erdős-Rényi random graphs. The strength of this result is that it shows that the fluctuations around the limit are $O(1)$. The statement is made complicated by the fact that H_n is always an integer so the support of $H_n - \log_\nu n$ changes with n.

The proof of Theorem 2.4.1 is done by a careful comparison of the growing cluster and a branching process. The proof is about 30 pages, so we will content ourselves to explain where the formula for the distribution of R_a comes from and refer the reader to the original paper for details. The total number of oriented edges is $\sim \mu n$ as $n \to \infty$, so for our sketch we will suppose that it is always equal to μn. Taking turns growing each cluster by one branching step,

$$P(H_n > k) \approx E \exp\left(- \sum_{i=2}^{k+1} \frac{Z^1_{\lceil i/2 \rceil} Z^2_{\lfloor i/2 \rfloor}}{\mu n} \right).$$

Here, one needs to look closely at the indices. We round up in the first case, and down in the second, so for $i = 2, 3, 4, \ldots$ we get $(\lceil i/2 \rceil, \lfloor i/2 \rfloor) =$

$(1, 1), (2, 1), (2, 2) \ldots$. When $k = 1$ there is only one term in the sum: $Z_1^1 Z_1^2/\mu n$. This gives the expected number of the Z_1^1 half-edges from 1 that are paired with one of the Z_1^2 half-edges from 2. The distance is > 1 if and only if the number of such pairings is 0. Since the events are almost independent, the number of pairings (conditional on the values of Z_1^1 and Z_1^2) is roughly Poisson and the probability of 0 is $E \exp(-Z_1^1 Z_1^2/\mu n)$.

Letting $\sigma_n = \lfloor \log_\nu n \rfloor$ and writing $n = \nu^{\log_\nu n} = \nu^{\sigma_n - a_n}$ we have

$$P(H_n > \sigma_n + j) \approx E \exp\left(-\mu \nu^{a_n + k} \sum_{i=2}^{\sigma_n + j + 1} \frac{Z^1_{\lceil i/2 \rceil} Z^2_{\lfloor i/2 \rfloor}}{\mu^2 \nu^{\sigma_n + j}} \right).$$

$\lceil i/2 \rceil + \lfloor i/2 \rfloor = i$ so using (2.51)

$$\frac{Z^1_{\lceil (\sigma_n + j + 1)/2 \rceil} Z^2_{\lfloor (\sigma_n + j + 1)/2 \rfloor}}{\mu^2 \nu^{\sigma_n + j - 1}} \to W_1 W_2,$$

and it follows that

$$\sum_{i=2}^{\sigma_n + j + 1} \frac{Z^1_{\lceil i/2 \rceil} Z^2_{\lfloor i/2 \rfloor}}{\mu^2 \nu^{\sigma_n + j}} \to \nu^{-1} W_1 W_2 \sum_{k=0}^{\infty} (1/\nu)^k = \frac{1}{\nu - 1}.$$

Combining our results gives

$$P(H_n > \sigma_n + j | H_n < \infty) \approx E(\exp(-\kappa \nu^{a_n + j} W_1 W_2) | W_1 W_2 > 0),$$

with $\kappa = \mu/(\nu - 1)$.

2.4.2 *Power Laws* $2 < \gamma < 3$

Consider a random graph with degree distribution $p_k \sim C_p k^{-\gamma}$ where $2 < \gamma < 3$. In this case, $q_{k-1} = k p_k/\mu \sim C_q k^{1-\gamma}$, so the mean is infinite and the tail of the distribution

$$Q_K = \sum_{k=K}^{\infty} q_k \sim C_Q k^{2-\gamma}, \tag{2.52}$$

where $C_Q = C_q/(\gamma - 2)$. Here we are keeping track of the values of the C's, a decision we will soon regret.

To study the average distance between two randomly chosen points on the giant component, we will first investigate the behavior of the branching process in order to figure out what to guess. The power $0 < \gamma - 2 < 1$, and q_k is concentrated on the nonnegative integers so q_k is in the **domain of attraction of a one-sided stable law with index** $\alpha = \gamma - 2$.

Theorem 2.4.2 *Let $X_1, X_2, \ldots$ be i.i.d. with distribution q_k which satisfies* (2.52) *with $2 < \gamma < 3$, and let $S_m = X_1 + \cdots + X_m$. As $m \to \infty$,*

$$S_m/(C_Q m)^{1/\alpha} \Rightarrow \Gamma_\alpha,$$

where $E(\exp(it\Gamma_\alpha) = \exp(\int_0^\infty (e^{itx} - 1)/x^{\alpha+1}\,dx)$

Proof To begin to understand the asymptotic behavior of S_m, let

$$N_m(a,b) = |\{k \le m : X_k/(C_Q m)^{1/\alpha} \in (a,b)\}|,$$

for $0 < a < b < \infty$. For each k the probability $X_k \in (a(C_Q m)^{1/\alpha}, b(C_Q m)^{1/\alpha})$ is

$$\sim \frac{1}{n}(a^{-\alpha} - b^{-\alpha}).$$

Since the X_k are independent, $N_m(a,b) \Rightarrow N(a,b)$ with a Poisson distribution with mean

$$a^{-\alpha} - b^{-\alpha} = \int_a^b \frac{\alpha}{x^{\alpha+1}}\,dx. \tag{2.53}$$

If we interpret $N(a,b)$ as the number of points in (a,b) the limit is a Poisson point process on $(0,\infty)$ with intensity $\alpha x^{-(\alpha+1)}$. In the limit there are finitely many points in (δ,∞) for $\delta > 0$ but infinitely many in $(0,\delta)$.

The last paragraph describes the limiting behavior of the random set

$$\mathcal{X}_m = \{X_k/(C_Q m)^{1/\alpha} : 1 \le k \le m\}.$$

To describe the limit of $S_m/(C_Q m)^{1/\alpha}$, we will "sum up the points." Let $\epsilon > 0$ and

$$I_m(\epsilon) = \{k \le m : X_k > \epsilon (C_Q m)^{1/\alpha}\},$$
$$\hat{S}_m(\epsilon) = \sum_{k \in I_m(\epsilon)} X_m \qquad \bar{S}_m(\epsilon) = S_m - \hat{S}_m(\epsilon).$$

$I_m(\epsilon)$ = the indices of the "big terms," i.e., those $> \epsilon n^{1/\alpha}$ in magnitude. $\hat{S}_m(\epsilon)$ is the sum of the big terms, and $\bar{S}_m(\epsilon)$ is the rest of the sum.

In the central limit theorem all the X_i make small contributions to the limit. In contrast, here the sum is determined by the big values, i.e., the contribution of $\bar{S}_m(\epsilon)$ is small if ϵ is. To do this we note that

$$E\left(\frac{X_k}{(C_Q m)^{1/\alpha}}; X_k \le \epsilon (C_Q m)^{1/\alpha}\right) \approx \frac{1}{(C_Q m)^{1/\alpha}} \sum_{k=1}^{\epsilon (C_Q m)^{1/\alpha}} k \cdot C_q k^{-1-\alpha}$$

$$\sim \frac{C_q}{(C_Q m)^{1/\alpha}} \cdot \frac{(\epsilon (C_Q m)^{1/\alpha})^{1-\alpha}}{1-\alpha} = C_\alpha \epsilon^{1-\alpha} m^{-1},$$

where C_α is a constant that only depends on α. Multiplying on each side by m gives

$$E(\bar{S}_m(\epsilon)/(C_Q m)^{1/\alpha}) \to C_\alpha \epsilon^{1-\alpha} \downarrow 0 \quad \text{as} \epsilon \downarrow 0. \tag{2.54}$$

If $Z = \text{Poisson}(\lambda)$ then

$$E(\exp(itaZ)) = \sum_{k=0}^{\infty} e^{-\lambda} \frac{e^{itak} \lambda^k}{k!} = \exp(\lambda(e^{ita} - 1)).$$

Dividing (ϵ, ∞) into small strips, using independence of the number of points in different strips, and passing to the limit gives

$$E \exp(it\hat{S}_m(\epsilon)/(C_Q m)^{1/\alpha}) \to \exp\left(\int_\epsilon^\infty (e^{itx} - 1) \frac{\alpha}{x^{\alpha+1}} \, dx \right). \tag{2.55}$$

Now $e^{itx} - 1 \sim itx$ as $t \to 0$ and $\alpha < 1$ so combining (2.54) and (2.55) and letting $\epsilon \to 0$ slowly we have

$$E(\exp(itS_m/(C_Q m^) 1/\alpha) \to \exp\left(\int_0^\infty (e^{itx} - 1) \frac{\alpha}{x^{\alpha+1}} \, dx \right),$$

which completes the proof of Theorem 2.4.2

Limiting behavior of the two-phase branching process. It is enough to understand the behavior of the branching process with the size-biased distribution. Since the original degree distribution is a power law with $\gamma \in (2, 3)$, this branching process has infinite mean. The next result and its proof come from Davies (1978). That paper is a little strange because the main focus is to prove the result with an assumption that is more general than $P(D > k) = x^{-\alpha} L(x)$ where $L(x)$ is slowly varying.

Theorem 2.4.3 *Consider a branching process with offspring distribution* ξ *with* $P(\xi > k) \sim C k^{-\alpha}$ *where* $\alpha = \gamma - 2 \in (0, 1)$. *As* $m \to \infty$, $\alpha^m \log(Z_m + 1) \to W$ *with* $P(W = 0) = \rho$ *the extinction probability for the branching process.*

Proof Copying from page 474 of the paper, if $Z_m > 0$ then

$$Z_{m+1} = \sum_{i=1}^{Z_m} \xi_{m,i},$$

where the $\xi_{m,i}$ are independent and have the same distribution as ξ. We can write

$$\log(Z_{m+1}+1) = \frac{1}{\alpha}\log(Z_m+1) + \log Y_m,$$
$$\text{where}\quad Y_m = \left(1+\sum_{i=1}^{Z_m} \xi_{m,i}\right)/(Z_m+1)^{1/\alpha},$$

and the $+1$'s are to avoid taking $\log 0$. Multiplying each side by α^n and iterating we have

$$\alpha^m \log(Z_{m+1}+1) = \alpha^m \log Y_m + \cdots + \alpha\log(Y_1) + \log(Z_1+1).$$

In our special case of Davies' result, Theorem 2.4.2 shows that as $m \to \infty$, Y_m converges to Γ_α, a one-sided stable law. To quote the previous edition, "straightforward but somewhat tedious estimates on the tail of the distribution of Y_n show that, see pages 474–477 of Davies (1978),"

$$E\left(\sum_{m=1}^{\infty} \alpha^m \log^+ Y_m\right) < \infty \quad\text{and}\quad E\left(\sum_{m=1}^{\infty} \alpha^m \log^- Y_m\right) < \infty.$$

This shows that $\lim_{m\to\infty} \alpha^m \log(Z_{m+1}+1) = W$ exists.

The limit can be 0 and will be when Z_n dies out. However when it does not, Markov chain theory tells us that $Z_n \to \infty$ and we have $W > 0$. The next step is to show that the limit W is nontrivial. Davies has shown that if $J(x) = P(W > x)$ then $P(J > 0) = \rho$ and

$$\lim_{x\to\infty} \frac{-\log(1-J(x))}{x} = 1,$$

i.e., the distribution of W is asymptotically exponential(1).

The double exponential growth of the branching process associated with the power law degree distribution with $2 < \gamma < 3$ suggests that the average distance between two randomly chosen members of the giant component will be $O(\log\log n)$. To determine the constant we note that our limit theorem says

$$\log(Z_t+1) \approx \alpha^{-t} W,$$

so $Z_t + 1 \approx \exp(\alpha^{-t} W)$. We replace $Z_t + 1$ by $n^{1/2}$ since we grow the branching processes from the two points up to this size to get them to intersect. Solving gives $(1/2)\log n = \alpha^{-t} W$. Discarding the W and writing $\alpha^{-t} = \exp(-t\log\alpha)$ we get

$$t \sim \frac{\log\log n}{\log(1/\alpha)}. \tag{2.56}$$

Notice that the 1/2 disappears so the time to grow the cluster to n^β is independent of β. The distance between the two vertices is

$$\approx 2t = \frac{2 \log \log n}{\log(1/\alpha)}. \tag{2.57}$$

van der Hofstad, Hooghiemstra, and Znamenski (2007) give a rigorous proof of (2.57). To state their result we need some notation

$$m_{\tau,n} = 2\left\lfloor \frac{\log \log n}{\log(\tau - 2)} \right\rfloor,$$
$$a_n = \left\lfloor \frac{\log \log n}{\log(\tau - 2)} \right\rfloor - \frac{\log \log n}{\log(\tau - 2)},$$
$$P(R_a > k) = E(\exp(-\kappa \nu^{a+k} W_1 W_2) | W_1 W_2 > 0),$$

where W_1 and W_2 are independent copies of W defined in (2.51).

Theorem 2.4.4 *If $\tau \in (2, 3)$ then there are random variables $R_{\tau,\alpha}$ so that as $n \to \infty$*

$$P(H_n = m_{\tau,n} + \ell | H_n < \infty) = P(R_{\tau,a_N} = \ell)$$

This trio of authors has also proved results in (2005) for $\tau \in (1, 2)$ but in this case the world is very tiny and the size-biased distribution is weird because it has infinite mass,

$$\lim_{n\to\infty} P(H_n = 2) = 1 - \lim_{n\to\infty} P(H_n = 3).$$

In general, we will ignore this case. To prove Theorem 2.4.4, HHZ start with Davies' result, then describe a coupling between the SPG (shortest path graph) and the branching process. The proof is nicely organized and carefully written so we refer the reader to the paper for the details.

2.4.3 *Chung-Lu Model*

It is easy to prove the result for the pairwise distances in their power-law model. See Chung-Lu (2002a), (2002b), (2003). This is an inhomogeneous random graph in which the edges are independent with the edge from i to j present with $p(i, j)$. These systems will be investigated more fully in Chap. 3, but for our purposes here it is enough to know that

$$p(i, j) = \frac{w_i w_j}{\text{vol}(G)},$$

where $w_i = (i/Bn)^{-1/(\gamma-1)}$ is the expected degree of i, and $\text{vol}(S) = \sum_{i \in S} w_i$.

Theorem 2.4.5 *Consider Chung and Lu's power-law graphs with $2 < \gamma < 3$. Then the distance between two randomly chosen vertices in the giant component H_n is asymptotically at most*

$$(2 + o(1)) \log\log n/(-\log(\gamma - 2)).$$

Proof Let $t = n^{1/\sqrt{\log\log n}} = \exp(\log n\sqrt{\log\log n})$. Note that t goes to ∞ slower than any power n and faster than any power of $\log n$. Consider the vertices $H_0 = \{i : w_i \geq t\}$. Looking at the definition of w_i, we have

$$w_i \geq t \quad \text{if and only if} \quad i \leq nBt^{1-\gamma}. \tag{2.58}$$

Recalling that $i, j \in H_0$ are connected with probability $\geq t^2/(\text{vol}(G))$ and when $\gamma > 2$, $\text{vol}(G) \sim \mu n$ we see that two vertices in H_0 are connected with probability p_0 where $np_0 \to \infty$ faster than $\log n$, so by Theorem 1.9.1 the probability H_0 is connected tends to 1. Using Theorem 1.9.4 we see that if $n' = |H_0|$ then the diameter of H_0 is

$$\sim \frac{\log n'}{\log(n'p)} \sim \frac{\log n}{(3-\gamma)\log t} = \Theta(\sqrt{\log\log n}).$$

To expand out from H_0 we let $H_k = \{i : w_i > t^{\alpha^k}\}$ for $k \geq 1$ and use the following easy result:

Lemma 2.4.6 *If $S \cap T = \emptyset$ and $vol(S)vol(T) \geq c\, vol(G)$ then the distance from S to T satisfies $P(d(S, T) > 1) \leq e^{-c}$.*

Proof

$$P(d(S,T) > 1) = \prod_{i\in S, j\in T}\left(1 - \frac{w_i w_j}{\text{vol}(G)}\right) \leq \exp\left(-\sum_{i\in S, j\in T}\frac{w_i w_j}{\text{vol}(G)}\right) \leq e^{-c},$$

since $1 - x \leq e^{-x}$.

Using (2.58), the volume of $\{i : w_i > r\}$ is asymptotically

$$(Bn)^{1/(\gamma-1)} \sum_{i=1}^{Bnr^{1-\gamma}} i^{-1/(\gamma-1)} \sim (Bn)^{1/(\gamma-1)} \cdot C(Bnr^{1-\gamma}(^{(\gamma-2)/(\gamma-1)} \sim C'nr^{2-\gamma}. \tag{2.59}$$

Using this result with S a single vertex gives the next result

Lemma 2.4.7 *Let $1 > \alpha > \gamma - 2$, $0 < \epsilon < \alpha - (\gamma - 2)$, and $r \geq (\log n)^{1/\epsilon}$. If $w_i \geq r^\alpha$ and $T = \{j : w_j > r\}$ then for large n*

$$P(d(\{i\}, T) > 1) \leq \exp(-Cr^{\alpha+(2-\gamma)}/2\mu) \leq n^{-2}.$$

Proof Plugging into Lemma 2.4.6 using $w_i \geq r^\alpha$, $\text{vol}(T) \sim rnCr^{1-\gamma}$, which comes from (2.59) $\text{vol}(G) \leq 2n\mu$, which holds for large n, gives the result.

This will allow us to prove the existence of connections from $i \in H_{k+1}$ to H_k as long as the degrees of the vertices are large enough. Suppose $t^{\alpha^k} \geq (\log n)^{1/\epsilon}$.

Using the previous lemma we see that if $j \in H_{k+1}$ then with probability $\geq 1 - n^{-2}$, j is connected to a point in H_k. Since H_0 is connected it follows that each H_k with $k \leq \ell \equiv \inf\{i : t^{\alpha^i} < (\log n)^{1/\epsilon}\}$ is connected. Let $m = (\log\log n)/(-\log\alpha)$. This is chosen so that

$$\frac{\alpha^m \log n}{\sqrt{\log\log n}} < 1,$$

and $t^{\alpha^m} < e$, so $\ell \leq m$.

H_0 has diameter $O(\sqrt{\log\log n})$, so at this point we have shown that H_ℓ is connected and has diameter smaller than $2m + O(\sqrt{\log\log n})$. To connect the remaining points we note that if n is large then $T = H_\ell$ has volume at least $(Cn/2)(\log n)^{(2-\beta)/\epsilon}$, so if S has volume $\geq n^{\alpha/\epsilon}$ then Lemma 2.4.6 implies

$$P(d(S,T) > 1) \leq \exp(-C(\log n)^{(\alpha-(\beta-2))/\epsilon}/2\mu) \leq n^{-2}.$$

Thus the component of a site i will connect to H_ℓ if it reaches size $(\log n)^{\alpha/\epsilon}$. It is easy to see that collisions in the growth of the cluster to size $(\log n)^{\alpha/\epsilon}$ and since the branching process grows doubly exponentially fast the amount of time required to reach this size is $O(\log\log\log n)$. □

2.5 First Passage Percolation

This topic has been extensively studied on $\mathbb{Z}^d$ since its introduction by Hammersley and Welsh (1965). Some of the important milestones in the development of the theory are the Springer Lecture Notes written by Smythe and Weirman (1978), Kesten's St. Flour Notes (1986), and Auffinger, Damron, and Hanson's (2016) survey *Fifty Years of First Passage Percolation*, available in book form from the AMS.

As in the well-studied case of first passage percolation on $\mathbb{Z}^d$, we think of FPP as describing fluid flow through a porous medium. One can also think of it as an SI model in which infecteds are never cured, though I personally prefer to instead think of that epidemic as modeling the spread of information or a rumor. On $\mathbb{Z}^d$ the main focus is on the asymptotic behavior of the wet region as a function of time t. There is a strong law of large numbers called the **shape theorem**. As this is being written the "central limit theorem" for the fluctuation of the shape is an important open problem. We put central limit theorem in quotes since the fluctuations do not have the usual $n^{1/2}$ and the limit is not Gaussian.

The topic of FPP on graphs is covered in great detail in Chap. 3 of van der Hofstad's notes for the 47th Summer School in Saint-Flour, or StFl for short. The chapter is 71 pages long, so we will only provide a summary of the main results and describe some ideas important for the proofs. FPP is constructed by assigning i.i.d. positive random variables Y_e to the edges. These can be though of as the time to traverse the edge, its length, or its weight. The main objects of study are what van der Hofstad calls

$$C_n(i,j) = \min_{\pi:i\to j} \sum_{e\in\pi} Y_e \quad \text{and} \quad H_n(i,j) = |\pi_n(i,j)|.$$

H_n is for **hopcount** is the length of the shortest path. I guess $\mathcal{C}$ is the cost of the path but since we use $\mathcal{C}$ for clusters and $\mathcal{T}$ for trees, we will denote the passage time by $t_n(i,j)$.

On our finite graphs we let U_1 and U_2 be randomly chosen vertices and examine the behavior of

$$H_n = H_n(U_1, U_2) \quad \text{and} \quad t_n = t_n(U_1, U_2),$$

which are analogues of the average pairwise distance. One can also study $\max_j t_n(i,j)$, the time until everyone knows the rumor and $max_{i,j} t_n(i,j)$.

Janson (1999) proved the following result on the complete graph

Theorem 2.5.1 *Let $\ell(n) = (\log n)/n$. If edge weights are independent exponential(1) then*

(i) for any fixed i and j $C_n(i,j)/\ell(n) \to 1$,
(ii) for any fixed i $\max_{1\le j\le n} C_n(i,j)/\ell(n) \to 2$,
(iii) $\max_{1\le i,j\le n} C_n(i,j)/\ell(n) \to 3$.

Almost all of the results in this section assume what StFl calls

Condition 1.6 (Regularity conditions for vertex degrees)

(a) The degree distributions D_n converge in distribution to a limit D.
(b) As $n \to \infty$, $ED_n \to ED$.
(c) As $n \to \infty$, $ED_n^2 \to ED^2$.

Obviously (b) follows from (a) and (c), but some results in StFl use (b) and not (c). In addition in all cases we will assume that the mean of the size-biased degree distribution $\nu = E[D(D-1)]/ED > 1$, so there is a giant component.

Here we will primarily be concerned with four papers written by Bhamidi, van der Hofstad and Hooghiemstra that we will refer to as BHH (2010a), (2010b), (2011), and (2017). In discussing these results we will follow the notation in the papers with a few exceptions.

2.5.1 *Exponential Passage Times, Finite Variance Degrees*

The discussion here follows BHH (2010b). We assume throughout this subsection that edge weights are i.i.d. exponential(1). When this holds the set of wet sites is a Markov process. As in Sect. 2.4, when we studied the distance between randomly chosen vertices, the key to computing passage times will be to run FPP from vertices i and j until they intersect.

In Theorem 3.7 of StFl it is assumed that

$$\nu = \frac{E[D(D-1)]}{ED} > 1,$$

$\nu_n = E[D_n(D_n - 1)]/ED_n = \nu + o(1/\log n)$, and

$$\text{there is an}\epsilon > 0\text{so that} \quad \lim_{n\to\infty} ED_n^{2+\epsilon} = ED^{2+\epsilon}. \tag{2.60}$$

Despite the $+\epsilon$, we will call this the finite variance case.

Theorem 3.1 of BHH (2010b) gives the following result. Throughout the discussion in this section we have replaced their power law exponent of τ by our traditional γ. Here the Gumbel distribution, also known as the double exponential, has cumulative distribution function

$$1 - \exp(-e^{-(x-\mu)/\beta}) = 1 - \exp(-c_1 e^{-c_2 x}).$$

Theorem 2.5.2 *Suppose that* $P(D \geq 2) = 1$, $\gamma > 3$ *and*

$$P(D \geq x) \leq cx^{-(\gamma-1)}.$$

(a) Then the hopcount H_n *satisfies the central limit theorem*

$$\frac{H_n - \alpha \log n}{\sqrt{\alpha \log n}} \Rightarrow \chi, \tag{2.61}$$

where χ *has a standard normal distribution and* $\alpha = \nu/(\nu - 1) > 1$.

(b) There is a random variable V *so that*

$$t_n - \frac{1}{\nu - 1} \log n \Rightarrow V. \tag{2.62}$$

The limiting random variable V *is*

$$V = -\frac{\log W_1}{\nu - 1} - \frac{\log W_2}{\nu - 1} + \frac{\Lambda}{\nu - 1} + \frac{\log(\mu(\nu - 1))}{\nu - 1}, \tag{2.63}$$

where W_1 *and* W_2 *are two independent copies of the limiting random variable of a continuous time branching process and* Λ *has a Gumbel distribution.*

This answer is remarkably explicit. On the two-dimensional lattice the passage time from $(0, 0)$ to $(n, 0)$ when edge weights are exponential(1) scales like γn, but the time constant γ can only be calculated by simulation. The simple explicit constant $1/(\nu - 1)$ reflects the fact that random graphs are locally tree-like

The reader will see the three terms in the definition of V several times, so we will now give our own intuitive explanation of the how they arise. If we let $\lambda = \nu - 1$ then the number of alive half-edges at time t in the shortest weight graph starting from i,

A_t^i, has $e^{-\lambda t} A_t^i \to W_i$. If we let $S_i = (-\log W_i)/\lambda$ then

$$A^i((1/2\lambda)\log n + S_i + s/2\lambda) \approx W_i \exp((1/2)\log n - \log W_i + s/2)$$
$$\approx e^{s/2} n^{1/2}.$$

The number of possible connections between the growing clusters at time s is $e^s n$. Since the probability of a connection between two vertices in this sparse graph is $O(1/n)$ the probability of no connection is $\approx \exp(-ce^s)$.

Main ideas of the proofs

The key idea is to grow the **shortest weight graph** from vertex 1, SWG_m^1, until it reaches size $\sqrt{n}$, which is order of magnitude of the size at which a connection is made. Then we grow the shortest weight graph from vertex 2, SWG_m^2, until they intersect. To define these growing graphs we let $SWG_0^i = \{i\}$ then repeatedly add the edge adjacent to SWG_m^i with minimal weight. The rules for the growth of SWG_m^2 are defined in the same way.

We know from results in Sect. 1.2 that the growing cluster starting from an individual is a tree while the number of vertices is $\leq n^{1/2-\epsilon}$. We now describe Construction 3.1 in StFl, which supposes that the graph is always a tree. In their analysis the degree of the ith unexplored vertex is deterministic, so that they can prove results in the setting of a limiting degree sequence described in Sect. 2.3. We will content ourself to study the process on graphs generated by the configuration model. In this case the d_i will be independent. d_1 has the original degree distribution, but the d_i for $i \geq 2$ have the size-biased degree distribution.

Here as in the proof of the Miller-Volz equations, which will be described in Sect. 4.5, we build the graph as we run the process. We start with d_i half-edges attached to each vertex $1 \leq i \leq n$. We find it convenient to talk about first passage percolation as an SI epidemic. If you do not know what this is then the next several lines will define it for you. Infected individuals are the ones who know the rumor. They infect neighboring individual at rate 1.

1. Start with the root, v_0, which is infected and has d_1 alive half edges (which BHH call stubs) attached. Immediately we choose one of the half-edges to pass on the infection.
2. After the first step, each alive half-edge lives for an independent exponential(1) amount of time before the infection crosses the edge.
3. When the infection crosses a half-edge for the mth time, it pairs with one chosen at random from the unpaired half-edges. The new vertex, called v_m, is infected, and we let d_{m+1} be the number of half-edges not counting the one used to make the connection.

An example drawn in Fig. 2.4 should help explain the definitions.

The number of alive half-edges on the ith step of the construction, when v_i is chosen.

$$s_i = d_1 + d_2 + \cdots + d_i - (i-1) = d_1 + B_2 + \cdots + B_i - (i-1),$$

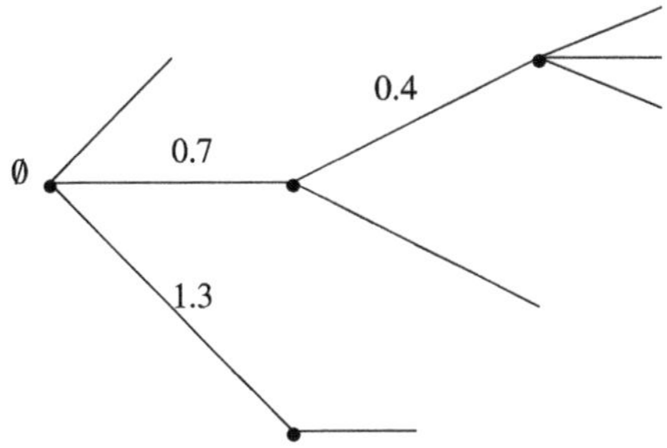

Fig. 2.4 $d_1 = 3, d_2 = 2, d_3 = 3, d_4 = 1$. $E_1/s_1 = 0.7$, $E_2/s_2 - 0.4$, $E_3/s_3 - 0.2$

where B is the size-biased version of d. The first formula is the notation used in BHH and StFl. In order to make formulas easier to write, we will use it here with the understanding that $d_i = B_i$ for $i \geq 2$

The next result is Proposition 3.8 in StFl (and Proposition 4.2 in BHH (2010a)).

Lemma 2.5.3 *Let v_m be the vertex added on the mth step of the construction.*

(a) The weight T_m of the shortest-weight path from the root to v_m, has distribution

$$T_m = \sum_{i=1}^{m} E_i/s_i,$$

where E_i are exponential(1) random variables.

(b) The generation number of v_m is equal in distribution to

$$G_m = \sum_{i=1}^{m} I_i,$$

where I_i are independent Bernoulli with $P(I_i = 1) = d_i/s_i$.

Remark 2.5.4 Note that since $m \to T_m$ is increasing, the distance to vertices added after time m are larger than T_m.

Proof To prove (a) note that at each time, the lack of memory property of the exponential implies that $T_m - T_{m-1}$ is exponential with rate s_M and is independent of T_i, $1 \leq i < m$.

(b) is proved by induction on m. The result is trivial for $m = 1$. A little thought reveals that

$$P(G_{m+1} = k) = \frac{d_{m+1}}{s_{m+1}} P(G_m = k - 1) + \left(1 - \frac{d_{m+1}}{s_{m+1}}\right) P(G_m = k).$$

The first term corresponds to choosing one of the d_{m+1} vertices that were added at time m and the second term to the complement of this event.

Having determined the distributions of the quantities of interest, it is fairly routine to use results about sums of independent random variables to show the following result. Here we are following the notation and formulation of Corollary 3.9 in StFl.

Lemma 2.5.5 *Let B_i be an i.i.d. sequence with $EB_i^{1+\epsilon} < \infty$ and let $\nu = EB_i$. Let $S_i = d_1 + B_2 + \cdots + B_{i-1} - (i-1)$.*

(a) Let $\lambda = \nu - 1$. There is a random variable X so that

$$T_m - \frac{1}{\lambda}\log m \Rightarrow X.$$

(b) If $\alpha = \nu/(\nu-1)$ then the generation number G_m has

$$\frac{G_m - \alpha \log m}{\sqrt{\alpha \log m}} \Rightarrow \chi,$$

where χ has a standard normal distribution.

Proof The first formula follows from the facts that $s_i/i \to \nu - 1$ almost surely var $(E_i/s_i|s_i) = 1/s_i^2$. The second from the Lindberg-Feller central limit theorem and $EB_i = \nu$.

We next identify the distribution of X,

$$T_m - \frac{1}{\lambda}\log m = \sum_{i=1}^{m} \frac{E_i - 1}{S_i} + \sum_{i=1}^{m}\left(\frac{1}{S_i} - \frac{1}{(\nu-1)i}\right) + \frac{1}{\nu-1}\left(\sum_{i=1}^{m}\frac{1}{i} - \log m\right).$$

Proof The key fact is the law of large numbers $S_i/i \to \nu - 1$. $M_m = \sum_{i=1}^m \frac{E_i-1}{S_i}$ is a martingale with EM_m^2 bounded so it converges almost surely. Using the law of large numbers, little analysis, see StFl for details, shows that the second series converges. Finally the last term converges to the Euler-Mascheroni constant.

Technicalities. If we were physicists we would be done with analyzing the growth of a single SWG. Theorem 1.2.3 implies that the expected number of collisions in an Erdős-Rényi random graph before the cluster reaches size n^α is $\le Cn^{2\alpha-1}$. That result generalizes easily to the current setting. Taking $\alpha = 1/2$, we expect only $O(1)$ collisions, which should only rarely alter the distance.

On the other hand, mathematicians who publish papers in journals are required to prove things, so BHH (2010b) create a coupling of the tree and $\mathbb{CM}$ versions, called Construction 4.4. Proposition 4.5 quantifies the differences between the two systems. Proposition 4.6 shows that the conclusions in Lemma 2.5.5 do not change as long as $m \le \bar{m}_n$, where $\log(\bar{m}_n/a_n) = o(\sqrt{\log n})$ where a_n is a constant that we will soon define.

Connection time. Having mentioned these technical problems, we will ignore them. To compute the distance from 1 to 2, we will grow SWG^1 until time $a_n = n^{1/2}$ and then grow SWG^2 until the time they first intersect (which BHH and STFL call $\mathcal{C}_n$):

$$\sigma_n = \min\{m \geq 0 : SWG^1_{a(n)} \cap SWG^2_m \neq \emptyset\}.$$

This is the first time the vertex added to SWG^2 is in $SWG^1_{a(n)}$. When we computed the distance in the finite variance case in Sect. 1.5, we grew each cluster until it had size $n^{1/2} + O(1)$. Part (a) of Proposition 4.9 in BHH (2010b) shows that

$$\log \sigma_n - \log a(n) = o_p(\sqrt{\log n}),$$

where the subscript p means in probability.

Passage time from 1 to 2, and hopcount for the minimum weight path. Here we are following Sect. 4.4 in BHH (2010b), which can be consulted for more detail. If v is the vertex in the intersection then the path from 1 to v in SWG^1 is the shortest weight path from 1 to v and the path from 2 to v in SWG^2 is the shortest weight path from 2 to v. To see that this is the shortest weight path from 1 to 2 see the remark after Lemma 2.5.3. It follows that weight of the path (or passage time) from 1 to 2 is

$$t_n = T^1_{a(n)} + T^2_{\sigma(n)}.$$

Let w be the vertex in $SWG_{\sigma(n)-1}$ that is connected to v. By the way the algorithm runs the vertex v is chosen with probability proportional to the number of live edges attached to it. This is the procedure used to add to SWG^1 so the hopcount in the shortest path from 1 to 2 is

$$H_n = G^1_{a(n)+1} + G^2_{\sigma(n)} - 1.$$

Since G^1 and G^2 are independent, $a_n = n^{1/2}$, and $\log(\sigma_2) \to 1/2$, part(b) of Theorem 1 follows. See (4.28) and the proof of Proposition 4.9 for more details.

Turning to the more delicate first conclusion, Proposition 4.9 shows that

$$(T^1_{a(n)} - (1/\lambda)\log a(n)), T^2_{\sigma(n)} - (1/\lambda)\log a(n)) \to (X_1, X_2),$$

with the X_i independent. The formula for the limit V given in (2.63) is derived in Appendix C, see (C.19).

Erdős-Rényi graphs. Theorems 2.1 and 2.2 in BHH (2011) give the analogue of Theorem 2.5.2 in the case $np \to \lambda$. They also consider what happens when $np \to \infty$. The absence of the assumption $P(D \geq 2) = 1$ causes some inconvenience, but due to the restriction to Erdős-Rényi graphs, the proofs are shorter and less technical.

2.5.2 Exponential Passage Times, Power Law Degrees

The next result is Theorem 3.31 in StFl. We will not discuss the case that the power $\gamma \in (1, 2]$, which is covered in BHH (2010a).

Theorem 2.5.6 *Suppose that* $P(D \geq 2) = 1$, *and for some* $\gamma \in (2, 3)$

$$c_1 x^{-(\gamma-1)} \leq P(D \geq x) \leq c_2 x^{-(\gamma-1)}.$$

(a) Then the hopcount H_n *satisfies the central limit theorem*

$$\frac{H_n - \alpha \log n}{\sqrt{\alpha \log n}} \Rightarrow \chi,$$

where χ *has a standard normal distribution and* $\alpha = 2(\gamma - 2)/(\gamma - 1) < 1$.

(b) There is a random variable V *so that*

$$t_n \Rightarrow V.$$

The limiting random variable V *is shown to be*

$$V = V_1 + V_2,$$

where V_1 *and* V_2 *are two independent copies of the explosion time of an infinite-mean Markovian continuous time branching process.*

This result is surprising since (i) the minimum weight path is $O(1)$ independent of n and has length $\sim \alpha \log n$ despite the fact that the distance between the two points is $\sim 2 \log \log n/(-\log(\gamma - 2))$. See Theorem 2.4.5 for a result for the Chung-Lu model. As the authors say "Thus it is worthwhile to make large detours to avoid high weight edges."

Main ideas of the proof
The proof is similar to the previous result. Indeed in BHH (2010b) the proofs of Theorems 2.5.2 and 2.5.6 are done in parallel. Let B_i be the sized biased version of d_i. Let

$$S_i = B_1 + \cdots + B_i.$$

Here (i) we have dropped the first degree which is not size-biased from the sum and (ii) since $\hat{S}_m / m^{1/(\gamma-2)}$ converges in distribution to a one-sided stable law with index $\gamma - 2$, we have dropped the $-(i - 1)$, which is insignificant.

Let v_m be the vertex added on the mth step of the construction. Lemma 2.5.3 extends immediately to the new setting, so our next step is to extend Lemma 2.5.5. Here

$$\hat{T}_m = \sum_{i=1}^{m} E_i/S_i \quad \text{and} \quad \hat{G}_m = \sum_{i=1}^{m} \hat{I}_i,$$

where $\hat{I}_i$ are independent Bernoulli with $P(\hat{I}_i = 1) = B_i/S_i$.

Lemma 2.5.7 *(a) There is a random variable X so that*

$$\hat{T}_m \to X \quad a.s. \text{ and } in L^1.$$

(b) The generation number $\hat{G}_m$ *has*

$$\frac{\hat{G}_m - \log m}{\sqrt{\log m}} \Rightarrow \chi,$$

where χ *has a standard normal distribution.*

Proof (a) Let $\eta \in (1, 1/(\gamma - 2))$. In order for $\hat{S}_m \le m^\eta$ we must have $B_i \le m^\eta$ for $i \le m$. Since $P(B_i > x) \le Cx^{2-\gamma}$

$$P(S_m \le m^\eta) \le (1 - Cm^{-\eta(\gamma-2)})^m \le \exp(-Cm^{1-\eta(\gamma-2)})$$

and we have $P(S_m \le m^\eta \text{ infinitely often}) = 0$. Since the E_i are exponential mean 1, $\max_{1\le i\le m} E_i \le C' \log m$, and $\Sigma \equiv \sum_{i=1}^{\infty} E_i/\hat{S}_i < \infty$ a.s. The bounds we have obtained imply that the convergence occurs in L^1 and hence that $E\Sigma < \infty$.

(b) Let $\mathcal{F}'_m$ be the σ-field generated by $\hat{S}_k$, $k \ge m$. Assuming $i_1 < \cdots < i_k$, let

$$H(i_1, i_2, \ldots i_m) = \{\hat{I}_{i_j} = 1 \text{ for } 1 \le j \le k\}.$$

Since the distribution of $B_1, \ldots B_{i_1}$ given $\mathcal{F}'_{i_1}$ is exchangeable

$$\begin{aligned} P(H(i_1, i_2, \ldots i_m)|\mathcal{F}'_{i_1}) &= E\left(\prod_{j=1}^{k} B_{i_j}/\hat{S}_{i_k}\right) \\ &= \frac{1}{i_1} \cdot E\left(\prod_{j=2}^{k} B_{i_j}|\hat{S}_{i_k}\right). \end{aligned}$$

It follows by induction that $P(H(i_1, i_2, \ldots i_m) = \prod_{j=1}^{k} 1/i_j$, so the events $\hat{I}_k$ are independent and have $P(I_{i_j} = 1) = 1/i_j$.

Connection time. As in the proof of Theorem 2.5.2, to compute the distance from 1 to 2, we will grow SWG^1 until time a_n and then grow SWG^2 until the time they first intersect (which BHH and StFl call C_n):

$$\sigma_n = \min\{m \geq 0 : SWG^1_{a(n)} \cap SWG^2_m \neq \emptyset\}.$$

However now $a_n = n^{(\gamma-2)/(\gamma-1)}$. To see why this value is chosen note that at time a_n the number of alive stubs in SWG^1 is of order $a_n^{1/(\gamma-2)}$ and we have

$$a_n \cdot a_n^{1/(\gamma-2)} = n^{(\gamma-2)/(\gamma-1)\cdot[1+(\gamma-2)/(\gamma-1)]} = n.$$

Again part (a) of Proposition 4.9 in BHH (2010b) implies

$$\log \sigma_n - \log a(n) = o_p(\sqrt{\log n}).$$

Passage time and hopcount from 1 to 2. As in the proof of Theorem 2.5.2 we have

$$t_n = \hat{T}^1_{a(n)} + \hat{T}^2_{\sigma(n)} \qquad H_n = \hat{G}^1_{a(n)+1} + \hat{G}^2_{\sigma(n)} - 1,$$

and Proposition 4.9 shows that

$$(T^1_{a(n)}, T^2_{\sigma(n)}) \to (X_1, X_2)$$

with the X_i independent. The fact that the limit $V = X_1 + X_2$ is proved in Sect. 6 of BHH (2010b).

2.5.3 General Continuous Passage Times

We will mainly discuss the case of finite variance degrees but will briefly mention the work of Baroni et al. (2017) on the power law case with $\gamma \in (2, 3)$ at the end. The result for the finite variance case here is from BHH (2017), and we will follow their notation, which differs slightly from StFl. The next result is their Theorem 1.2.

Theorem 2.5.8 *Suppose that the edge weights have a continuous distribution F_ξ. For the graph we assume weak convergence of the distribution of vertex degrees, convergence of their second moments, and*

$$ED_n^2 \log^+(D_n) \to ED^2 \log^+(D) < \infty. \tag{2.64}$$

Then there are constants α, λ, $\beta \in (0, \infty)$, and $\alpha_n \to \alpha$, $\gamma_n \to \gamma$ so that

$$\left(\frac{H_n - \gamma_n \log n}{\sqrt{\beta \log n}}, t_n - \frac{1}{\alpha_n} \log n \right) \Rightarrow (\chi, Q),$$

where χ and Q are independent. χ has a standard normal distribution, while Q has a continuous distribution that will be described in Theorem 2.5.9.

Again we will approximate the number of active half-edges by a branching process. Since passage times are no longer exponential, the approximating branching process is more complex. Let F be the degree distribution, and define the distribution function of the size-biased degrees D^* by

$$F^*(x) = E[D1_{(D\le x)}]/ED.$$

Now let $BP^*(t)$ denote the following branching process.

(a) At time $t = 0$ we start with one individual we refer to as the original ancestor or the root of the branching process.
(b) Each individual v lives for an amount of time that has distribution F_ξ and then dies. At the time of its death it gives birth to $X_v^* = D^* - 1$ children.
(c) Each of the children starts an independent copy of the original process.

The process just described is often called an **age-dependent branching process** but it is sometimes called a **Bellman-Harris** process in honor of the pioneering work of Bellman and Harris (1948).

The expected number of children $EX_v^* = \nu > 1$. The **Malthusian parameter** gives the exponential growth rate of the branching process BP^*. It is the unique solution $\alpha \in (0, \infty)$ of the equation

$$\nu \int_0^\infty e^{-\alpha t} dF_\xi(t) = 1. \tag{2.65}$$

When $\nu > 1, \alpha > 0$.

Assumption (2.64) in the theorem implies that $EX_v^* \log X_v^* < \infty$. "Standard theory" implies that there is a random variable W^* so that

$$e^{-\alpha t}|BP^*(t)| \Rightarrow W^* \quad \text{a.s. and in} L^1, \tag{2.66}$$

and $W^* > 0$ when the branching process does not die out.

By (2.65) we can define a **stable age distribution** by

$$\bar{F}_\xi(x) = \nu \int_0^x e^{-\alpha y}\, dF_\xi(y). \tag{2.67}$$

As the name suggests, it gives an approximation to the age of a randomly chosen individual which becomes exact as time $t \to \infty$. Let

$$\bar{\nu} = \int_0^\infty x e^{-\alpha x}\, d\bar{F}_\xi(x) \quad \text{and}$$

$$\bar{\sigma}^2 = \int_0^\infty (x - \bar{\nu})^2 e^{-\alpha x}\, d\bar{F}_\xi(x)$$

be the mean and variance of the stable age distribution $\bar{F}_\xi$.

Theorem 2.5.9 *The parameter α is the Malthusian growth rate defined in* (2.65), *α_n is the solution with ν replaced by ν_n while*

$$\gamma = \frac{1}{\alpha\bar{\nu}} \qquad \gamma_n = \frac{1}{\alpha_n \bar{\nu}_n} \quad \textit{and} \quad \beta = \frac{\bar{\sigma}^2}{\bar{\nu}^3\alpha}. \tag{2.68}$$

Further Q can be identified (see (3.5.7) in StFl) as

$$Q = \frac{1}{\lambda}\left(-\log(W^{(1)}) - \log(W^{(2)}) - \Lambda + c\right),$$

where $P(\Lambda \le x) = \exp(-e^{-x})$ has the Gumbel distribution. $W^{(1)}$ and $W^{(2)}$ are two independent copies of the random variable in (2.66), *are independent of Λ and*

$$c = \log[ED(\nu - 1)^2/(\nu\alpha\bar{\nu})]. \tag{2.69}$$

Power law degrees $2 < \gamma < 3$. Baroni, van der Hofstad, and Komathy (2017) considered this case. To a large extent the difficulties are the union of those in Sects. 2.5.2 and 2.5.3 above. One feature not seen in Sect. 2.5.2 is that the age-dependent branching process does not always explode. There is still universality in that all of those that are explosive have the same behavior (Theorem 4). A simple nonexplosive case occurs when the passage times are bounded below, e.g., $Y = 1 + X$ where $X \ge 0$. In this situation (Theorem 5) the weight and hopcount of the optimal path have the same behavior as the distance. For details, see the paper or StFl.

2.6 Critical Regime

Dhara, van der Hofstad, van Leeuwarden and Sen (2017), hereafter referred to as Dhara et al., studied the case in which the third moment of the degree distribution is finite, and hence the size-biased distribution has a finite second moment. In this case the critical behavior is the same as in the Erdős-Rényi case as described in Sect. 1.8, i.e., rescaled cluster sizes converge to the multiplicative coalescent. They assumed what we will call **condition** A_3, where the subscript 3 indicates we are assuming a finite third moment

(i) If D_n is the degree of a randomly chosen vertex then as $n \to \infty$, D_n converges in distribution to a limit D with $E[D^3] < \infty$.
(ii) $E(D_n^3) \to E(D^3)$.
(iii) For some $t \in \mathbb{R}$

$$\nu_n = \frac{\sum_{i=1}^n d_i(d_i - 1)}{\sum_{i=1}^n d_i} = 1 + tn^{-1/3} + o(n^{-1/3}).$$

(iv) $P(D = 1) > 0$.

To state their results we need some notation. Let $m_r = E[D^r]$,

$$\mu = m_1, \qquad \eta = m_3 m_1 - m_2^2, \quad \text{and} \quad \beta = 1/\mu.$$

Let $B^\lambda_{\mu,\eta}(s)$ be a nonhomogeneous Brownian motion given by

$$B^\lambda_{\mu,\eta}(s) - \frac{\sqrt{\eta}}{\mu} B(s) + \lambda s - \frac{\eta s^2}{2\mu^3},$$

where B is a standard Brownian motion, and define the reflected version by

$$W^\lambda(s) = B^\lambda_{\mu,\eta}(s) - \min_{0 \le t \le s} B^\lambda_{\mu,\eta}(s).$$

We say that $\gamma = (\ell, r)$ is an excursion interval of $W^\lambda(s)$ if $W^\lambda(\ell) = W^\lambda(r) = 0$ and $W^\lambda(s) > 0$ for $s \in (\ell, r)$. Let γ_j^λ, $j \ge 1$ be the ordered excursion lengths of $W^\lambda(s)$. Their Theorem 3.2 states

Theorem 2.6.1 *Fix any real number λ. Under assumption A_3*

$$\{n^{-2/3}|\mathcal{C}_j|, j \ge 1\} \Rightarrow \{\gamma_j^\lambda, j \ge 1\},$$

where ⇒ indicates convergence in distribution.

Power-laws $3 < \gamma < 4$. Dhara et al. (2020) studied the situation in which the tail of the degree distribution

$$P(d_i > k) = k^{-\gamma+1} L_0(k), \tag{2.70}$$

where $3 < \gamma < 4$ and L_0 is a **slowly varying function** defined by the property that for any x, $L_0(tx)/L_0(t) \to 1$ as $t \to \infty$. Functions that converge to positive constants are slowly varying, as are the functions $K(t) = \log t$ or $\log\log t$. When (2.70) holds the size-biased distribution $\hat{d}_i$ has

$$P(\hat{d}_i > k) = k^{-\gamma+2} L_0(k),$$

so sums of $\hat{d}_i$ are in the domain of a stable law with index $\gamma - 2 \in (1, 2)$ and asymmetric $\kappa = 1$.

To state their results we need some notation (again their τ is our γ)

$$\alpha = 1/(\gamma - 1),\ \rho = (\gamma - 2)/(\gamma - 1),\ \eta = (\gamma - 3)/(\gamma - 1),$$
$$a_n = n^\alpha L(n), \qquad b_n = n^\rho / L(n), \qquad c_n = n^\eta / L^2(n).$$

Not much is lost if you assume that the slowly varying function is a constant.

Generalizing a definition from the discussion of Aldous' theorem in Sect. 1.8.3,

$$\ell^p_\downarrow = \left\{ \mathbf{x} = (x_1, x_2, \ldots : x_1 \ge x_2 \ge \cdots \quad \text{and} \quad \sum_{i=1}^\infty x_i^p > \infty \right\}.$$

Assumption 2.6.2 Let $\mathbf{d} = (d_1, d_2, \ldots d_n)$ be the degree sequence written in decreasing order.

(i) $d_i/a_n \to \theta_i$ where $\theta = (\theta_1, \theta_2, \ldots) \in \ell^3_\downarrow \setminus \ell^2_\downarrow$.
(ii) Let D_n be the degree of a randomly chosen vertex. Then $D_n \Rightarrow D$,

$$\frac{1}{n}\sum_{i=1}^n d_i \to \mu = ED, \qquad \frac{1}{n}\sum_{i=1}^n d_i^2 \to E[D^2],$$

and $\lim_{K\to\infty} \limsup_{n\to\infty} a_n^{-3} \sum_{i>K} d_i^3 = 0$.
(iii) For some λ

$$\nu_n(\lambda) \equiv \frac{\sum_{i=1}^n d_i(d_i - 1)}{\sum_{i=1}^n d_i} = 1 + \lambda c_n^{-1} + o(c_n^{-1}).$$

(iv) $P(D = 1) > 0$.

Theorem 2.6.3 *Under assumption 1, if C_j is the jth largest cluster*

$$\{b_n^{-1}|C_j|, j \ge 1\} \Rightarrow \{\tilde{L}^\lambda_j, j \ge 1\},$$

where $\tilde{L}_j(\lambda)$ is the jth largest excursion of a process $\tilde{S}^\lambda_\infty$ which we will now define.

Let $\theta = (\theta_1, \theta_2, \ldots) \in \ell^3_\downarrow \setminus \ell^2_\downarrow$ be the sequence from (i) in Assumption 1. Let $\mathcal{I}_j(s) = 1_{\{\xi_j \le s\}}$ where ξ_j are independent with distribution exponential(θ_i/μ). Let

$$\tilde{S}^\lambda_\infty(t) = \sum_{i=1}^\infty \theta_i(\mathcal{I}_j(s) - (\theta_i/\mu)t) + \lambda t,$$
$$\text{refl}(\tilde{S}^\lambda_\infty(t) = \tilde{S}^\lambda_\infty(t) - \min_{0\le u\le t} \tilde{S}^\lambda_\infty(u).$$

Recall that if B_t is a one-dimensional Brownian motion starting at 0 then

$$\text{refl}(B_t) = B_t - \min_{0 \le s \le t} B_s$$

is a reflecting Brownian motion that has the same distribution as $|B_t|$. If this is not a familiar fact, note that $\text{refl}(B_t) \ge 0$ and that $\min_{0 \le s \le t} B_s$ only increases when $\text{refl}(B_t) = 0$.

Remarks. Theorem 2.6.3 is analogous to Theorem 1.8.4. As in that result, the key is to prove convergence of the exploration process to a limit. One can also prove a version of Theorem 1.8.5, which describes the complexity of the components. For more about the scaling limit see the paper "Novel scaling limits for critical inhomogeneous random graphs" by Bhamidi, van der Hofstad, and van Leeuwarden (2012), and the paper by Norros and Reittu (2006).

Open Problem 2.6.4 Prove (or disprove) that if we do not change the ordering of the vertices then the limit of the rescaled exploration process is a stable process with index $\gamma - 2$ and asymmetric parameter $\kappa = 1$.

2.7 Percolation

Percolation is one of the simplest processes to study on random graphs. If we take a random graph and independently delete edges with probability $1 - p$ then we have another random graph. Here, we will content ourselves to compute the degree distribution for the thinned graph and use the two-phase branching process to compute the critical value p_c and to determine the asymptotics of the probability of a giant component as $p \downarrow p_c$. For those who want a fully rigorous treatment, see Fountoulakis (2007) or Janson (2009), who also considers phase transition results for the k-core.

Suppose that the original degree distribution was p_k. If we keep edges with probability p then the probability a vertex chosen at random will have j neighbors after thinning is

$$\widehat{p}_j = \sum_{k=j}^{\infty} p_k \binom{k}{j} p^j (1-p)^{(k-j)}, \tag{2.71}$$

and the mean degree is $\hat{\mu} = p\mu$. As we explore the cluster containing a vertex, individuals in the first and subsequent generations will have k neighbors (excluding their parent) with probability $q_k = (k+1)p_{k+1}/\mu$ for $k \ge 0$. The probability that they will have j children after thinning is

$$\widehat{q}_j = \sum_{k=j}^{\infty} q_k \binom{k}{j} p^j (1-p)^{(k-j)}. \tag{2.72}$$

Some readers may wonder: what would happen if we took the size-biased version of the thinned distribution instead? Using $\bar{q}$ for this distribution

$$\bar{q}_{j-1} = \frac{j\widehat{p}_j}{\mu p} = \frac{j}{\mu p}\sum_{k=j}^{\infty} p_k \frac{k!}{(k-j)!j!} p^j (1-p)^{(k-j)}$$

$$= \sum_{k=j}^{\infty} \frac{kp_k}{\mu} \frac{(k-1)!}{(k-j)!(j-1)!} p^{j-1}(1-p)^{(k-j)}.$$

Setting $m = k + 1$ the above is

$$= \sum_{m=j-1}^{\infty} q_m \binom{m}{j-1} p^{j-1}(1-p)^{(m+1-j)} = \widehat{q}_{j-1}.$$

If ν is the mean of q, then the mean of $\widehat{q}$ is $\hat{\nu} = \nu p$, and there is a giant component if

$$\nu p > 1. \tag{2.73}$$

Let G_0 be the generating function of p_k and G_1 be the generating function of q_k. Let $\widehat{G}_0$ be the generating function of $\widehat{p}_k$ and $\widehat{G}_1$ be the generating function of $\widehat{q}_k$.

$$\widehat{G}_0(z) = \sum_{j=0}^{\infty} wide\hat{p}_j z^j = \sum_{j=0}^{\infty}\sum_{k=j}^{\infty} p_k \binom{k}{j} p^j (1-p)^{k-j} z^j$$

$$= \sum_{k=0}^{\infty} p_k \sum_{j=0}^{k} \binom{k}{j} (pz)^j (1-p)^{k-j} = G_0(pz + (1-p)).$$

An almost identical calculation gives $\widehat{G}_1(z) = G_1(pz + (1-p))$.

The results for the giant cluster in percolation are the same as the ones for the configuration model in Theorem 2.1.3.

Theorem 2.7.1 *The condition for the existence of a giant component is $\hat{\nu} = \nu p > 1$. In this case the fraction of vertices in the giant component is asymptotically $1 - \widehat{G}_0(\hat{\rho}_1)$, where $\hat{\rho}_1$ is the smallest fixed point of $\hat{G}_0$ in* $[0, 1]$.

In the remainder of this section we will apply this result to various examples, most of which have power law degree distributions $p_k \sim Ck^{-\gamma}$.

	p_c	$1 - \widehat{G}_0(\hat{\rho}_1)$
finite variance	> 0	$\sim C(\theta - 1)$
$3 < \gamma < 4$	> 0	$\sim C(\theta - 1)^{1/(\gamma-3)}$
$\gamma = 3$	$= 0$	$\exp(-(1 + o(1)/cp)$
$2 < \gamma < 3$	$= 0$	$\sim cp^{(\gamma-2)/(3-\gamma)}$

The graph in Fig. 2.5 is probably easier to understand than the table.

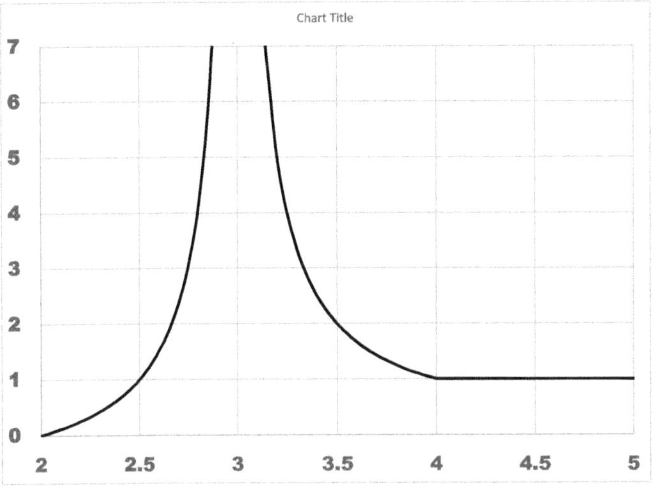

Fig. 2.5 Critical exponent for size of percolation cluster as a function of the power law exponent γ

2.7.1 *When q Has Finite Variance*

The computation in this case is essentially the same as the one done for the Erdős-Rényi graph in Sect. 1.9.1. Consider a branching process with offspring distribution r_k with mean $\theta > 1$ and finite second moment. Let ϕ be the generating function of r_k and ρ be the probability the system dies out, which is the unique solution of

$\phi(\rho) = \rho$ in $[0, 1)$,

$$\phi(1) = 1, \qquad \phi'(1) = \sum_{k=0}^{\infty} kr_k = \theta, \qquad \phi''(1) = \sum_{k=0}^{\infty} k(k-1)r_k \equiv \theta_2.$$

If θ is close to 1 then ρ will be close to 1. If x is close to 1, expanding ϕ in power series around 1 gives

$$\phi(1-x) \approx 1 - \theta x + \theta_2 x^2/2 + \cdots$$

For a fixed point at x we want $1 - x = \phi(1-x)$, which means (dropping the $+\cdots$)

$$(\theta - 1)x = \theta_2 x^2/2 \quad \text{or} \quad x = 2(\theta - 1)/\theta_2. \tag{2.74}$$

Applying this result on the branching process with the size-biased distribution.

$$1 - \rho_1 \sim C_1(\theta - 1) \quad \text{where} \theta = \hat{\nu} - \mu p.$$

This implies

$$1 - \widehat{G}_0(\rho_1) \approx 1 - \widehat{G}_0(1 - C_1(\theta - 1)) \sim \widehat{G}_0'(1) \cdot C_1(\theta - 1). \qquad (2.75)$$

This calculation generalizes easily to all of the other examples: the asymptotic behavior of $1 - \widehat{G}_0(\rho_1)$ is the same as that of $1 - \rho_1$

2.7.2 *Power Law Distributions* $3 < \gamma < 4$

When $3 < \gamma < 4$ the mean of q_k is finite but the variance of q_k is infinite. To understand the behavior of the generating function as $z \to 1$, we will use the following Tauberian theorem, which is Theorem 5 in Sect. XIII.5 of Feller, Volume 2. To state this result we need a definition: $L(t)$ is **slowly varying** (at infinity) if for any $0 < x < \infty$, $L(tx)/L(t) \to 1$. This obviously holds if $L(t)$ converges to a positive limit. $L(t) = (\log t)^a$ is a more interesting example.

Theorem 2.7.2 *Let* $q_n \geq 0$ *and suppose that* $Q(s) = \sum_{n=0}^{\infty} q_n s^n$ *converges for* $0 \leq s < 1$. *If* L *varies slowly at* ∞ *and* $0 \leq \alpha < \infty$ *then each of the two following relations implies the other*

$$Q(s) \sim (1 - s)^{-\alpha} L(1/(1 - s)) \qquad \text{as } s \uparrow 1,$$

$$q_0 + \cdots + q_n \sim n^{\alpha} L(n)/\Gamma(\alpha + 1) \qquad \text{as } s \uparrow 1.$$

Note that the result only gives us conclusions for the sum of the q's. To see that one cannot hope to get a result for the q's rather than their sums, define $\tilde{q}_{2n} = q_{2n-1} + q_{2n}$, $\tilde{q}_{2n-1} = 0$, and note that

$$\tilde{Q}(s) \leq Q(s) \leq \tilde{Q}(s)/s,$$

so the two generating functions have the same behavior at $s = 1$.

The Tauberian theorem requires that we work with quantities that diverge as $s \uparrow 1$ so we start from

$$G_1''(z) = \sum_k k(k - 1)(k - 2) p_k z^{k-3}/\mu.$$

When $2 < \gamma < 3$ we start with $G_1'(z)$. The reader might want to look at the simpler argument first. If $p_k \sim ck^{-\gamma}$ with $\gamma < 4$ then

$$\sum_{k=1}^{n} k(k - 1)(k - 2) p_k/\mu \sim cn^{4-\gamma},$$

so our Tauberian theorem implies

$$G_1''(z) \sim C(1-z)^{\gamma-4}.$$

Integrating and allowing the C to change from line to line gives

$$G_1(1) - G_1(1-x) \sim \int_0^x Cu^{\gamma-3}\,du = C'x^{\gamma-2}.$$

It follows that

$$(\theta - 1)x \approx Cx^{\gamma-2},$$

so $x \approx (\theta-1)^{1/(\gamma-3)}$. Using the last computation we see that

$$1 - \rho_1 \sim (\theta-1)^{1/(\gamma-3)}.$$

Using (2.75) the same result holds for $1 - \widehat{G}_0(\rho_1)$.

Theorem 2.7.3 *Suppose $p_k \sim C_p k^{-\gamma}$ as $k \to \infty$ with $3 < \gamma < 4$ and let $\theta = p\nu$. Then the percolation probability for the type one branching process satisfies*

$$1 - \widehat{G}_0(\rho_1) \sim C(\theta-1)^{1/(\gamma-3)}.$$

Skipping over the borderline case $\gamma = 3$ for the moment.

2.7.3 Scale-Free Networks: $2 < \gamma < 3$

Theorem 2.7.4 *Given a degree distribution with $p_k \sim ck^{-\gamma}$ with $2 < \gamma \le 3$, the critical percolation probability is 0. If $2 < \gamma < 3$ the size of the giant component $\sim cp^{(\gamma-2)/(3-\gamma)}$ as $p \to 0$.*

Theorem 2.7.2 requires that we work with quantities that diverge as $s \uparrow 1$ so we consider

$$G_1'(z) = \sum_k k(k-1)p_k z^{k-2}/\mu.$$

If $p_k \sim ck^{-\gamma}$ with $2 < \gamma < 3$ then

$$\sum_{k=1}^n k(k-1)p_k/\mu \sim cn^{3-\gamma},$$

so our Tauberian theorem implies

$$G_1'(z) \sim c(1-z)^{\gamma-3}.$$

Integrating gives

$$G_1(1) - G_1(1-x) \sim \int_0^x cu^{\gamma-3}\,du = c'x^{\gamma-2}.$$

To solve $G_1(1-p+px) = x$ we write

$$1 - x = G_1(1) - G_1(1-p+xp) \sim c(p(1-x))^{\gamma-2}.$$

Rearranging we have

$$(1-x) \sim cp^{(\gamma-2)/(3-\gamma)}.$$

This gives the asymptotics for the survival probability $1-\rho_1$ for the homogeneous branching process. Using (2.75) the same result holds for $1-\widehat{G}_0(\rho_1)$.

2.7.4 Borderline Case $\gamma = 3$

Theorem 2.7.5 *If* $\gamma = 3$ *the size of the giant component is* $\exp(-(1+o(1))/cp)$.

Proof If $p_k \sim ck^{-3}$ then

$$\sum_{k=1}^{n} k(k-1)p_k/\mu \sim c\log n,$$

so our Tauberian theorem implies

$$G_1'(z) \sim -c\log(1-z).$$

Integrating gives

$$G_1(1) - G_1(1-x) \sim -\int_0^x c\log u\,du \sim cx\log(1/x).$$

To solve $G_1(1-p+px) = x$ we write

$$1-x = G_1(1) - G_1(1-p+xp) \sim cp(1-x)\log(1/p(1-x)).$$

Rearranging we have $1/cp = \log(1/p) + \log(1/(1-x))$. Since $\log(1/p) \ll 1/p$ we have

$$1-x = \exp(-(1+o(1))/cp).$$

This give the asymptotics for the survival probability $1-\rho_1$. Using (2.75) the same result holds for $1-\widehat{G}_0(\rho_1)$.

References

Auffinger A, Damron M, Hanson J (2016) Fifty years of first passage percolation. arXiv:5110.3262v2. Available as a book from American Math Society, Providence RI

Barabási AL, Albert R (1999) Emergence of scaling in random networks. Science 286:509–512

Barabási AL, Jeong H, Néda Z, Ravasz E, Schubert A, Vicsek T (2002) Evolution of the social network of scientific collaborations. XXPhys A 311:590–614

Baroni E, van der Hofstad R, Komathy J (2017) Nonuniversality of weighted random graphs with infinite variance degrees. J Appl Prob 54:146–164

Bellman R, Harris TE (1948) On the theory of age-dependent stochastic branching processes. Proc Natl Acad Sci 34:601–604

Bhamidi S, van der Hofstad R, van Leeuwaarden JSH (2012) Novel scaling limits for critical inhomogeneous random graphs. Ann Probab 40:2299–2361

Bhamidi S, van der Hofstad R, Hooghiemstra G (2010) Extreme value theory, Poisson-Dirichlet distributions, and first passage percolation on random networks. Adv Appl Prob 42(3):706–738

Bhamidi S, van der Hofstad R, Hooghiemstra G (2010) First passage percolation on random graphs with finite mean degrees. Ann Appl Probab 20:1907–1965

Bhamidi S, van der Hofstad R, Hooghiemstra G (2011) First passage percolation on the Erdős-Rényi random graph. Comb Probab Comput 20:683–707

Bhamidi S, van der Hofstad R, Hooghiemstra G (2017) Universality for first passage percolation on sparse random graphs. Ann Probab 45:2568–2630

Bollobás B (1980) A probabilistic proof of an asymptotic formula for the number of labelled regular graphs. Eur J Comb 1:311–318

Chung F, Lu L (2002) Connected components in random graphs with a given expected degree sequence. Ann Comb 6:125–145

Chung F, Lu L (2002) The average distance in a random graph with given expected degrees. Proc Natl Acad Sci 99:15879–15882

Chung F, Lu L (2003) The average distance in a random graph with given expected degrees. Internet Math 1:91–114

Davies PL (1978) The simple branching process: a note on convergence when the mean is infinite. J Appl Prob 15:466–480

de Price DJS (1965) Networks of scientific papers. Science 149:510–515

Dhara S, van der Hofstad R, van Leeuwarden JSH, Sen S (2020) Heavy-tailed configuration models at criticality. Ann Inst Henri Poincaré 56:1515–1558

Dhara S, van der Hofstad R, van Leeuwarden JSH, Sen S (2017a) Critical window for the configuration model: finite third moment degrees. Electron J Probab 22, paper 16

Fountoulakis N (2007) Percolation on sparse random graphs with given degree sequence. Internet Math 4:329–356

Gut A (2005) Probability: a graduate course. Springer, New York

Hammersley J, Welsh D (1965) First-passage percolation, subadditive processes, stochastic networks, and generalized renewal theory. In: Proceedings of the 1965 Berkeley symposium. Springer, New York,

Ibragimov R, Sharakhmetov S (2001) The best constant in the Rosenthal inequality for nonnegative random variables. Statist Probab Lett 55:367–376

Janson S (2008) The largest component in a subcritical random graph with a power law degree distribution. Ann Appl Probab 18:1651–1668

Janson S (2009) On percolation in random graphs with given vertex degrees. Electron J Probab 14:87–118, paper 5

Janson S (1999) One, two, and three times $(\log n)/n$ for paths in a complete graph with random weights. Comb Prob Comput 8:347–361

Janson S, Luczak M (2009) A new approach to the giant component problem. Rand Struct Alg 34:197–216

Kesten H (1986) Aspects of first passage percolation. École d'été de probabilités de Saint-Flour, XIV-1984 Springer Lecture Notes in Math 1180

Liljeros F, Edling CR, Amaral LAN, Stanley HE, Aberg Y (2001) The web of human sexual contacts. Nature 411:907–908

Miller JC (2011) A note on a paper by Erik Volz: SIR dynamics in random networks. J Math Biol 62:349–358

Molloy M, Reed B (1995) A critical point for random graphs with a given degree sequence. Rand Struct Alg 6:161–179

Molloy M, Reed B (1998) The size of the giant component of a random graph with a given degree sequence. Comb Probab Comput 7:295–305

Newman MEJ, Strogatz SH, Watts DJ (2001) Random graphs with arbitrary degree distributions and their applications. Phys Rev E 64, paper 026118

Newman MEJ (2003) The structure and function of complex networks. SIAM Rev 45(2):167–256

Norros I, Reittu H (2006) On a conditionally Poisson graph process. Adv Appl Probab 38:59–75

Smythe RT, Weirman JC (1978) First passage percolation on the square lattice. Springer Lecture Notes in Math 671

van der Hofstad R, Hooghiemstra G, Znamenski D (2005) Distances in random graphs with infinite mean degrees. Extremes 8:111–141

van der Hofstad R, Hooghiemstra G, Znamenski D (2007) Distances in random graphs with finite mean and infinite variance degrees. Electron J Probab 12:703–766

van der Hofstad R, Hooghiemstra G, van Mieghem (2005) Distances in random graphs with finite variance degrees. Rand Struct Alg 26:76–123

Chapter 3
Inhomogeneous Random Graphs

In maximum generality an inhomogeneous random graph is one in which an edge between vertices i and j is present with probability $p(i, j)$. However, it is not possible to prove anything significant in this level of generality. Bollobás, Janson, and Riordan (BJR) (2007) have found a framework that is very general and allows for a rich theory to be developed. The paper filled an entire issue of *Random Structures and Algorithms*. In the special case that we will consider, the vertices $\{1, 2, \dots, n\}$ are embedded in $(0, 1]$ by the map $x \to x/n$. The interval is equipped with the Borel sets and Lebesgue measure, and there is a kernel κ defined on $(0, 1] \times (0, 1]$ (with properties to be specified later) so that

$$p(i, j) = \kappa(i/n, j/n)/n.$$

In Sect. 3.1 we will consider the situation in which there are only a finite number of different types of vertices. This is the simplest possible example, but it is important, because results for this case are a first step for proving things in general. These graphs can be called **multitype Erdős-Rényi graphs**, since the number of edges that connect type j vertices to type k has a Poisson distribution. Let $m_{j,k}$ be the mean number of vertices of type k that are neighbors of a vertex of type j. Generalizing results from the single type case, there is a giant component when the largest eigenvalue of $m_{j,k}$ is larger than 1, and the fraction of neighbors of type j in the giant component is given by the positive fixed point of the generating function of the offspring distribution. Finally, in the subcritical phase the largest cluster is $O(\log n)$.

In Sect. 3.2 we introduce a number of examples to show that the BJR framework is useful. Three examples will be our constant companions as we develop the theory.

- **Dubins' model** has several identities. It can be a graph with connection probabilities $p(i, j) = c/(i \vee j)$ on $1, \dots, n$ or on the positive integers, $\mathbb{N}$. In the latter setting all vertices in the random graph have infinite degree, so work focuses on

R. Durrett, *Dynamics on Graphs*, Probability Theory and Stochastic Modelling 110,
https://doi.org/10.1007/978-3-032-21908-4_3

the question: is the graph connected. Dubins' model is a close relative of the **uniformly grown random graph**, which was called the CHKNS (chickens) model in my 2007 book. On the nth step of the construction a new vertex labeled n is added, and with probability δ a new edge is added between two vertices chosen at random from the graph (so it is very unlikely that n is one of the endpoints).

- **The square root model** is an inhomogeneous graph on $\{1, \ldots, n\}$ with $p(i, j) = c/\sqrt{ij}$. It is called a mean-field version of the **preferential attachment** (PA) model because in PA the probability of an edge connecting i and j is asymptotically $p(i, j)$ (see Lemma 3.2.6). The "mean-field" in the name comes from the fact that in the PA model the presence the edges are not independent.
- **The Chung-Lu model** is defined by assigning a weight w_k to vertex k and declaring that an edge connects i and j with probability $p(i, j) = (w_i w_j / \sum_k w_k) \wedge 1$, and edges are independent. We will be most interested in the version that has a power-law degree distribution $p_k \sim Ck^{-\gamma}$. In this case $\kappa(x, y) = c\psi(x)\psi(y)$ with $\psi(x) = x^{-1/(\gamma-1)}$. So the square root model corresponds to $\gamma = 3$.

In Sect. 3.3 we will describe the set-up and the main results. The key to the analysis is a linear integral operator, which in our special case is

$$(T_\kappa f)(x) = \int_0^1 \kappa(x, y) f(y)\, dy$$

with norm $\|T_\kappa\|_2 = \sup\{\|T_\kappa f\|_2 : f \geq 0, \|f\|_2 \leq 1\}$. The system is supercritical if $\|T_\kappa\|_2 > 1$ and subcritical is $\|T_\kappa\|_2 < 1$. The survival probability is the maximum fixed point of the operator

$$\Phi_\kappa = 1 - \exp(-T_\kappa f).$$

This is a generalization of the answer in the one and finite type cases but now we have to compute a function.

In Sect. 3.4 we will introduce some theory that will allow us to compute in Sect. 3.5 the survival probabilities σ for our three examples. The answers are exotic in the first two cases

$$\sigma(\delta) \sim c \exp(-\pi(\delta - 1/8)^{1/2}) \quad \text{CHKNS}$$
$$\sigma(c) \sim 2e^{1-\gamma} \exp(-1/(2c)) \quad \text{square root}$$

where this time $\gamma = 0.57721566\ldots$ is Euler's constant

In the Chung-Lu model we have an answer that matches the percolation probabilities for the power law configuration model graphs discussed in Sect. 2.7. Finally in Sect. 3.6 we give results for cluster sizes in the subcritical regime. The results for the Chung-Lu model are again similar to the ones for the power law case in Sect. 2.3. However, the fact that edges are independent in the Chung-Lu model allows for a much simpler proof.

3.1 Finitely Many Types

For each $1 \leq k \leq m$, there n_k individuals of type k, and there is an edge from an inrdividual of type j to one of type k with probability $a(j,k)/n$ where $n = \sum_k n_k$. If we let $\ell(x)$ be the type of vertex x then

$$p(x, y) = \frac{a(\ell(x), \ell(y))}{n}.$$

It is natural to arrange the indices so that the first n_1 vertices are of type 1, the next n_2 are of type 2, etc. and to assume that $n_k/n \to \pi_k$. Let V_k be the set of vertices of type k. When there is only one type and $a(1, 1) = \lambda$, we have a Erdős-Rényi graph. In the limit $n \to \infty$ vertices have degrees that are Poisson with mean λ. In Sect. 1.2 we compared the growth of clusters to a branching process in which individuals have a Poisson(λ) number of offspring and concluded that

(i) there is a giant component (with size of order n) if and only if $\lambda > 1$.
(ii) if $\lambda > 1$ then as $n \to \infty$ the fraction of vertices in the giant component converges to the survival probability of the branching process.

In the multitype case, if we let $N_{j,k}$ be the number of neighbors of type k for a vertex of type j then (i) the $N_{j,k}$ have a distribution that is

$$\text{binomial}(n_k, a(j,k)/n) \approx \text{Poisson}(a(j,k)\pi_k),$$

and (ii) $N_{j,1}, \ldots, N_{j,m}$ are independent.

Our first goal is to generalize results for the Erdős-Rényi case and show that the phase transition can be understood by studying a **multitype branching process**. Let $\mu(j,k) = EN_{j,k}$. If there is ℓ so that if $\mu^{\ell}_{jk} > 0$ for all j, k (in which case Athreya and Ney (1972) say the branching process is **positive regular**) then the **Perron-Frobenius theorem** gives the following asymptotic result.

Theorem 3.1.1 *As* $n \to \infty$

$$\mu^n_{j,k} \sim \alpha^n u_j v_k, \tag{3.1.1}$$

where α *is the largest eigenvalue of* μ, *and* v, u *are the associated left* $(1 \times m)$ *and right* $(m \times 1)$ *eigenvectors normalized so that* $u \cdot v = 1$.

The maximum eigenvalue is often denoted by ρ but in this book that letter is reserved for the extinction probability of the branching process.

It may seem strange to have the right eigenvector on the left in the limit in (3.1.1), but this is dictated by the fact that uv is an $m \times m$ matrix. As a further check, note that in the case $\mu(j,k)$ is a Markov chain transition probability (i.e., the rows sum to 1), the maximum eigenvalue $\alpha = 1$, the left eigenvector v is the stationary distribution while the right eigenvector u is a vector of all 1's, so $u_j v_k$ is a matrix in which each row is the stationary distribution.

3.1.1 Phase Transition

It follows easily from (3.1.1) that

Theorem 3.1.2 *If* $\alpha < 1$ *then the branching process is subcritical and* $\sup_i E|\mathcal{C}_i| < \infty$.

This conclusion holds for the random graph since the branching process provides an upper bound on cluster growth. The key to studying the case $\alpha > 1$ is a result about supercritical multitype branching process.

Theorem 3.1.3 *(Kesten and Stigum (1966)) If* $\alpha > 1$ *then there is a one-dimensional random variable* W *so that with probability 1*

$$Z_n^j/\alpha^n \to W\, v,$$

where v *is the left eigenvector. One has either (i)* $E(W|Z_n = e_i) = u_i$ *or (ii)* $W = 0$ *with probability one. (i) holds if and only if*

$$E(Z_1^j \log^+ Z_1^j | Z_0 = e_i) < \infty \quad \textit{for all } 1 \le i, j \le n.$$

Here the positive part is to make $\log^+ 0 = 0$. This result gives a necessary and sufficient condition for the limit to be nontrivial. In the random graph with finitely many types, the number of offspring N_{ij} is binomial so $E(Z_1^j)^2 < \infty$.

Theorem 3.1.4 *If* $\alpha > 1$ *then the inhomogeneous random graph with finitely many types has a giant component. In the limit as* $n \to \infty$ *the fraction of vertices of type* k *in this component is* ρ_k*, the probability the multitype branching process starting from a single type* k *individual does not die out.*

Proof This time we cannot reduce the cluster growth process to a random walk, so we will instead run the branching process using a **breadth-first search**. That is, we first generate the children of the ancestor at time $t = 1$, let the number be Y_1, and write them in an ordered list. We generate the offspring of the first individual in the list, then the second, and so on until we have generated all the children of the first generation. We then order the Y_2 individuals in the second generation and generate their offspring one at a time. Note that there are two time scales here. The time t counts the number of individuals that have given birth, $m \to Y_m$ gives the size of the mth generation, which is known at time T_m on the birth counting time scale.

Lemma 3.1.5 *Let* δ *be chosen so that* $\bar{\alpha} = \alpha(1-\delta) > 1$*. At the times* T_m *when the growing cluster has* $\le \delta n_k$ *vertices in* V_k *for all* k *the process* Y_m *dominates a multitype branching process with mean matrix* $(1-\delta)\mu_{j,k}$,

Proof Suppose that at the tth step of the construction we are generating the offspring of a type j individual. If the growing cluster does not yet contain δn_k vertices in V_k for any k, we can pick sets $G_{t,k} \subset V_k$ that have $|G_{t,k}| = (1-\delta)n_k$ and $G_{t,k}$ does not contain any sites of the growing cluster. We allow the vertex x of type j to give birth onto each $y \in G_k$ with probability $a(j,k)/n$ independent of everything that has happened so far in the construction. This completes the proof of the lemma.

There is no need to do the comparison at time $t = 1$ but if we do then Y_m is exactly a multitype branching process with mean matrix $(1-\delta)\mu(j,k)$ and in addition the number of individuals of type k of an individual of type j is binomial$(n_k, a(j,k)/n)$. We say that the process Y_m survives if the total number of offspring reaches $n^{2/3}$. Note that if n is large it follows that the growing cluster does not yet contain δn_k vertices in V_k for any k.

If the ancestor has type k then the probability the branching process survives converges to ρ_k as $n \to \infty$. Using the limit theorem for supercritical branching processes in Theorem 3.1.3, it is easy to see that if the total progeny reaches $n^{2/3}$ at time M then the exponential growth in Theorem 3.1.3 implies that with high probability $Y_M > n^{0.6}$. Reasoning as in Step 3 in Sect. 1.3 (and noting that there is an ℓ so that we have $\mu^\ell(j,k) > 0$) it follows that two supercritical branching processes that survive will intersect with high probability.

To complete the proof now we note that part (i) of Theorem 3.1.9 below, there is a $\beta > 0$ so that if the multitype branching process has $\beta \log n$ particle total then the probability that it survives is $\geq n^{-2}$ independent of the actual collection of types that are present. □

3.1.2 Computing the Survival Probability

The basic data for a multitype branching process is

$$p^j(i_1, \dots, i_m),$$

the probability an individual of type j gives birth to i_1 individuals of type 1, i_2 individuals of type 2, …, i_m individuals of type m. Here we are following Chap. V of the book of Athreya and Ney (1972) so we will use their notation. In particular the superscript j here and in $f^j(s)$ below are superscripts and not powers. Define the multivariate generating function by

$$f^j(s) = \sum_i p^j(i_1, \dots i_m) s_1^{i_1} \cdots s_m^{i_m}, \tag{3.1.2}$$

where $s = (s_1, \dots, s_m)$ and combine these into a vector

$$f(s) = (f^1(s), \dots f^m(s)).$$

Example 3.1.6 (Random graphs) Since the numbers of neighbors $N_{j,k}$ are independent binomial$(n_k, a(j,k)/n)$ and the generating function of binomial(m, r) is

$$\phi(z) = \sum_{\ell=0}^{m} z^\ell \binom{m}{\ell} r^\ell (1-r)^{m-\ell} = (1 - r + rz)^m,$$

the generating function defined in (3.1.2) is

$$f^j(z) = \prod_{k=1}^{m} \left(1 - \frac{a(j,k)}{n}(1-z)\right)^{n_k} \approx \exp\left(-\sum_{k=1}^{m} a(j,k)\pi_k(1-z)\right).$$

By analogy with the one type case we should expect

Lemma 3.1.7 *If f_k is the generating function of $(Z_k^1, \ldots, Z_k^m)$ then $f_{k+\ell}(s) = f_k(f_\ell(s))$.*

Throughout this section, we will regard facts about multitype branching processes as well-known and not provide proofs. The reader can consult Athreya and Ney for details, although in this case one must go back to Harris' (1963) book to find the proof.

In the one-type case we have to exclude the situation in which each individual has exactly one offspring. In the multitype case we say that Z_n is **singular** if this holds.

Lemma 3.1.8 *If Z_n is positive regular and nonsingular then*

$$P(Z_n = j \text{ infinitely often}) = 0$$

for any $j \geq 0$ with $j \neq 0$.

To prove this it is enough to show that for each $j \geq 0$ with $j \neq 0$ the probability the process dies out is $\geq \epsilon_j > 0$ for then the result follows from Lévy's 0-1 law, Theorem 4.6.9 in PTE5.

Let q^i be the probability of extinction for the multitype branching process starting from one individual of type i, and let $q = (q^1, \ldots, q^m)$.

Theorem 3.1.9 *Assume Z_n is positive regular and nonsingular and let α be the maximum eigenvalue of the mean matrix μ.*

(i) If $\alpha \leq 1$ then $q^i = 1$ for all i. If $\alpha > 1$, $q^i < 1$ for all i.
(ii) $\lim_{n\to\infty} f_n(s) = q$ if $s \in \mathcal{Q} = \{s : 0 \leq s < 1\}$.
(iii) q is the only solution in $\mathcal{Q}$ of $f(q) = q$.

(iii) follows immediately from (ii). In the one type case we start from $s = 0$ and iterate to find q.

3.1.3 Subcritical Regime

The goal of this section is to show that

Theorem 3.1.10 *If* $\alpha < 1$, *there are constants* $\theta, \epsilon > 0$ *so that*

$$P(|\mathcal{C}_x| \geq k) \leq e^{\theta(1-\epsilon k)}.$$

Proof To prove this we will find the cluster by revealing the neighbors of the vertices one at a time as in Sect. 1.3. Let A_t^i be the number of active vertices of type i at time t. Let v_i be the right-eigenvector of the mean matrix and let $Y_t = \sum_i A_t^i v_i$. Let r_t be the type of the vertex whose neighbors are revealed at time $t+1$. Since $\sum_j \mu(i, j)v_j = \lambda v_i$ if we ignore collisions then Y_t has negative drift

$$E(Y_{t+1} - Y_t | r_t = i) = (\lambda - 1)v_i.$$

The assumption of no collisions is valid since with high probability the number of sites involved will be $O(\log n)$. Y_t is not a random walk but it only has finitely many different types of jumps, so we can find an exponential martingale. Let $\mathcal{F}_t$ be the information known at time t and let $\epsilon > 0$. On $\{r_t = i\}$

$$E(e^{\theta[Y_{t+1}+\epsilon(t+1)]}|\mathcal{F}_t) = e^{\theta[Y_t+\epsilon t]} E\left(\exp\left(\epsilon - \theta v_i + \theta \sum_j N_{i,j} v_j \right) \middle| \mathcal{F}_t \right). \quad (3.1.3)$$

Ignoring the $e^{\epsilon - \theta v_i}$, the conditional expectation on the right-hand side is

$$\prod_j \left(1 - \frac{a(i, j)}{n} + \frac{a(i, j)}{n} e^{\theta v_j} \right)^{n_j} \approx \exp\left(\mu(i, j)[e^{\theta v_j} - 1] \right), \quad (3.1.4)$$

since $\mu(i, j) = a(i, j)\pi_j$.

The inequality $e^{\theta v_j} - 1 \geq \theta v_i$ goes the wrong way for finding a supermartingale so we need a complimentary upper bound. The next result is routine and the proof is easy but we include the details for completeness

Lemma 3.1.11 *Let* $\eta > 0$. *If* δ *is small enough then for all* $0 \leq x \leq \delta$, $e^x - 1 \leq (1 + \eta)x$.

Proof Using the power series expansion for e^x we see that if $0 \leq x \leq \delta$

$$e^x - 1 = \sum_{m=1}^{\infty} \frac{x^m}{m!} \leq x + x \left(\frac{\delta}{2!} + \frac{\delta^2}{3!} + \cdots \right) \leq x \left(1 + \frac{\delta}{2(1 - \delta)} \right)$$

since $k! \geq 2$ for $k \geq 2$. Choosing δ small proves the result.

Using (3.1.3) and (3.1.4) with the Lemma, we see that if θ is chosen small enough so that $0 \le \theta v_i \le \delta$ for all i then on $\{r_t = i\}$

$$E(e^{\theta[Y_{t+1}+\epsilon(t+1)]}|\mathcal{F}_t) \le e^{\theta[Y_t+\epsilon t]} \exp\left(\epsilon - \theta v_i + \theta(1+\eta)\sum_j \mu(i,j)v_j\right).$$

The last quantity on the right

$$\epsilon - \theta v_i + \theta(1+\eta)\sum_j \mu(i,j)v_j = \epsilon - \theta v_i(1 - \lambda(1+\eta)).$$

If we pick η small enough so that $\lambda(1+\epsilon) < 1$ and then ϵ small then $\exp(\theta[Y_t + \epsilon t])$ is a supermartingale. Let τ be the time that the exploration process terminates and recall that if we start with one active site x then $\tau = |\mathcal{C}_x|$. Using the stopping theorem for nonnegative supermartingales, Theorem 4.8.4 in PTE5, we have

$$e^{\theta} \ge Ee^{\theta\epsilon\tau} \ge e^{\theta\epsilon k}P(|\mathcal{C}_x| \ge k),$$

which proves the desired result. □

3.2 Motivating Examples

To motivate the machinery that needs to be developed, we will first show that it covers some interesting examples. The three sections here introduce three examples that will be our constant companions as we explain the theory.

3.2.1 *Dubins' Model*

Kalikow and Weiss (1988) considered a random graph on the positive integers $\mathbb{N}$ in which there is an edge between i and j with probability

$$p_{ij} = \left(\frac{\lambda}{\max\{i,j\}}\right) \wedge 1,$$

where $a \wedge b = \min\{a,b\}$. The existence of an infinite component is trivial: for any $\lambda \in (0,1]$, $p_{1,j} = \lambda/j$, so $\sum_j p_{1,j} = \infty$ and the component containing 1, $\mathcal{C}_1$ is always infinite. Because of this, they considered the question: Is $\mathbb{N}$ connected with positive probability? They showed that the answer was yes if $\lambda > 1$ and no if $\lambda < 1/4$. We call this Dubins' model, since Dubins had earlier conjectured that the critical value $\lambda_c = 1$. Shepp (1989) later proved that the answer to the question was yes if $\lambda > 1/4$ and no if $\lambda \le 1/4$, so the critical value $\lambda_c = 1/4$.

Callaway et al. (2001) or CHKNS for short, introduced the following simple version of a randomly grown graph. Start with one vertex labelled 1 and no edges. At each time $n \ge 2$, we add one vertex labelled n and with probability δ add one edge between two randomly chosen vertices $1 \le i < j \le n$. Note that the newly added vertex is not necessarily an endpoint of the added edge and when n is large, it is likely not to be.

CHKNS began their analysis by examining $N_k(t) =$ the expected number of components of size k at time t. Ignoring terms of $O(1/t^2)$, which come, for example, from picking the same cluster twice:

$$N_1(t+1) = N_1(t) + 1 - 2\delta\,\frac{N_1(t)}{t}, \tag{3.2.1}$$

$$N_k(t+1) = N_k(t) - 2\delta\,\frac{kN_k(t)}{t} + \delta \sum_{j=1}^{k-1} \frac{jN_j(t)}{t} \cdot \frac{(k-j)N_{k-j}(t)}{t}. \tag{3.2.2}$$

To explain the first equation, note that at each discrete time t one new vertex is added, and a given isolated vertex becomes the endpoint of an added edge with probability $\approx 2\delta/t$. For the second equation, note that the probability an edge connects to a given cluster of size k is $\approx 2\delta k/t$, while the second term corresponds to mergers of clusters of size j and $k-j$. There is no factor of 2 in the last term since we sum from 1 to $k-1$.

Theorem 3.2.1 *As $t \to \infty$, $N_k(t)/t \to a_k$ where $a_1 = 1/(1+2\delta)$ and for $k \ge 2$*

$$a_k = \frac{\delta}{1+2\delta k} \sum_{j=1}^{k-1} ja_j \cdot (k-j)a_{k-j}.$$

To solve for the a_k, which gives the limiting number of clusters of size k per site, CHKNS used generating functions. Let

$$h(x) = \sum_{k=1}^{\infty} x^k a_k \quad \text{and} \quad g(x) = \sum_{k=1}^{\infty} x^k k a_k.$$

Multiplying the formulas for a_k in Theorem 3.2.1 by $(1+2\delta k)x^k$, recalling the formula for a_1 is different from the others, and summing gives

$$h(x) + 2\delta g(x) = x + \delta g^2(x). \tag{3.2.3}$$

Since $h'(x) = g(x)/x$ differentiating (3.2.3) gives $g(x)/x + 2\delta g'(x) = 1 + 2\delta g(x)g'(x)$. Rearranging we have $2\delta g'(x)(1-g(x)) = 1 - g(x)/x$

$$g'(x) = \frac{1}{2\delta x} \cdot \frac{x - g(x)}{1 - g(x)}. \tag{3.2.4}$$

Let $b_k = ka_k$ be the number of vertices that belong to clusters of size k. $g(1) = \sum_{k=1}^{\infty} b_k$ gives the fraction of vertices that belong to finite components. $1 - g(1)$ gives the fraction of sites that belong to clusters whose size continues to grow in time. Even though it is not known that the missing mass in the limit belongs to a single cluster, it is common to call $1 - g(1)$ the fraction of sites that belong to the giant component. The next result gives the mean size of finite components. Here and throughout the motivating examples we state results without proving them (Fig. 3.1).

Using the differential equation (3.2.4) one can show

Lemma 3.2.2 *(i) If* $g(1) < 1$ *then* $\sum_{k=1}^{\infty} kb_k = g'(1) = 1/2\delta$.

(ii) If $g(1) = 1$ *then* $g'(1) = (1 - \sqrt{1 - 8\delta})/4\delta$.

Theorem 3.2.3 *The critical value* $\delta_c = \sup\{\delta : g(1) = 1\} = 1/8$ *and hence*

$$\sum_k kb_k = \begin{cases} (1 - \sqrt{1 - 8\delta})/4\delta & \delta \le 1/8, \\ 1/2\delta & \delta > 1/8. \end{cases}$$

Note that this implies that the mean cluster size $g'(1)$ is always finite, but is discontinuous at $\delta = 1/8$, since the value there is 2 but the limit for $\delta \downarrow 1/8$ is 4.

Durrett (2003) modified the model so that a Poisson mean δ number of vertices are added at each step. This version is nicer since in the Poisson case if we let $A_{i,j,k}$ be the event no (i, j) edge is added at time k then $P(A_{i,j,k}) = \exp\left(-\delta/\binom{k}{2}\right)$ for $i < j \le k$ and these events are independent.

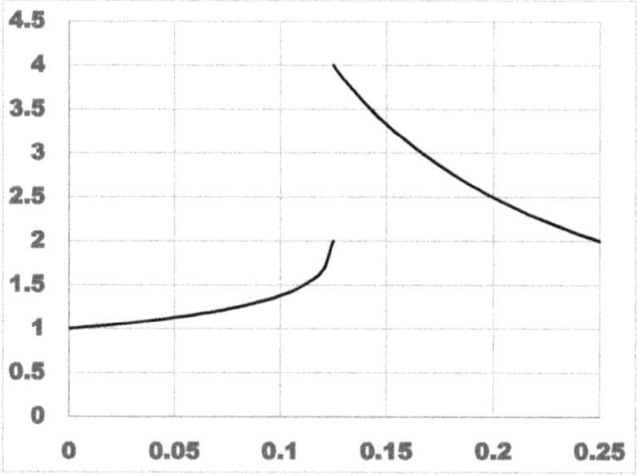

Fig. 3.1 Graph of the result in Theorem 3.2.3.

$$P(\cap_{k=j}^{n} A_{i,j,k}) = \prod_{k=j}^{n} \exp\left(-\frac{2\delta}{k(k-1)}\right)$$
$$= \exp\left(-2\delta\left(\frac{1}{j-1} - \frac{1}{n}\right)\right) \geq 1 - 2\delta\left(\frac{1}{j-1} - \frac{1}{n}\right). \qquad \#1$$

The last formula is not simple, so he also considered two approximations

$$\approx 1 - 2\delta\left(\frac{1}{j} - \frac{1}{n}\right), \qquad \#2$$
$$\approx 1 - \frac{2\delta}{j}. \qquad \#3$$

Referring to these three models by their numbers, he showed that

Theorem 3.2.4 *In models #1, #2, or #3, the critical value for the existence of a component of order n is* $\delta_c = 1/8$.

In addition Durrett (2003) obtained some results about the decay of connection probabilities at the critical value and in the subcritical regime. See Sect. 7.3 in *Random Graph Dynamics*, the previous version. Bollobás et al. (2005) were not aware of Durrett's work when they wrote their paper. They did an independent analysis of the CHKNS model, proving a remarkable result about the supercritical phase: the size of the giant component is infinitely differentiable at the critical value. To put their work into context we need to describe previous work.

To investigate the size of the giant component, CHKNS integrated the differential equation (3.2.4) near $\delta = 1/8$. Letting $S(\delta) = 1 - g(1)$ the fraction of vertices in the infinite component they plotted $\log(-\log S))$ vs $\log(\delta - 1/8)$ and concluded that

$$S(\delta) \sim \exp(-\alpha(\delta - 1/8)^{-\beta}),$$

where $\alpha = 1.132 \pm 0.008$ and $\beta = 0.499 \pm 0.001$. Based on this they conjectured that $\beta = 1/2$. Inspired by their conjecture Dorogovstev, Mendes, and Samukhin (2000) calculated (using methods that were not completely rigorous but turned out to be very accurate) that as $\delta \downarrow 1/8$,

$$S \equiv 1 - g(1) \approx c\exp(-\pi/\sqrt{8\delta - 1}). \qquad (3.2.5)$$

This result shows $\beta = 1/2$ and $\alpha = \pi/\sqrt{8} = 1.1107$. For a look at the awesome power of physics to study problems of this type see the survey article by Dorogovstev and Mendes (2002). BJR (2005) used $c = 2\delta$ as their parameter. They showed that if $\eta > 0$

$$\exp\left(-\frac{\pi + \eta}{2\sqrt{\epsilon}}\right) \leq C_1(G_n(1/4 + \epsilon)) \leq \exp\left(-\frac{1 - \eta}{2\sqrt{\epsilon}}\right).$$

Riordan (2005) improved the result to give the sharp constant

$$\mathcal{C}_1(G_n(1/4+\epsilon)) \approx \exp\left(-\frac{\pi}{2\sqrt{\epsilon}} + O(\log(1/\epsilon))\right),$$

which agrees with (3.2.5)

3.2.2 The Square Root Model

This can be succinctly described as a mean-field version of the preferential attachment model but it will take a while to explain what that means. Barabási and Albert (1999) introduced a model in which at every time step, we add a new vertex with m edges that link the new vertex to m different vertices already present in the system. To incorporate preferential attachment, we assume that the probability π that a new vertex will be connected to a vertex i depends only on the degree of that vertex.

To have a precise definition, we will suppose that the process starts at time 1 with two vertices linked by m parallel edges, so that the total degree at any time t is $2mt$. When we add a new vertex we will add edges one a time, with the second and subsequent edges doing preferential attachment using the updated degrees. This scheme has the desirable property that a graph of size n for a general m can be obtained by running the $m = 1$ model for nm steps and then collapsing vertices $km, km-1, \ldots, (k-1)m+1$ to make vertex k.

They did simulations that suggested a power law distribution with power $\gamma = 2.9 \pm 0.1$ and gave the following argument for $\gamma = 3$. If we consider the degree of i, k_i, to be a continuous variable then

$$\frac{\partial k_i}{\partial t} = m \cdot \frac{k_i}{\sum_j k_j} = \frac{k_i}{2t},$$

since $\sum_j k_j = 2mt$. The solution to this differential equation is

$$k_i(t) = m(t/t_i)^{1/2},$$

where t_i is the time the vertex was introduced. From this we see that

$$P(k_i(t) > k) = P(t_i < tm^2/k^2) = m^2/k^2,$$

since vertices are added uniformly on $[0, t]$. Differentiating

$$P(k_i(t) = k) = 2\,m^2/k^3.$$

Dorogovstev et al. (2000) took a different approach, using what they call the "master equation," which can be used to obtain rigorous asymptotics for the mean number of vertices of degree k, $N(k,t)$. By considering what happens on one step, in which vertices of degree k can be created from those of degree $k-1$ or lost by becoming degree $k+1$, we see that

$$N(k,t+1) - N(k,t) = \frac{m(k-1)}{2mt} N(k-1,t) - \frac{mk}{2mt} N(k,t) + \delta_{k,m}. \tag{3.2.6}$$

Here, we ignore the possibility of more than one edge being attached to one vertex, and the updating of edges that occurs as the m edges are added. The last term takes care of the fact that when $k = m$ we add one vertex of degree m at each time. To make the equation correct for $k = m$ we suppose $N(j,t) = 0$ for $j < m$.

Letting $p_k = \lim_{t\to\infty} N(k,t)/t$, and using (3.2.6) we find that $p_m = 2/(m+2)$ and

$$p_k = \frac{(k-1)p_{k-1}}{2} \cdot \frac{2}{k+2} = p_{k-1}\frac{k-1}{k+2} \quad \text{for } k > m.$$

The solution to this recursion is

$$p_k = \frac{2m(m+1)}{k(k+1)(k+2)}. \tag{3.2.7}$$

Bollobás and Riordan (2004) have performed a rigorous analysis of the diameter of the graphs produced by the Barabási-Albert model. To explain their result we first consider the case $m = 1$. They inductively define a sequence of directed random graphs G_1^t on $\{i : 1 \le i \le t\}$. Start with G_1^1, the graph with one vertex and one loop. Given G_1^{t-1} form G_1^t by adding the vertex t together with a directed edge from t to I where I is chosen randomly with

$$P(I = i) = \begin{cases} d_i^{t-1}/(2t-1) & 1 \le i \le t-1, \\ 1/(2t-1) & i = t, \end{cases} \tag{3.2.8}$$

where d_i^{t-1} = the degree of i in G_1^{t-1}. In words, we consider the outgoing edge from t when we consider where to attach the other end of the edge. Note that each vertex will have out degree 1. To extend the definition to $m > 1$, use the definition above to define random graphs G_1^{mt} on $\{v_i : 1 \le i \le mt\}$ and then combine the vertices $v_{(k-1)m+1}, \ldots, v_{km}$ to make vertex k.

Their main result is:

Theorem 3.2.5 *Let $m \ge 2$ and $\epsilon > 0$. Then with probability tending to 1, G_m^n is connected and*

$$(1-\epsilon)\log n/(\log\log n) \le \textit{diameter}(G_m^n) \le (1+\epsilon)\log n/(\log\log n).$$

The case $m = 1$, in which the graph is a tree, is excluded from the result. The upper bound is false in this case.

In order to prove the lower bound they consider G_1^N with $N = nm$ and compare G_1^N with a random graph in which an edge from i to j is present with probability $c/\sqrt{ij}$. Let g_j be the vertex to which j sends an edge when it is added to the graph.

Lemma 3.2.6 *(a) If* $1 \le i < j$ *then* $P(g_j = i) \le C_1(ij)^{-1/2}$.
(b) If $1 \le i < j < k$ *then* $P(g_j = i, g_k = i) \le C_2 i^{-1}(jk)^{-1/2}$.

This result motivates the consideration of the inhomogeneous graph with

$$p(i, j) = c/\sqrt{ij}, \tag{3.2.9}$$

which is often called the mean-field version of preferential attachment because edges are independent. There is a giant component in this graph for all $c > 0$. Riordan (2005) has shown for the square root model defined in (3.2.9) that there is a function $f(c)$ so that with high probability the largest component has $(f(c) + o(1))n$ vertices as $n \to \infty$ where

$$f(c) \sim 2e^{1-\gamma} \exp(-1/(2c) \qquad \text{as } c \to 0. \tag{3.2.10}$$

3.2.3 Chung-Lu (2002)

In this model weights w_i are assigned to vertices and we define the probability of an edge between i and j to be

$$p(i, j) = \frac{w_i w_j}{\sum_k w_k} \wedge 1. \tag{3.2.11}$$

The weight w_i is the expected degree of i since (ignoring the $\wedge 1$ and including the $j = i$ (self-loop) term in the sum.)

$$\sum_j \frac{w_i w_j}{\sum_k w_k} = w_i.$$

If we let

$$w_i = (i/Cn)^{-1/(\gamma-1)}, \tag{3.2.12}$$

then $w_i > w$ if and only if $i/n < Cw^{-(\gamma-1)}$ so the graph has a power law degree distribution $P(w_i > w) \sim Cw^{-(\gamma-1)}$. If $\gamma > 2$ then $Ew_i < \infty$ so $\sum_k w_k \sim \mu n$. To get the Chung-Lu model we take $a = 1/(\gamma - 1)$, $\psi(x) = x^{-a}$ and $\kappa(x, y) = c\psi(x)\psi(y)$. In this case

$$p(i,j) = \kappa(i/n, j/n)/n = \frac{c(i/n)^{-a}(j/n)^{-a}}{n}.$$

3.3 Welcome to the Machine

The description of the framework used in BJR (2007) will require a number of definitions. We will restrict ourselves to a level of generality that is enough to cover the examples we are interested in.

A **ground space** is a pair $(\mathcal{S}, \mu)$ where $\mathcal{S}$ is a separable metric space and μ is a Borel probability measure. For us this will usually be $\mathcal{S} = (0, 1]$ and $\mu =$ Lebesgue measure, i.e., $d\mu(x) = dx$

A **vertex space** $\mathcal{V}$ is a triple $(\mathcal{S}, \mu, \{x^n\}_{n\geq 1})$ where $(\mathcal{S}, \mu)$ is a ground space and $x^n = (x_1^n, \ldots, x_n^n)$ is a random sequence of points in S so that as $n \to \infty$

$$\nu_n(A) = \#\{i \leq n : x_i^n \in A\}/n \to \mu(A) \tag{3.3.1}$$

for every μ-continuity set, i.e., $\mu(\partial A) = 0$ where ∂A is the boundary of A. In most of our examples $x_i^n = i/n$. We will not consider **generalized vertex spaces** where the number of vertices is random.

A **kernel** κ on a ground space $(\mathcal{S}, \mu)$ is a symmetric non-negative Borel measurable function on $\mathcal{S} \times \mathcal{S}$.

Given a vertex space and a kernel, we can define a **random graph** $G^{\mathcal{V}}(n, \kappa)$ in which an edge from i to j is present with probability

$$p_{ij}^n = \kappa(x_i^n, x_j^n)/n \wedge 1.$$

To avoid taking the minimum with 1 one can define

$$p_{ij}^n = 1 - \exp(-\kappa(x_i^n, x_j^n)/n).$$

A kernel κ is **graphical** if (i) κ is continuous almost everywhere on $\mathcal{S} \times \mathcal{S}$, (ii) $\kappa \in L^1(\mathcal{S} \times \mathcal{S}, \mu \times \mu)$ and (iii) if $e(G)$ is the number of edges in G then

$$\frac{1}{n} Ee(G^{\mathcal{V}}(n, \kappa)) \to \frac{1}{2} \iint \kappa(x, y)\, d\mu(x)\, d\mu(y).$$

(ii) and (iii) imply that the expected number of edges is $O(n)$.

If $\kappa \in L^1(\mathcal{S} \times \mathcal{S}, \mu \times \mu)$ then $\int_{\mathcal{S}} \kappa(x, y)\, d\mu(y) < \infty$ for μ-a.e. x. It is convenient to assume that

$$\int_{\mathcal{S}} \kappa(x, y)\, d\mu(y) < \infty \quad \text{for all } x \in \mathcal{S}. \tag{3.3.2}$$

Since we will consider only examples on (0, 1] we will restrict our attention to kernels that are **piecewise continuous**. That is, there are $0 = a_0 < a_1 < a_2 < \cdots < a_k = 1$ so that κ is continuous on each (a_{i-1}, a_i). In addition we require that on each interval (a_{i-1}, a_i) κ converges to a finite limit as either endpoint is approached, unless $a_{i-1} = 0$, in which case we can have $\kappa \to \infty$.

It is useful to have some concrete examples to consider. In all these cases $\mathcal{S} = (0, 1]$, μ is Lebesgue measure and $x_i^n = i/n$.

Erdős-Rényi random graph. $\kappa(x, y) = \lambda$.

Finitely many types. For $1 \le k \le K$ we have intervals $I_k = (a_{k-1}, a_k]$ with $0 = a_0 < a_1 < \cdots < a_K = 1$. For $x \in I_k$, $y \in I_\ell$, $\kappa(x, y) = a(k, \ell)$.

Dubins' model. $\kappa(x, y) = c/\max\{x, y\}$

The next two are examples of **rank-one systems** which have $\kappa(x, y) = c\psi(x)\psi(y)$.

Square root model. $\kappa(x, y) = c/\sqrt{xy}$. In this case $\kappa \notin L^1$. As we will see this implies that the critical value $c_c = 0$.

Chung and Lu (2002). $\kappa(x, y) = c\psi(x)\psi(y)$ with $\psi(x) = x^{-1/(\gamma-1)}$.

3.3.1 Phase Transition

To begin we define a linear integral operator

$$(T_\kappa f)(x) = \int_{\mathcal{S}} \kappa(x, y) f(y)\, d\mu(y). \tag{3.3.3}$$

The norm $\|T_\kappa\| = \sup\{\|T_\kappa f\|_2 : f \ge 0, \|f\|_2 \le 1\}$. In the finite type case this operator is essentially the mean matrix of the branching process, and the norm is the same as the maximum eigenvalue. To be precise, if we let $g(x) = w_k$ for $x \in I_k$ (defined in the finitely many types example above) then

$$(T_\kappa g)(x) = \sum_k a(j, k) w_k \quad \text{for } x \in I_j.$$

In addition we define the nonlinear operator

$$\Phi_\kappa f = 1 - \exp(-T_\kappa f). \tag{3.3.4}$$

The survival probability $\rho_\kappa(x)$ is the maximum fixed point of Φ. Recall that in the Erdős-Rényi and configuration models the extinction probability is defined in terms of the smallest fixed point of generating functions. In the Erdős-Rényi case when the mean degree is $\lambda > 1$ the extinction probability satisfies

$$z = \exp(-\lambda(1 - z)),$$

so $\rho = 1 - z$ satisfies $\rho = 1 - \exp(-\lambda\rho)$. To make the connection with (3.3.4) note that $\kappa(x, y) = \lambda$ for Erdős-Rényi .

To state the next result we need a definition. κ is **reducible** if there is an A with $0 < \mu(A) \le \mu(\mathcal{S})$ so that $\kappa(x, y) = 0$ on $A \times (\mathcal{S} - A)$. Otherwise it is **irreducible**.

Theorem 3.3.1 (i) *If* $\|T_\kappa\| < 1$ *then the largest component* $\mathcal{C}_1$ *is* $o(n)$ *while if* $\|T_\kappa\| > 1$ *it is* $\Theta(n)$.
(ii) *If* κ *is irreducible then*

$$(1/n)|\mathcal{C}_1| \to \rho = \int_0^1 \rho_\kappa(x)\, d\mu(x).$$

The reasoning is similar to that for Erdős-Rényi. x is part of the giant component if the branching process started at x does not die out.

Rank-one case. When $\kappa(x, y) = c\psi(x)\psi(y)$ the operator T has

$$T_\kappa f = c\left(\int f\psi\, d\mu\right) \cdot \psi.$$

Since T maps all functions to multiples of ψ, ψ is the only possible eigenvector. It has eigenvalue $c \int \psi(x)^2\, dx$ and critical value

$$c_c = 1/\int \psi(x)^2\, dx.$$

Square root model. $\psi(x) = cx^{-1/2}$ which has $\int \psi^2\, dx = \infty$, so $c_c = 0$.

Chung-Lu model. $\psi(x) = x^{-1/(\gamma-1)}$ so $c_c = 0$ if $\gamma \le 3$ while it is > 0 if $\gamma > 3$.

Dubins' model. BJR observe in their Sect. 16.1 that

$$U : f \to e^{-x/2} f(e^{-x})$$

is an isometry from $L^2(0, 1)$ onto $L^2(0, \infty)$. The operator $\tilde{T}_\kappa = UT_\kappa U^{-1}$ has kernel

$$\tilde{\kappa}(x, y) = e^{-x/2}\kappa(e^{-x}, e^{-y})e^{-y/2} = e^{-x/2-y/2+x\wedge y} = e^{-|x-y|/2}.$$

Thus $\tilde{T}_\kappa$ is the restriction to $(0, \infty)$ of the convolution with $h(x) = ce^{-|x|/2}$. "Taking Fourier transforms we find $\|T_\kappa\| = 4c$." Since it is not hard to find the critical value of the model directly, will not go into the details.

3.3.2 Degree Distribution

In the special case $\mathcal{S} = (0, 1]$ with μ Lebesgue measure, we define

$$\lambda(x) = \int_0^1 \kappa(x, y)\, dy$$

to be the **expected degree** of x.

Theorem 3.13 in BJR (2007). Writing Z_k for the number of vertices of degree k,

$$Z_k/n \to P(D = k) = \int_0^1 e^{-\lambda(x)} \frac{\lambda(x)^k}{k!}\, dx.$$

In words, the degree D of a randomly chosen vertex has the mixed Poisson distribution

$$\int_0^1 \text{Poisson}(\lambda(x))\, dx.$$

Let Λ be the expected degree of a randomly chosen vertex. That is, $\Lambda = \lambda(U)$ where U is uniform on $(0, 1]$.

Corollary 3.13 in BJR (2007). Suppose that

$$P(\Lambda > t) = \mu\{x : \lambda(x) > t\} \sim at^{-\alpha}$$

for some $a > 0$ and $\alpha > 1$. Let $Z_{\geq k}$ be the number of vertices with degree $\geq k$ then

$$Z_{\geq k}/n \to P(D \geq k) \sim ak^{-\alpha}.$$

Proof By Theorem 3.13 it suffices to show that $P(D \geq k) \sim ak^{-\alpha}$. To do this we note that if $Z = \text{Poisson}(\lambda)$ then $(Z - \lambda)/\lambda^{1/2}$ converges to a normal distribution as $\lambda \to \infty$ so if $1/2 < b < 1$

$$P(\lambda - \lambda^b < \text{Poisson}(\lambda) < \lambda + \lambda^b) \to 1.$$

Using this we conclude that if k is large and $\epsilon > 0$

$$P(\text{Poisson}(\lambda(x)) \geq k) \approx \begin{cases} 1 & \text{if } \lambda(x) \geq (1 + \epsilon)K, \\ 0 & \text{if } \lambda(x) \leq (1 - \epsilon)K. \end{cases}$$

If we suppose for simplicity that $\lambda(x)$ is decreasing we have

$$P(D \geq k) \approx \mu\{x : \lambda(x) \geq k\},$$

which gives the desired result.

Rank-one model. $\lambda(x) = c\psi(x) \int_0^1 \psi(y)\, dy$. If $\psi(x) = x^{-a}$ and $a < 1$ then

$$\lambda(x) = cx^{-a} \int_0^1 y^{-a}\, dy = cx^{-a}/(1-a).$$

Letting $B = c/(1-a)$ we have $\lambda(x) > t$ if $Bx^{-a} > t$ or $x < (t/B)^{-1/a}$. If $t > B$ then

$$P(\Lambda > t) = (t/B)^{-1/a}.$$

In the Chung-Lu model $a = 1/(\gamma - 1)$, so $P(\Lambda > t) \sim Ct^{-\gamma+1}$ so the degree distribution has $p_k \sim C'k^{-\gamma}$.

3.4 Computing the Survival Probability

The fact that the survival probability $\rho_\kappa(x)$ is the maximum fixed point of

$$\Phi_\kappa f = 1 - \exp(-T_\kappa f)$$

is not really useful since we have to compute a function. Because of this we will follow Sect. 5 of Riordan (2005) who found survival probabilities for some of the examples we are most interested in by using branching process with a type space that is $(0, 1]$. To define the branching process we note that the children of y are a **Poisson process with mean measure** $\kappa(y, z)\, dz$ on $(0, 1]$. That is, if we let $N(A)$ be the number of points in A then

(i) If $A_1, \ldots, A_n$ are disjoint $N(A_1), \ldots, N(A_k)$ are independent.
(ii) $N(A)$ has Poisson $\left(\int_A \kappa(y, z)\, dz\right)$ distribution.

The next result will make this definition more explicit.

Lemma 3.4.1 *If a particle in X_k has type a and m children they will have joint distribution*

$$f_m(a, b_1, \ldots b_m) = \exp\left(-\int_0^1 \kappa(a, b)\, db\right) \frac{1}{m!} \prod_{i=1}^m \kappa(a, b_i).$$

Proof To see this suppose we have points $1/n, 2/n, \dots, 1$ and a random graph with

$$p(i, j) = \kappa(i/n, j/n)/n.$$

The probability that j connects to $k_1, \dots, k_m$ and to no other vertices is

$$= \prod_{k \neq k_i} (1 - \kappa(j/n, k/n)/n) \prod_{i=1}^{m} \kappa(j/n, k_i/n) \cdot \frac{1}{n}.$$

Letting $a = j/n$ nd $b_i = k_i/n$, the first term is

$$\approx \exp\left(- \int_0^1 g(\alpha, \beta)\, d\beta \right).$$

If we think of b_i as an interval of length $1/n$ and $g(a, b)$ as a density function the second term is

$$\approx \prod_{i=1}^{m} g(a, b_i).$$

If we get rid of the ordering $\beta_1 < \cdots < \beta_m$ we need to divide by $m!$

Our next task is to define the **success probability**. In Riordan (2005)'s approach we let

$$F(g)(a) = \sum_{m \geq 0} \int_{b_1, \dots b_m \in \mathcal{S}} f_m(a, b_1, \dots b_m) \left(1 - \prod_{i=1}^{m} (1 - g(b_i)) \right) db_n \cdots db_1.$$

If we assign to each child b an event called success that has probability $g(b)$ then $F(g)(a)$ is the probability of success of at least one child of a. If we let $S_\infty(b)$ be the probability that the branching process starting with one individual of type b does not die out then $S_\infty = F(S_\infty)$.

This already gives us enough to use Riordan's approach. For completeness we derive a formula for $\Phi_\kappa g$. This is from Sect. 5 of BJR (2007), which has results that enable one to conclude that under natural assumptions there are only two solutions 0 and ρ.

Lemma 3.4.2 *Consider the random offspring of a single particle of type x and let N be the total number of children. If $0 \leq g \leq 1$ is measurable*

$$(\Phi_\kappa g)(x) = 1 - E \prod_{i=1}^{N} (1 - g(\xi_i)) = 1 - e^{-(T_\kappa g)(x)}. \tag{3.4.1}$$

Proof Let $d\nu(y) = \kappa(x, y)\,dy$ and $\nu' = \nu/\nu(\mathcal{S})$ Given $N = n$, if we list the locations in random order $\xi_1, \xi_2, \ldots, \xi_n$ the locations are independent with distribution ν' so

$$\prod_{i=1}^{n} E(1 - g(\xi_i)) = (E[1 - g(\xi_1)])^n$$
$$= \left(1 - \int_0^1 g(y)\,d\nu'(y)\right)^n = \left(1 - \frac{(T_\kappa g)(x)}{\nu(S)}\right)^n.$$

Using $P(N = n) = e^{-\nu(\mathcal{S})}\nu(\mathcal{S})^n/n!$ taking expectation, and writing $h(x) = (T_\kappa g)(x)$ to simplify the formulas gives

$$E\prod_{i=1}^{N}(1 - g(\xi_i)) = \sum_{n=0}^{\infty} e^{-\nu(\mathcal{S})}\frac{\nu(\mathcal{S})^n}{n!} \cdot \left(1 - \frac{h(x)}{\nu(S)}\right)^n$$
$$= e^{-h(x)} \sum_{n=0}^{\infty} e^{-\nu(\mathcal{S})+h(x)} \frac{[\nu(\mathcal{S}) - h(x)]^n}{n!} = e^{-h(x)},$$

which proves the second equality.

To approximately calculate S_∞, Riordan uses (see his page 912):

Lemma 3.4.3 *If* $\phi : (0, 1] \to [0, 1]$ *satisfies* $F(\phi) \geq \phi$ *then* $S_\infty \geq \phi$.

Lemma 3.4.4 *If* $\phi : (0, 1] \to [0, 1]$ *satisfies* $F(\phi) \leq \phi$ *and* $F(\phi)$ *is not* $\equiv 0$ *then* $S_\infty \leq \phi$.

3.5 Survival Probabilities for Examples

3.5.1 Square-Root Model

We now move on to Sect. 6 in Riordan (2005). Section 6.1 couples the random graph to the branching process to show that the size of the giant component is close to the survival probability of the branching process. We skip that detail and move on to Sect. 6.2 where he computes the survival probability of the branching process. If the initial type is a then the expected number of first generation children that survive is

$$\mu = \int_0^1 \frac{c}{\sqrt{ab}} S_\infty(b)\,db.$$

The set of offspring is a Poisson process on $(0, 1]$. We thin the process by keeping only the points that survive. Survival events are independent of each other and independent of the Poisson process so the number of survivors has a Poisson distribution.

If the number that survive is Poisson with mean μ the probability at least one survives is $1 - e^{-\mu}$ so

$$S_\infty(a) = 1 - \exp\left(- \int_0^1 \frac{c}{\sqrt{ab}} S_\infty(b)\, db \right). \tag{3.5.1}$$

If we let $A_c = \int_0^1 (c/\sqrt{b}) S_\infty(b)\, db$ then

$$S_\infty(a) = 1 - \exp(-A_c/\sqrt{a}),$$

so using (3.5.1), equating the exponentials, and multiplying by $\sqrt{a}$ we have

$$A_c = c \int_0^1 \frac{1}{\sqrt{b}} [1 - \exp(-A_c/\sqrt{b})]\, db. \tag{3.5.2}$$

We want to solve for A_c, or at least determine the behavior of A_c as $c \to 0$. Let

$$g(x) = \int_0^1 \frac{1}{\sqrt{b}} [1 - \exp(-x/\sqrt{b})]\, db.$$

With this notation $A_c = cg(A_c)$.

Lemma 3.5.1 *$g(x) \sim 2x \log(1/x)$ as $x \to 0$.*

Proof The first step is to change variables $z = \sqrt{b}$, $dz = (1/2)b^{-1/2} db$ to get

$$g(x) = 2 \int_0^1 dz\, (1 - e^{-x/z}).$$

Noting that

$$1 - e^{-x/z} \approx \begin{cases} x/z & z \gg x \\ 1 & z \ll x \end{cases}$$

suggests an approach to approximating the integral

$$2 \int_0^{Kx} dz\, (1 - e^{-x/z}) \in [0, 2Kx].$$

If K is large

$$2\int_{Kx}^{1} dz\,[1 - e^{-x/z}] \approx 2x \int_{Kx}^{1} \frac{dz}{z} = -2x \ln Kx = 2x \log(1/x) - 2x \log(K).$$

Taking $K = \log\log(1/x)$ now gives the desired result.

To compute the asymptotics for the critical value now we note that

$$c = \frac{A_c}{g(A_c)} \sim \frac{1}{2\log(1/A_c)} \quad \text{or} \quad \log(1/A_c) \sim 1/2c,$$

which implies $A_c = \exp(-(1 + o(1))/2c)$. Riordan says one can get more precise results by using

$$g(x) = 2 - 2e^{-x} + 2x E_1(x) \quad \text{where} \quad E_1(x) = \int_1^x \exp(-xt)/t\, dt.$$

This leads to the result given earlier in (3.2.10)

$$f(c) \sim 2e^{1-\gamma} \exp(-1/2c) \qquad \text{as } c \to 0.$$

3.5.2 *Chung-Lu Model (2002)*

This was an undergraduate project for Alice Ding at Duke. The finished product, Ding and Durrett (2023), is available on the arXiv. Our first step is to repeat the argument above for a general rank-1 model

$$\kappa(x, y) = c\psi(x)\psi(y).$$

If the initial type is a then the expected number of first generating children that survive is

$$\mu = \int_0^1 c\psi(a)\psi(b)S_\infty(b)\, db.$$

If the number that survive is Poisson with mean μ the probability at least one survives is $1 - e^{-\mu}$ so

$$S_\infty(a) = 1 - \exp\left(-\int_0^1 c\psi(a)\psi(b)S_\infty(b)\, db\right). \tag{3.5.3}$$

If we let $A_c = \int_0^1 c\psi(b)S_\infty(b)\,db$ then

$$S_\infty(a) = 1 - \exp(-A_c\psi(a)),$$

so we have

$$A_c = c\int_0^1 \psi(b)[1 - \exp(-A_c\psi(b))]\,db. \tag{3.5.4}$$

Again if we let

$$g(x) = \int_0^1 \psi(b)[1 - \exp(-x\psi(b))]\,db, \tag{3.5.5}$$

then with this notation $A_c = cg(A_c)$.

$\gamma > 3$

In the Chung-Lu model $\psi(x) = x^{-1/(\gamma-1)}$. When $\gamma > 3$

$$\int_0^1 \psi(x)^2\,dx < \infty, \tag{3.5.6}$$

so $c_c = 1/\int_0^1 \psi(x)^2\,dx > 0$. If (3.5.6) holds then as $x \to 0$

$$g(x) \sim x\int_0^1 \psi(y)^2\,dy,$$

so, in contrast to the square root model, g has a finite slope at 0.

$$g'(x) = \int_0^1 \psi^2(b)\exp(-x\psi(b))\,d\beta,$$

so $g'(x)$ decreases as x increases i.e., x is concave.

Since $g(0) = 0$, $g'(0) = \int_0^1 \psi(x)^2\,dx$, and (3.5.5) implies that as $x \to \infty$

$$g(x) \to \int_0^1 \psi(b)\,db \quad \text{and hence} \quad g(x)/x \to 0.$$

Thus we will have a solution of $x/c = g(x)$ only if $1/c < \int_0^1 \psi(x)^2\,dx$, i.e., $c > c_c$. Expanding g in a power series about 0

$$g(x) = xg'(0) + \frac{x^2}{2}g''(0) + \cdots,$$

so dropping the three dots (and leaving it to the reader to justify this) we have

$$A_c = cg(A_c) \approx cg'(0)A_c + g''(0)\frac{A_c^2}{2}.$$

Solving and recalling $c_c = 1/g'(0)$ we have

$$A_c = \frac{2(cg'(0) - 1)}{-g''(0)} = \frac{2g'(0)}{-g''(0)}(c - c_c).$$

If the negative sign in the constant looks weird recall that $g''(0) < 0$ since g is concave.

$$S_\infty(a) = 1 - \exp(-A_c\psi(a)) \sim \psi(a)A_c,$$

as $A_c \to 0$ so as $c \downarrow c_c$

$$S_\infty(a) \sim C\psi(a)(c - c_c)^\beta \quad \text{with } \beta = 1.$$

Thus in the Chung-Lu model with $\gamma > 3$ the critical behavior of the survival probability has the same behavior in the Erdős-Rényi case. This is the behavior one should expect since the degree distribution has finite second moment.

When $\gamma = 3$ we have $\psi(x) = x^{-1/2}$ and this is essentially the square root model, so we move on to

$2 < \gamma < 3$

In this case

$$g(x) = \int_0^1 \frac{1}{b^{1/(\gamma-1)}}[1 - \exp(-x/b^{1/(\gamma-1)})]\,db$$

Imitating the proof for the square-root model we change variables

$$z = b^{1/(\gamma-1)}, \quad \text{and} \quad dz = \frac{1}{\gamma - 1} b^{1/(\gamma-1)-1},$$

which means

$$db = (\gamma - 1)b^{-1/(\gamma-1)+1}\, dz = (\gamma - 1)z^{-1+(\gamma-1)}\, dz = (\gamma - 1)\frac{dz}{z^{2-\gamma}}.$$

Thus we have

$$g(x) = (\gamma - 1)\int_0^1 \frac{1}{z^{3-\gamma}}(1 - \exp(-x/z))\, dz.$$

To deal with this formula we change variables again $z = x/y$, $dz = -(x/y^2)dy$ to get

$$g(x) = x^{\gamma-2}\int_x^\infty y^{1-\gamma}(1 - e^{-y})\, dy.$$

When $\gamma > 2$ we have $1 - \gamma < -1$ so the integral is convergent at ∞. Near 0 we have $1 - e^{-y} \sim y$. The exponent $2 - \gamma < -1$ when $\gamma < 3$ so the integral is convergent at 0 and the above is

$$\sim C_\gamma x^{\gamma-2} \quad \text{where} \quad C_\gamma = \int_0^\infty y^{1-\gamma}(1 - e^{-y})\, dy.$$

As $x \to 0$, $x/g(x) \sim x^{3-\gamma}/C_\gamma$. Using $c = A_c/g(A_c)$ we have $c \sim A^{3-\gamma}/C_\gamma$ and

$$A_c \sim (C_\gamma c)^{1/(3-\gamma)}.$$

Recall that $A_c = \int_0^1 cb^{-1/(\gamma-1)} S_\infty(b)\, db$ so this is the survival probability starting from a random point chosen according to a measure that is finite when $\gamma > 2$. From this we conclude that the critical exponent $\beta = 1/(3 - \gamma)$. It may look strange that $\beta = \infty$ when $\gamma = 3$ but that value corresponds to the square-root model.

3.6 Component Sizes in the Subcritical Case

3.6.1 Cases Where the Max Is $O(\log n)$

Theorem 3.12 in BJR shows that if

$$\sup_{x,y} \kappa(x, y) < \infty \tag{A}$$

then in the subcritical regime the largest component is $O(\log n)$.

The proof given in Sect. 12 of their paper (see Theorem 12.5) is more than a little mysterious (to me at least) but if one is willing to assume instead

(A') There is a subcritical kernel $\bar{\kappa}$ for an inhomogeneous graph with finitely many types that has $\bar{\kappa}(x, y) \geq \kappa(x, y)$

then the conclusion follows easily from the reasoning in Sect. 3.1.3.

Turova (2011) For the rest of the section we will concentrate on the **rank-one case** because in that situation we can define a cluster exploration process that is a random walk. For simplicity, we will work with the limiting object in which the children of y in the branching process approximation are a Poisson process with mean measure $\kappa(y, z)\, dz$ on $(0, 1]$. See Sect. 3.4 for a little more detail. In the rank-1 case $\kappa(y, z) = c\psi(y)\psi(z)$ and hence

$$c_c = 1/\int_0^1 \psi(y)^2 dy.$$

In order to have a subcritical phase we will suppose $\int \psi(y)^2 dy < \infty$ Using the notation of Turova (2011), we let

$$M = \int \psi(u)\, du < \left(\int \psi(y)^2 dy\right)^{1/2} < \infty.$$

When $M < \infty$ the set of offspring of the vertex y can be constructed in two steps.

(i) The total number of offspring is $N(y) =$ Poisson with mean

$$\int \kappa(y, z)\, dz = c\psi(y)M.$$

(ii) Conditional on $N(y) = m$, the labels of the m offspring are distributed according to $\bar{\psi}(z) = \psi(z)/M$.

Given this procedure we can construct the exploration process as follows. If we start from $X_0 = x_0$ then the number of offspring $N(x_0) =$ Poisson with mean $c\psi(x_0)M$. Since $A_0 = 1$ we have

$$A_1 = N(x_0).$$

For all $t \geq 1$ the vertex X_t whose neighbors are added on the tth step is distributed according to $\bar{\psi}(z)$, and

$$A_{t+1} = A_t - 1 + N(X_t).$$

The distribution of $N(X_t)$ is a mixture

$$\int_0^1 \bar{\psi}(y) \,\text{Poisson}(c\psi(y)M)\, dy.$$

At time $\tau = \inf\{t : A_t = 0\}$ we have found the τ members of the cluster.

The process is easier to analyze if it is homogeneous in time so we will suppose that X_0 is distributed according to $\bar{\psi}(z)$. To simplify the typing let $Y_t = -1 + N(X_t)$. Recalling the formula for the Poisson and dropping the subscript t, the moment generating function of Y is

$$\phi(\theta) \equiv Ee^{\theta Y} = e^{-\theta} \int_0^1 \bar{\psi}(y) e^{c\psi(y)M\cdot(e^\theta - 1)}\, dy.$$

$W_t = \exp(\theta A_t)/\phi(\theta)^t$ is a martingale for $t \le \tau$. To choose θ we note that

$$\phi'(\theta) = -\phi(\theta) + e^{-\theta} \int_0^1 \frac{\psi(y)}{M} e^{c\psi(y)M\cdot(e^\theta - 1)} \cdot c\psi(y)Me^\theta\, dy. \tag{3.6.1}$$

Taking $\theta = 0$ we have

$$\phi'(0) = -1 + \int_0^1 c\psi(y)^2 dy < 0,$$

since we are supposing $c < c_c$.

Since $\phi(0) = 1$ and $\phi'(0) < 0$, if we choose $\theta_0 > 0$ small enough then $\phi(\theta_0) < 1$. Using the optional stopping theorem for the nonnegative supermartingale W_t at time τ which has $A_\tau = 0$

$$e^{\theta_0} \ge E(\phi(\theta_0)^{-\tau}).$$

If we let $e^\alpha = 1/\phi(\theta_0) > 1$ then we have $Ee^{\alpha\tau} \le e^{\theta_0}$ and

$$P(\tau \ge k) \le e^{\theta_0} e^{-\alpha k}. \tag{3.6.2}$$

The last result is for the cluster size when we start from a random vertex chosen according to $\bar{\psi}(y)$. If we start from a fixed vertex x then $A_1 = N(x)$ and X_1 has distribution $\bar{\psi}(y)$ so by (3.6.2)

$$P(\tau \ge k + 1 | X_0 = x, N(x) = m) \le e^{\theta_0 m} e^{-\alpha k}.$$

Here the martingale starts when $A_1 = m$ instead of $A_0 = 1$. Multiplying by $P(N(x) = m)$ and summing over m

$$P(\tau \geq k + 1 | X_0 = x) \leq e^{-\alpha k} E e^{\theta_0 N(x)} = e^{-\alpha k} \exp(c\psi(x) M(e^{\theta} - 1)).$$

To bound this it is natural to assume that

$$C_a = \int_0^1 \exp(a\psi(x))\,dx < \infty \quad \text{for some } a > 0. \tag{A1}$$

To be able to relate the integral to $\sum_{i=1}^n \exp(a\psi(i/n))$ we will assume that

$$x \to \psi(x) \text{ is decreasing.} \tag{A2}$$

(Together these two assumptions are a little stronger than Turova's Assumption 1.1.) If θ_0 is small then $c_c M(e^{\theta_0} - 1) \leq a$, so

$$\frac{1}{n} \sum_{i=1}^{n} P(\tau \geq k + 1 | X_0 = i/n) \leq e^{-\alpha k} \int_0^1 e^{a\psi(x)}\,dx,$$

and we have

$$\sum_{i=1}^{n} P(\tau \geq k + 1 | X_0 = i/n) \leq n C_a e^{-\alpha k}.$$

If we take $k = ((1 + \epsilon)/\alpha) \log n$ then the probability that the largest cluster $> (1 + \epsilon/\alpha) \log n$ is $\leq C_a/n^{\epsilon} \to 0$, so

$$P\left(\max_{1 \leq i \leq n} |C_i| > \frac{1 + \epsilon}{\alpha} \log n \right) \to 0. \tag{3.6.3}$$

Here we have gotten an upper bound on component sizes that are $O(\log\, n)$. Turova (2011) obtained results that were asymptotically sharp

Theorem 3.6.1 *(Theorem 1.3. in Turova (2011).) Let $\kappa(x, y) = c\psi(x)\psi(y)$ and define $r(c)$ by*

$$c^{cr} = \left(\int \psi^2(x)\,dx \right)^{-1}.$$

As $n \to \infty$ we have

$$\frac{C_1(n, \kappa)}{\log n} \to \frac{1}{\log r(c)}.$$

Her Theorem 1.4 gives an explicit formula for $r(c)$. Intuitively this is done by optimizing the θ in the bound that comes from the moment generating function, but it takes more than a little work to carry out the plan announced in the previous sentence. We refer the reader to her paper for details.

3.6.2 Cases Where the Max Is a Power of n

We will consider the Chung-Lu model introduced in Sect. 3.2.3. $\kappa(x, y) = c\psi(x)\psi(y)$ where $\psi(x) = x^{-a}$ with $a = 1/(\gamma - 1)$ and we assume $\gamma > 3$. To make it easier to do computations we will have self-loops at i with probability $p(i, i)$ and assume they add only 1 to the degree of i instead of 2.

Degree distribution. The probability of an edge between i and j is

$$p(i, j) = \kappa(i/n, j/n)/n = ci^{-a}j^{-a}n^{2a-1}.$$

Lemma 3.6.2 *As* $n \to \infty$

$$E(d_i) \sim \frac{c}{1-a} \cdot \frac{n^a}{i^a}.$$

Proof The expected degree of i

$$E(d_i) = \frac{c}{i^a} n^{2a-1} \sum_{j=1}^{n} \frac{1}{j^a}.$$

To evaluate the sum, we think about Riemann approximating sums for the integral of $1/x^a$ when the end points of the intervals are integers. Since $1/x^a$ is decreasing,

if we evaluate at the left end point

$$\sum_{j=1}^{n} \frac{1}{j^a} \geq \int_1^{n+1} x^{-a}\, dx = \frac{(n+1)^{1-a} - 1}{1-a}.$$

To get a bound in the other direction note that if we evaluate at the right endpoint

$$\sum_{j=2}^{n} \frac{1}{j^a} \leq \int_1^{n} x^{-a}\, dx \leq \frac{n^{1-a}}{1-a},$$

which proves the desired result.

Given the formula for the mean degree in Lemma 3.6.2 the next result should not be surprising, but to prove it we need estimates of the deviations of the degrees from their means.

Lemma 3.6.3

$$P\left(d_1 = \max_{1 \le i \le n} d_i\right) \to 1. \tag{3.6.4}$$

$d_1/Ed_1 \to 1$ as $n \to \infty$ and hence

$$\max_{1 \le i \le n} d_i/Ed_1 \to 1. \tag{3.6.5}$$

Proof Our first step is to compute the variance.

$$\text{var}\,(d_i) = \sum_{j=1}^{n} p(i,j)(1-p(i,j)) \le \sum_{i=1}^{n} p(i,j) = Ed_i.$$

Chebyshev's inequality says that if $E(Y^2) < \infty$ then

$$P(|Y - EY| > y) \le \frac{\text{var}\,(Y)}{y^2}, \tag{3.6.6}$$

so taking $y = (Ed_i)^{2/3}$ we have

$$P(|d_i - Ed_i| > (Ed_i)^{2/3}) \le \frac{Ed_i}{(Ed_i)^{4/3}} \to 0.$$

From this it follows that $d_i/Ed_i \to 1$ in probability, i.e., for any fixed i if $\delta > 0$

$$P(1 - \delta < d_i/Ed_i < 1 + \delta) \to 0, \tag{3.6.7}$$

as $n \to \infty$. Since $Ed_i = Ed_1/i^a$, we can conclude that for any fixed I

$$P\left(d_1 = \max_{1 \le i \le I} d_i\right) \to 1. \tag{3.6.8}$$

The argument cannot be used to prove (3.6.4) because the error bound from Chebyshev's inequality is not good enough. To get a better bound we turn to the moment generating function. The analogue of Chebyshev's inequality is:

Lemma 3.6.4 *If $\theta > 0$ and the moment generating function $\phi(\theta) = Ee^{\theta Y} < \infty$ then*

$$e^{\theta y} P(Y \ge y) \le \phi(\theta). \tag{3.6.9}$$

Proof This holds since $e^{\theta Y} \ge e^{\theta y}$ when $Y \ge y$.

Since oriented edges from i to j are independently present with probability $p(i, j)$

$$Ee^{\theta d_i} = \prod_{j=1}^{n} \left(1 - p(i, j) + p(i, j)e^{\theta}\right) = \prod_{j=1}^{n} \left(1 + p(i, j)[e^{\theta} - 1]\right), \qquad (3.6.10)$$

since the m.g.f. of a sum of independent random variables is the product of the m.g.f. We need to simplify the last expression. To do this we are guided by the fact that the sum of independent Bernoulli random variables is approximately Poisson, while if Z is Poisson with mean μ the mgf is

$$Ee^{\theta Z} = \sum_{k=0}^{\infty} e^{-\mu} \frac{\mu^k}{k!} e^{\theta k} = \exp(\mu[e^{\theta} - 1]).$$

Using $1 + x \leq e^x$ in (3.6.10) and letting $\mu_i = \sum_j p(i, j)$, we have

$$Ee^{\theta d_i} \leq \exp(\mu_i[e^{\theta} - 1]),$$

so using (3.6.9), if $c > 1$

$$P(d_i \geq c\mu_i) \leq \exp(\mu_i[e^{\theta} - 1 - \theta c]).$$

This is a very powerful inequality but we have to manipulate it to be useful. A more systematic approach is to optimize over θ, but that can get messy. To simplify our inequality we set $\theta = \log 2$ to get

$$P(d_i \geq c\mu_i) \leq \exp(\mu_i[1 - c \log 2]).$$

We have already taken care of indices $1 \leq i \leq I$ so we can assume $i > I$. Taking $c = \mu_1/(2\mu_i)$ we have

$$P(d_i \geq \mu_1/2) \leq \exp\left(\mu_i \left[1 - \frac{\mu_1}{2\mu_i} \log 2\right]\right).$$

The ratio $\mu_1/\mu_i = i^a$. If $(\mu_1/2\mu_I) \log 2 > 2$ then for $i > I$ we have $(\mu_1/4\mu_i) \log 2 > 1$, and hence

$$P(d_i \geq \mu_1/2) \leq \exp\left(\mu_i \left[-\frac{\mu_1}{4\mu_i} \log 2\right]\right) = \exp\left(-\frac{\mu_1}{4} \log 2\right).$$

It follows that

$$\sum_{i=I+1}^{n} P(d_i \geq \mu_1/2) \leq n \exp\left(-\frac{\mu_1}{4} \log 2\right),$$

as $n \to \infty$ since $\mu_1 \sim Cn^{1-a}$. Combining this with (3.6.8) proves (3.6.4). (3.6.5) now follows from (3.6.7).

Subcritical cluster sizes. As in Sect. 3.6.1, we use the exploration process. If we start from $X_0 = x_0$ then the number of neighbors $N(x_0) =$ Poisson with mean $c\psi(x_0)M$ where $\psi(x) = x^{-1/(\gamma-1)}$, $M = \int_0^1 \psi(z)\,dz$. Since initially the active set $A_0 = 1$ we have

$$A_1 = N(x_0).$$

For all $t \geq 1$ the index of the vertex X_t whose neighbors are revealed on the tth step is distributed according to $\bar{\psi}(z) = \psi(z)/M$, and

$$A_{t+1} = A_t - 1 + N(X_t). \tag{3.6.11}$$

The distribution of $N(X_t)$ is a mixture

$$\int_0^1 \bar{\psi}(y)\,\text{Poisson}(c\psi(y)M)\,dy.$$

The expected value is

$$EN(X_t) = c\int_0^1 \psi(y)^2\,dy < 1 \qquad \text{if } c < c_c. \tag{3.6.12}$$

Using the reasoning from Sect. 3.2.3, when we compute the degree distribution we can think of Poisson$(c\psi(y)M)$ as being a point mass at the mean. $c\psi(y)M \geq x$ when

$$y^{-1/(\gamma-1)} \geq x/cM \qquad \text{or} \qquad y \leq (x/cM)^{-(\gamma-1)}.$$

Letting $z = (x/cM)^{-(\gamma-1)}$ we have

$$P(N(X_t) \geq x) = \int_0^z \frac{\psi(y)}{M}\,dy = \frac{C}{M}x^{1-1/(\gamma-1)} = C_5 x^{-(\gamma-1)+1}, \tag{3.6.13}$$

where the two C's are constants whose values are not important. The sudden appearance of the subscript 5 is to make it easier to compare with Sect. 2.3. Taking the derivative with respect to y we see that, as in the configuration model, the distribution of the number of vertices added is Cy times the original degree distribution.

Our next goal is to show that in the subcritical regime of the Chung-Lu model with $\gamma > 3$ the size of the largest cluster is of the same order as the largest degree.

Theorem 3.6.5 *In the Chung-Lu model with $\gamma > 3$ then for any $c < c_c$ there is a constant $C_{\gamma,c}$ so that*

$$P\left(\max_v |\mathcal{C}(v)| \geq C_{\gamma,c} n^{-1/(\gamma-1)}\right) \to 0.$$

By analogy with Janson's result for the configuration model, Theorem 2.3.1, it is natural to expect that

$$C_{\gamma,c} = Ed_1/(1-\nu)$$

where $\nu = EN(X_t)$.

Proof We proved a version of Theorem 3.6.5 for the configuration model in Sect. 2.3. To be able to more easily compare with the previous proof we rewrite the recursion (3.6.11) to look like the one in Sect. 2.3

$$S_{t+1} = S_t - 1 + \xi_t,$$

where $S_t = A_t$ and $\xi_t = N(X_t)$. The new proof is almost the same as before but easier. We do not need the first part of the argument where we develop a comparison of the ξ_i with independent random variables X_i. If $|\mathcal{C}(x_0)| \geq M$ then $\tau \geq M$ and

$$S_M = d(x_0) + \sum_{i=1}^{M} (\xi_i - 1). \tag{3.6.14}$$

As noted in (3.6.12) we have $E\xi_i = \nu < 1$.

For some A large and $\delta < 1/(\gamma - 1)$ small, to be chosen later, let

$$M = An^{1/(\gamma-1)} \quad \text{and} \quad M_1 = n^{1/(\gamma-1)-\delta}. \tag{3.6.15}$$

Note that M is the desired upper bound on $\mathcal{C}_1$. If we define the truncated random variables $Y_i = \xi_i 1_{(\xi_i \leq M_1)}$ then $EY_i \leq E\xi_i = \nu$. If $|\mathcal{C}(x_0)| > M$ then (3.6.14) holds and so using $\xi_i \leq \Delta = \max_i d_i$ and computing as in (3.2.11)

$$\begin{aligned} M &\leq d(x_0) + \sum_{i=1}^{M} \xi_i \leq \Delta + \sum_{i=1}^{M} \xi_i 1_{(\xi_i \leq M_1)} + \sum_{i=1}^{M} \xi_i 1_{(\xi_i > M_1)} \\ &\leq \sum_{i=1}^{M} Y_i + \Delta\left(1 + \sum_{i=1}^{M} 1_{(\xi_i > M_1)}\right) \qquad (3.6.16) \\ &\leq \nu M + \sum_{i=1}^{M} (Y_i - EY_i) + \Delta\left(1 + \sum_{i=1}^{M} 1_{(\xi_i > M_1)}\right). \end{aligned}$$

Thus if $\nu < 1 - 2\epsilon$ (which holds for $\epsilon = (1-\nu)/3$)

$$P((|\mathcal{C}(x_0)| \geq M) \leq P\left(\sum_{i=1}^{M} |Y_i - EY_i| > \epsilon M\right) + P\left(\sum_{i=1}^{M} 1_{(X_i > M_1)} > \epsilon \frac{M}{\Delta} - 1\right). \tag{3.6.17}$$

For the first sum fix a number $r \geq \gamma$ so that $r\delta > 2$. Rosenthal's inequality for nonnegative random variables given in (2.3.13) states

$$E\left(\sum_{k=1}^{n} Z_i\right)^r \leq C_r \max\left(\left(\sum_{k=1}^{n} EZ_i^2\right)^{r/2}, \sum_{k=1}^{n} EZ_i^r\right).$$

Letting $Z_i = |Y_i - EY_i|$ we have

$$E\left(\sum_{k=1}^{n} |Y_i - EY_i|\right)^r \leq C_r\left((nE(Y_i - EY_i)^2)^{r/2} + nE|Y_i - EY_i|^r\right).$$

Taking $n = M$ we have

$$\begin{aligned} E\left|\sum_{i=1}^{M} |Y_i - EY_i|\right|^r &\leq C_6 M^{r/2}(E|Y_1 - EY_1|^2)^{r/2} + C_7 M E|Y_1 - EY_1|^r \\ &\leq C_6 M^{r/2}(EY_1^2)^{r/2} + C_8 M EY_1^r. \end{aligned} \tag{3.6.18}$$

To go from the first line to the second we note that $E(X-c)^2$ is minimized when $c = EX$. For the second term writing $E(Y; A)$ for the integral of Y over A

$$\begin{aligned} E|X - EX|^r &= E(|X - EX|^r; X \geq EX) + E(|X - EX|^r; X < EX) \\ &\leq E(|X|^r; X \geq EX) + |EX|^r P(X < EX) \leq 2E|X|^r, \end{aligned}$$

where in the second line we have used the fact that $X \geq 0$ and then Jensen's inequality to put the rth power inside the expected value

We estimate the second moment of Y_1 using (3.6.13)

$$\begin{aligned} EY_1^2 &= \int_0^\infty 2xP(Y_1 > x)\,dx = \int_0^{M_1} 2xP(\xi_1 > x)\,dx \\ &\leq 1 + C_5 \int_1^{M_1} 2x^{3-\gamma}\,dx \leq C_9 M_1, \end{aligned} \tag{3.6.19}$$

since $\gamma > 3$. Trivially $EY_1^r \le M_1^r$. Using Markov's inequality, (3.6.18), the second moment estimate in (3.6.19), $Y_1 \le M_1$ and $M_1/M = A^{-1}n^{-\delta}$ we get (here $C_{10} = c_6\epsilon^{-r}C_6$ and $C_{11} = c_6\epsilon^{-r}C_8$ but ϵ is fixed)

$$\begin{aligned} P\left(\sum_{i=1}^{M} |Y_i - EY_i| > \epsilon M\right) &\le (\epsilon M)^{-r} E\left|\sum_{i=1}^{M} |Y_i - EY_i|\right|^r \\ &\le C_{10}(EY_1^2/M)^{r/2} + C_{11}M^{1-r}EY_1^r \qquad (3.6.20) \\ &\le C_{12}(M_1/M)^{r/2} + C_{13}M(M_1/M)^r \\ &\le C_{14}n^{-r\delta/2} + C_{15}n^{1-r\delta} = o(n^{-1}), \end{aligned}$$

since $r\delta > 2$.

For the second sum in (3.6.17) write $I_i = 1_{(\xi_i > M)}$ and following (2.3.17) conclude that

$$\begin{aligned} P\left(\sum_{i=1}^{M} I_i \ge L\right) &\le \binom{M}{L} P(I_k = 1,\ 1 \le k \le L) \\ = \binom{M}{L} P(I_1 = 1)^L &= \binom{M}{L} P(X > \xi_1)^L \le (MP(\xi_1 > M_1))^L. \qquad (3.6.21) \end{aligned}$$

Now by the tail bound on ξ_i in (3.6.13) and the choices of M and M_1 in (3.6.15)

$$\begin{aligned} MP(\xi_1 > M_1) &\le MC_5M_1^{-(\gamma-1)+1} \\ &\le C_5An^{1/(\gamma-1)+(2-\gamma)[1/(\gamma-1)-\delta]} = CAn^{(\gamma-2)\delta-(\gamma-3)/(\gamma-1)}. \end{aligned}$$

Choose $\delta > 0$ so that $\delta_1 = (\gamma-3)/(\gamma-1) - (\gamma-2)\delta > 0$. Then $MP(\xi_1 > M_1) = O(n^{-\delta_1})$ and

$$P\left(\sum_{i=1}^{M} I_i \ge L\right) = O(n^{-L\delta_1}). \qquad (3.6.22)$$

If we choose $L > 1/\delta_1$ so that the last error term is $o(n^{-1})$, then (3.6.5) and Lemma 3.6.2 imply that

$$\Delta/Ed_1 \to 1 \qquad Ed_1 \sim cN^a/(1-a),$$

where $a = 1/(\gamma-1)$. By (3.6.15) $M = An^{1/(\gamma-1)}$ so if A is chosen large enough

$$\epsilon\frac{M}{\Delta} - 1 > L \qquad \text{for large } n. \qquad (3.6.23)$$

Combining this with (3.6.22) and (3.6.20) we have shown that the right-hand side of (3.6.17) is $o(n^{-1})$, which gives the conclusion of Theorem 3.6.5.

To show that all of our constants can be chosen the way we want them, we note that $\epsilon = (1 - \nu)/3$ where $\nu = E\xi_i$, δ was chosen small enough so that

$$\frac{\gamma - 3}{\gamma - 1} - \delta(\gamma - 2) > 0,$$

$r \geq \gamma$ was chosen so that $r\delta \geq 2$, and finally A was chosen to give (3.6.23).

References

Athreya KB, Ney PE (1972) Branching processes. Springer, New York

Barabási AL, Albert R (1999) Emergence of scaling in random networks. Science 286:509–512

Bollobás B, Janson S, Riordan O (2005) The phase transition in the uniformly grown random graph has infinite order. Rand Struct Alg 26:1–36

Bollobás B, Janson S, Riordan O (2007) The phase transition in inhomogeneous random graphs. Rand Struct Alg 31:3–122

Bollobás B, Riordan O (2004) The diameter of a scale-free random graph. Combinatorica 24:5–34

Callaway DS, Hopcroft JH, Kleinberg JM, Newman MEJ, Strogatz SH (2001) Are randomly grown graphs really random? Phys Rev E 64, paper 041902

Chung F, Lu L (2002) Connected components in random graphs with a given degree expected sequence. Ann Comb 6:125–145

Ding A, Durrett R (2023) The phase transition in Chung-Lu random graphs. arXiv:2305.08203

Dorogovstev SN, Mendes JFF (2002) Evolution of networks. Adv Phys 51:1079–1187

Dorogovstev SN, Mendes JFF, Samukhin AN (2000) Structure of growing networks with preferential linking. Phys Rev Lett 85:4633–4636

Durrett R (2003) Rigorous results for the CHKNS random graph model. In: Proceedings of discrete random walks 2003. https://doi.org/10.46298/dmtcs.3345

Harris TE (1963) The theory of branching processes. Springer, New York

Kalikow S, Weiss B (1988) When are random graphs connected? Israel J Math 62:257–268

Kesten H, Stigum B (1966) A limit theorem for multidimensional Galton-Watson processes. Ann Math Stat 37:1211–1223

Riordan O (2005) The small giant component in scale-free random graphs. Comb Probab Comput 14:897–938

Shepp LA (1989) Connectedness of certain random graphs. Israel J Math 67:23–33

Turova TS (2011) The largest component in subcritical inhomogeneous random graphs. Comb Probab Comput 20:131–154

Chapter 4
Epidemics

In this chapter we will consider four epidemic models whose names are abbreviated as SI, SIR, SIS, and SIRS. At any time each vertex v is in state $S =$ susceptible, $I =$ infected, or $R =$ removed (is immune to infection). Common to all four dynamics is the infection step: $S - I$ edges become $I - I$ at rate λ. Once an individual becomes infected they stay infected for an amount of time T. These times for the various vertices are independent. We will primarily be concerned with the following situations:

- $T = \infty$ is the **SI mode**l, a depressing scenario in which individuals never recover, so we prefer to think of it as the spread of a rumor or other information. This type of spread can also be modeled by first passage percolation, which we considered in Sect. 2.5.
- The simplest system with $T < \infty$ is the **SIR model**. If T is a fixed constant we have a simple process with connections to percolation, which was considered in Sect. 2.7. If T is exponential with rate γ, i.e., $P(T > t) = e^{-\gamma t}$, then the system has the Markov property.
- In the SIR system, when the infected period is over the individual enters the removed state. In the **SIS model**, which is a close relative of the contact process considered in Chap. 5, when the infected period is over the individual is again susceptible to the disease.
- Finally there is the **SIRS model**, discussed in Sect. 4.7.3, in which immunity in the removed state is only temporary. This is the case for the flu since the dominant strain of influenza changes from year to year.

We begin in Sect. 4.1 with the classical theory of epidemics in a homogeneously mixing population (or on a complete graph). Infection times are exponential so models are Markov chains and lead in the limit to ODEs. In Sect. 4.2 we investigate spatial epidemics in which the infection time is constant, which is the **Reed-Frost model**. There is a close relationship between the network of infections that occur in a

R. Durrett, *Dynamics on Graphs*, Probability Theory and Stochastic Modelling 110,
https://doi.org/10.1007/978-3-032-21908-4_4

homogeneously mixing population and the edges of an Erdős-Rényi random graph. This makes it easy to compute the probability of a large epidemic and the fraction of people infected in one (which are equal).

In Sect. 4.3 we take the important step of allowing the infection times to have general distributions. If we let $I_{x,y}$ be the event that x infects neighbor y then $I_{x,y}$ and $I_{x,z}$ are independent when the infection time is constant. This breaks down in general, but since attempted infections from different sites are independent it is still possible to use branching processes to compute the probability of a large epidemic. However, that probability is no longer the same as the fraction of sites infected in a large epidemic. The good news is that since the criterion for a supercritical epidemic is that the mean number of infecteds > 1, we can compute the threshold from the individual $P(I_{x,y})$.

In Sect. 4.4 we discuss the Miller-Volz equations, a set of differential equations that describe the spread of an epidemic on a graph generated by the configuration model. When the original paper of Volz (2008) was published, I believe that these equations were regarded as heuristics that produced good approximations. Indeed the papers take pains to say that "calculations closely match simulations." Somewhat surprisingly, as we explain in Sect. 4.5, they turn out to be exact in the large graph limit. Several proofs of this have been given. We follow the approach of Janson, Luczak, and Windridge (2014) that proves the result under minimal assumptions.

In Sect. 4.6 we describe an epidemic model that incorporates the important feature that humans are organized into households. It maintains simplicity by assuming that the relationship between the houses is that of the complete graph. We explain the work of Ball, Mollison, and Scalia-Tomba (1997) and use their results to explore a pressing problem from the summer of 2020 that was the focus of an REU project: How should Duke University prepare to welcome students back to campus in the Fall in order to minimize the impact of the covid epidemic?

To study this question we investigated the **dorm model**, which is a two-level version of the household model. Dorms have rooms, and rooms can have one or two students. Our analysis showed that epidemic spread was reduced if all students had single rooms. Duke did implement this idea for the 2020–2021 academic year, but we cannot claim credit for Duke making this decision.

Our final topic, pursued in Sect. 4.7, is epidemics on $\mathbb{Z}^2$. Here the main result is of academic rather than applied interest: if we start with one infected individual and the epidemic does not die out then the epidemic grows linearly and has an asymptotic shape. At the end of the section we consider the **SIRS epidemic**. A result of Durrett and Neuhauser (1991) shows that if the SIR epidemic is supercritical and the $R \to S$ transition occurs at a positive rate then there is an equilibrium in which the epidemic persists.

4.1 On the Complete Graph

On the complete graph each individual interacts equally with all the members of the population, so the system is often referred to as a *homogeneously mixing population*. Let β be the total rate at which an individual sends out germs so that $\lambda = \beta/N$ is the infection rate per edge. To simplify the arithmetic we forget about the fact that there are only $N - 1$ edges.

4.1.1 SI Model

In Bailey's classic book *The Mathematical Theory of Infectious Diseases* (first edition 1957, second 1975) the SI model is called the **simple epidemic**. Since there are only susceptibles and infecteds we have $S = N - I$ and when the population size is large we have one differential equation

$$\frac{dS}{dt} = -\beta\frac{SI}{N} = -\frac{dI}{dt}.$$

Bailey rescales time to make $\beta/N = 1$, takes $N = n + 1$, $S(0) = n$, $I(0) = 1$ to arrive at

$$S(t) = \frac{n(n+1)}{n + e^{(n+1)t}} \qquad I(t) = \frac{n+1}{1 + ne^{(n+1)t}}.$$

If we rescale time $t = r/(n+1)$ and let $s = S/(n+1)$, $i = I/(n+1)$ then we have

$$s(r) = \frac{n}{n + e^r}, \qquad i(r) = \frac{1}{1 + ne^r}.$$

The rate at which new cases occur $i'(r) = ne^{-r}/(1 + ne^{-r})^2$ (Fig. 4.1).

4.1.2 SIR Model

In this case we assume that an infected individual stays infected for an exponentially distributed time T with mean $1/\gamma$, i.e., $P(T > t) = e^{-\gamma t}$. The following differential equations hold when the population size is large:

$$\begin{aligned} \frac{dS}{dt} &= -\beta\frac{SI}{N} \\ \frac{dI}{dt} &= \beta\frac{SI}{N} - \gamma I \\ \frac{dR}{dt} &= \gamma I. \end{aligned} \tag{4.1.1}$$

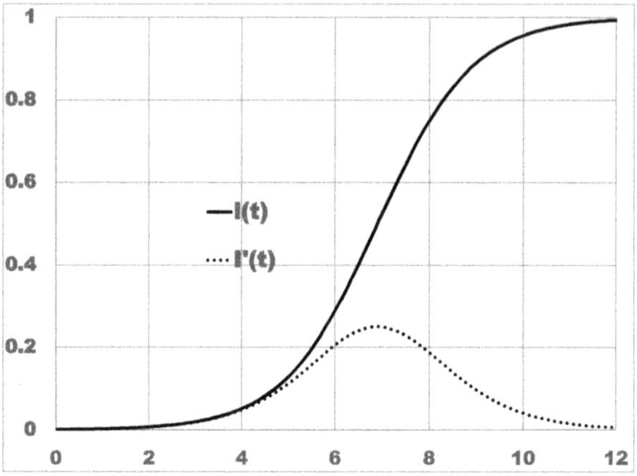

Fig. 4.1 SI epidemic curve with $n = 1000$. Solid line is $i(t)$, dotted line is $i'(t)$, which gives the number of new infecteds

In the SIR epidemic if most of the individuals are susceptible, an infected person creates new infecteds at rate $\approx \beta$ and becomes healthy at rate γ. A basic fact about the exponential distribution is that if $S = \text{exponential}(\beta)$ and $T = \text{exponential}(\gamma)$ then

$$P(S < T) = \frac{\beta}{\beta + \gamma}. \tag{4.1.2}$$

This implies that the number of individuals Y infected by one infected in an otherwise susceptible population has distribution

$$P(Y = k) = \left(\frac{\beta}{\beta + \gamma}\right)^k \frac{\gamma}{\beta + \gamma} \qquad \text{for } k = 0, 1, 2, \ldots \tag{4.1.3}$$

This is the **shifted geometric distribution** with parameter $p = \gamma/(\beta + \gamma)$, which gives the distribution of the number of failures before the first success in a sequence of independent trials. The ordinary geometric distribution, which gives the number of trials to get the first success has mean $1/p$, so the shifted one has mean

$$\frac{1}{p} - 1 = \frac{\beta + \gamma}{\gamma} - 1 = \frac{\beta}{\gamma}.$$

β/γ is called the **basic reproduction number** and is usually denoted by R_0, although in this book we will simply write R. No matter what notation is used, it is the number of secondary infections caused by one infection in the initial phase of the epidemic when only a small fraction of the individuals are removed.

Asymptotic behavior. Suppose $\beta > \gamma$ so that $R > 1$. Letting $s = S/N$, $i = I/N$, and $r = R/N$, the equations in (4.1.1) become

$$\begin{aligned}\frac{ds}{dt} &= -\beta si \\ \frac{di}{dt} &= \beta si - \gamma i \\ \frac{dr}{dt} &= \gamma i.\end{aligned} \tag{4.1.4}$$

We cannot solve these equations explicitly, but we can extract some useful information about the limiting behavior. Using the first two equations

$$\frac{di}{ds} = \frac{di/dt}{ds/dt} = \frac{\beta si - \gamma i}{-\beta si} = -1 + \frac{\gamma}{\beta s}.$$

Solving we have

$$i = -s + \frac{\gamma}{\beta} \ln s + C \quad \text{where } C = i(0) + s(0) - (\gamma/\beta)\ln(s(0)). \tag{4.1.5}$$

For initial conditions, we choose $i(0) = \epsilon$, $s(0) = 1 - \epsilon$. We have set $I(0) = \epsilon N$ since the ODE is not accurate when the number of infecteds is small. For this initial condition

$$C = 1 - (\gamma/\beta)\ln(1 - \epsilon),$$

so using (4.1.5) we have

$$i(t) + s(t) = 1 + \frac{\gamma}{\beta} \ln(s(t)/(1 - \epsilon)).$$

As $t \to \infty, i(t) \to 0, s(t) \to s(\infty)$. The limit of $s(t)$ exists since $s(t)$ is decreasing. If we let $\epsilon \to 0$ then

$$s(\infty) = 1 + \frac{\gamma}{\beta} \ln(s(\infty)).$$

Rearranging we have

$$s(\infty) = \exp(-(\beta/\gamma)[1 - s(\infty)]). \tag{4.1.6}$$

Extinction probability. If we think of the people infected by an individual as their children then in the initial stages of the epidemic we have a branching process. As noted in (4.1.3) the probability of k children is

$$p_k = \left(\frac{\beta}{\beta + \gamma}\right)^k \frac{\gamma}{\beta + \gamma} \qquad \text{for } k - 0, 1, 2, \ldots$$

The generating function of this distribution is

$$\phi(x) = \sum_{k=0}^{\infty} \left(\frac{\beta x}{\beta+\gamma}\right)^k \frac{\gamma}{\beta+\gamma}$$
$$= \frac{\gamma}{\beta+\gamma}\left(1 - \frac{\beta x}{\beta+\gamma}\right)^{-1} = \frac{\gamma}{\gamma+\beta(1-x)}.$$

$\phi(x) = x$ is a quadratic equation:

$$0 = -\gamma + \gamma x + \beta x - \beta x^2 = (\beta x - \gamma)(1-x), \tag{4.1.7}$$

so if $\beta > \gamma$ then the extinction probability

$$\rho = \gamma/\beta. \tag{4.1.8}$$

For a simpler proof, note that the first event is either the death of the ancestor with probability $\gamma/(\gamma+\beta)$ or the birth of a second particle with probability $\beta/(\beta+\gamma)$, so

$$x = \frac{\gamma}{\gamma+\beta} \cdot 1 + \frac{\beta}{\gamma+\beta} \cdot x^2,$$

since when we have two particles the probability both of their families die out is x^2. Rearranging we have the quadratic in (4.1.7).

Note that the probability that the epidemic dies out given in (4.1.8) is not the same as the limiting number of susceptibles since $s(\infty) = \gamma/\beta$ does not solve the equation in (4.1.6). To explain this suppose that for each ordered pair of individuals with probability $\lambda/(\gamma+\lambda) \approx \lambda/\gamma$ (recall $\lambda = \beta/N$) we draw an arrow from x to y to indicate that if x is infected it will infect y. To see if y will become infected we can work backwards from y crossing arrows in the direction opposite their orientation. In the limit as $n \to \infty$ the result is a branching process with a Poisson distribution with mean $n\lambda/\gamma = \beta/\gamma$. The probability that this **backward branching process** dies out is the solution of (4.1.6).

4.2 Fixed Infection Times

During the 1920s, mathematician Lowell Reed and physician Wade Hampton Frost developed a stochastic model for disease propagation, used in their biostatistics and epidemiology classes at Johns Hopkins University. They did not publish their work but it has been the subject of a large number of papers. See Jacquez (1987) for references to early work. In 1950 their model was the subject of a television program entitled *Epidemic theory: What is it?*, which is available on YouTube.

The basic version of their model is as follows: A set of n individuals is given, indexed by $i \in \{1, \dots, n\}$. At time 0 a single individual is infected. Once infected the individual is infectious for one unit of time, after which it enters the removed state. While infectious it will succeed in infecting a susceptible individuals with probability p, with the attempts independent for all target individuals and all infectious individuals. Thus if there are $I(t)$ infected individuals at time t the probability a susceptible is infected at time $t + 1$ is

$$1 - (1 - p)^{I(t)},$$

so the transition probability for the number of infecteds is

$$P(I(t+1) = m | I(t) = k) = \binom{S(t)}{m} \left(1 - (1 - p)^k\right)^m (1 - p)^{k(S(t)-m)}.$$

The susceptibles and infecteds are updated by

$$S(t + 1) = S(t) - I(t) \qquad R(t + 1) = R(t) + I(t).$$

Barbour and Mollison (1990) noticed that there was a close relationship between the Reed-Frost epidemic model and the Erdős-Rényi random graph. The key observation is that each edge e will be $S - I$ (or $I - S$) only once, so if we let $\eta(e)$ be independent, 1 with probability p, and 0 with probability $1 - p$ then we can use the η to determine what infections will occur in the Reed-Frost model. If we use the same variables to determine which edges are present in the Erdős-Rényi graph then the set of individuals infected in the Reed-Frost model coincides with C_x, the cluster containing x in the Erdős-Rényi graph.

In the Reed-Frost epidemic we start with $G = K_n$, the complete graph on n vertices. We can also start with some fixed graph G which represents the social structure of the population. For example, we could take G to be Erdős-Rényi$(n, \mu/n)$. In this case, if the time until the infection spreads across the edge S is exponential(λ) and the time until the infected vertex becomes removed $T \equiv 1$, then the infection will be transmitted to the susceptible with probability

$$\tau_f = P(S \le 1) = 1 - e^{-\lambda}. \tag{4.2.1}$$

Here, the subscript 'f' is for "fixed time." The last formula generalizes easily to other distributions of T, see (4.3.1) below, but only in the fixed time case are the infections for different edges independent.

Due to the last observation, we can delete edges with probability $e^{-\lambda}$ and the connected components of the resulting graph will give the epidemic sizes when one member of the cluster is infected. Thus, we can have a large epidemic if and only if the reduced graph has a giant component. In the physics literature this idea is attributed to Grassberger (1983), and in complex networks to Newman (2002). Results for percolation on random graphs generated by the configuration model were given in

Sect. 2.7, so there is not much to say about the critical value for a large epidemic. Since the epidemic is equivalent to percolation, the probability of a large epidemic and its size are the same. Using results for the Erdős-Rényi graph proved in Chapter 1 it is routine to show.

Theorem 4.2.1 *If the original graph is Erdős-Rényi with mean degree μ, then the reduced graph is Erdős-Rényi with mean degree $\mu\tau_f$. Thus, a large epidemic occurs with positive probability if $\mu\tau_f > 1$. If z_0 is the fixed point smaller than 1 of the generating function*

$$G(z) = \exp(-\mu\tau_f(1-z)), \tag{4.2.2}$$

then $1 - z_0$ gives both the limiting probability an infected individual will start a large epidemic, and the fraction of individuals who will become infected when a large epidemic occurs.

This result can also be applied to Reed-Frost model. If $p = c/n$ with $c > 1$ and we let σ be the solution of $1 - \sigma = \exp(-c\sigma)$ then σ gives the fraction of susceptibles that remain after a large epidemic. This is equation (2) in Barbour and Mollison (1990). Other results can be obtained for the Reed-Frost epidemic by computing the corresponding quantity for the Erdős-Rényi graph. If you are interested in techniques for analyzing this model you should take a look at Martin-Löf (1986).

4.3 General Infection Times

We now turn to the case in which the duration of the infection, T, is random. By conditioning on the value of T we see that the probability an infected site fails to infect a given susceptible neighbor is

$$P(S < T) = \int_0^\infty dt\ f_T(t)e^{-\lambda t} = Ee^{-\lambda T}. \tag{4.3.1}$$

This implies that the probability that the infection is transmitted in the model is

$$\tau_g = E(1 - e^{-\lambda T}) \tag{4.3.2}$$

(g is for general). Using the reasoning for (4.3.1), the probability two neighbors both escape infection is

$$\int_0^\infty dt\ f_T(t)(e^{-\lambda t})^2 = E(e^{-2\lambda T}) > (Ee^{-\lambda T})^2$$

where in the last step we have used $E(X^2) > (EX)^2$ with $X = e^{-\lambda T}$. Hence if y_1 and y_2 are neighbors of x then the events "x infects y_1" and "x infects y_2"are not independent. Infections on edges out of x are positively correlated events since they are more likely to be present if T_x is large.

Construction and connection with percolation. We split each edge $\{x, y\}$ in the graph into two oriented edges (x, y) and (y, x). Let T_x be the duration of the infection at x and for each oriented edge (x, y) let $I_{(x,y)}$ be independent infection times that are exponential with rate λ. To define a graph we say that the

$$\text{oriented edge } (x, y) \text{ is present if } T_x > I_{(x,y)}. \tag{4.3.3}$$

As noted earlier, two edges coming from the same vertex are not independent. In contrast, edges out of different vertices are.

The goal of this section is to compute the probability of a large epidemic and the fraction of individuals that will be infected if one occurs. When the duration of the infection is constant these quantities are the same (see Theorem 4.2.1), but not if the duration is random. Following the oriented edges out from x, the set of all sites that can be reached, C_x, is the **outward cluster**. We have used C since this is the usual percolation cluster. If we want to know whether y will be infected in the epidemic, we start at y and define a cluster by crossing edges in the direction opposite their orientation. The set of points that can be reached (the **inward cluster**), $\mathcal{D}_y$, is the set of initial infections that will lead to an infection of y.

Suppose for the moment, that we have a graph generated by the configuration model in which p_k is the degree distribution, $q_k = (k+1)p_{k+1}/\mu$ is the size-biased distribution (as usual we do not count the edge that brought the infection to the vertex). If ν is the mean of the size-biased distribution then the condition for a giant component is

$$\nu\tau_g > 1. \tag{4.3.4}$$

To simplify computations, we will first ignore the fact that the first individual has a different distribution, i.e., we pretend that the epidemic is a branching process with an offspring distribution that is a thinned version of q_k, i.e., we only retain edges that the infection crosses. To prepare for the real computation that comes later, we define the generating functions

$$G_0(z) = \sum_{k=0}^{\infty} p_k z^k \quad \text{and} \quad G_1(z) = \sum_{k=0}^{\infty} q_k z^k.$$

Outward infection degree. Since G_1 is the generating function for q_k, the generating function of the number of neighbors that will become infected (i.e., are part of the outward cluster) is

$$G_1^o(z) = \int_0^\infty dt\, f_T(t) \sum_{j=0}^\infty z^j \sum_{k=j}^\infty q_k \binom{k}{j} (1 - e^{-\lambda t})^j e^{-(k-j)t\lambda}. \tag{4.3.5}$$

Interchanging the order of summation, and using the binomial theorem, we have

$$\begin{aligned}
&= \int_0^\infty dt\, f_T(t) \sum_{k=0}^\infty q_k \sum_{j=0}^k z^j \binom{k}{j} (1 - e^{-\lambda t})^j e^{-(k-j)t\lambda} \\
&= \int_0^\infty dt\, f_T(t) \sum_{k=0}^\infty q_k (e^{-\lambda t} + z(1 - e^{-\lambda t}))^k \\
&= \int_0^\infty dt\, f_T(t)\, G_1(e^{-\lambda t} + z(1 - e^{-\lambda t})) \\
&= \int_0^\infty dt\, f_T(t)\, G_1(1 - 1 + e^{-\lambda t} + z(1 - e^{-\lambda t})) \\
&= EG_1(1 + (z - 1)[1 - e^{-\lambda T}]).
\end{aligned}$$

Inward infection degree. Since the edges that come in to a site y come from different sites, the infection crosses them with probability τ_g and the events are independent. In this case the computation is simpler since we do not have to integrate over t,

$$\begin{aligned}
G_1^i(z) &= \sum_{j=0}^\infty z^j \sum_{k=j}^\infty q_k \binom{k}{j} \tau_g^j (1 - \tau_g)^{k-j} \\
&= \sum_{k=0}^\infty q_k (\tau_g z + (1 - \tau_g))^k = G_1(z\tau_g + 1 - \tau_g).
\end{aligned} \tag{4.3.6}$$

Comparison. If $p_k > 0$ for some $k \geq 3$ then $q_k > 0$ for some $k \geq 2$ and the generating function G_1 is strictly convex. Using $E[1 - e^{-\lambda T}] = \tau_g$ and Jensen's inequality

$$\begin{aligned}
G_1^o(z) &= EG_1(1 + (z - 1)[1 - e^{-\lambda T}]) \\
&> G_1(1 + (z - 1)\tau_g) = G_1^i(z).
\end{aligned} \tag{4.3.7}$$

Thus if we let ρ_i^a be the fixed point of G_i^a in $[0, 1)$ then $\rho_1^o = G_1^o(\rho_1^o)$ and (4.3.7) implies $\rho_1^o > G_1^i(\rho_1^o)$ so we have

$$\rho_1^i < \rho_1^o. \tag{4.3.8}$$

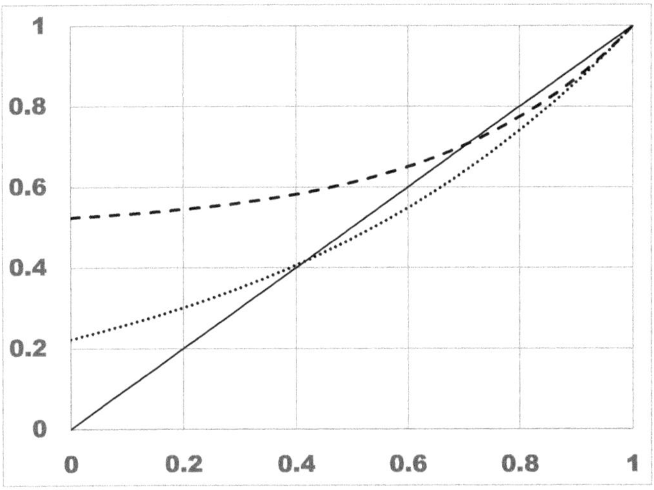

Fig. 4.2 $G^i(z)$ (dotted line) versus $G^o(z)$ (dashed line) for the concrete example with $\mu = 3$

To see the last conclusion it is useful to look at Fig. 4.2. In words independence decreases the extinction probability. Note also that correlation does not affect the expected value of the sum so both distributions have mean $\nu\tau_g$.

To define G_0^o and G_0^i replace q_k by p_k in (4.3.5) and (4.3.6) and the subscript 1 in by 0. By the reasoning above $G_0^o(z) \geq G_0^i(z)$. Using the usual branching process arguments

Theorem 4.3.1 *Suppose* $\mu\tau_g > 1$. *In this case the probability of a large epidemic is*

$$\sigma_o = 1 - G_0^o(\rho_1^o).$$

Theorem 4.3.2 *Suppose* $\mu\tau_g > 1$. *In this case the expected fraction of vertices infected is*

$$\sigma_i = 1 - G_0^i(\rho_1^i).$$

Thus in the case of nonconstant infection times, the final size of a large epidemic is not the same as the probability of a large epidemic. Since $G_0^o(z) > G_0^i(z)$ and the G_0's are increasing the survival probability $\sigma_0 < \sigma_i$. Again, it is increased by independence.

Concrete Example. It is not easy to compute $G_i^0(z)$ so, we will take a rather extreme infection time distribution $P(\tau = \infty) = p$ and $P(\tau = 0) = 1 - p$. In this case an infected node either infects no one with probability $1 - p$ or all of its neighbors with probability p. To further simplify assume that the degree distribution is Poisson with mean μ, so the branching process is homogeneous and we can drop the subscripts 0 and 1.

In our concrete example the offspring distribution of the backwards branching process is Poisson(μp) and has $\bar{G}^i(z) = \exp(-\mu p(1-z))$, which is like bond percolation while

$$G^o(z) = (1-p) + pG(z) = 1 - p + p\exp(-\mu(1-z))$$

is like site percolation.

4.4 Miller-Volz Equations

In 2008 Erik Volz published a remarkable paper deriving differential equations for the spread of epidemics on random graphs with specified degree distributions. In 2011 Miller reduced Volz's system of ODE to a single equation. Although the networks Miller and Volz considered were undirected in the sense that either of two neighbors can transmit the infection to the other, it is useful to keep track of who infects who, so we replace each undirected edge $\{a, b\}$ by two directed edges (a, b) and (b, a), which will be referred to as **arcs**. To quote Volz, "the first element in the ordered pair will be called the **ego** and the second the **alter**."

Let $\mathcal{A}_{X,Y}$ be the set of arcs with ego in X and alter in Y, $\mathcal{A}_X$ be the set of arcs with ego in X, and let $|\mathcal{A}|$ be the total number of arcs in the graph. Let

$$M_{X,Y} = \frac{|\mathcal{A}_{X,Y}|}{|\mathcal{A}|} \quad \text{and} \quad M_X = \frac{|\mathcal{A}_X|}{|\mathcal{A}|}$$

be the fraction of arcs in $\mathcal{A}_{X,Y}$ and in $\mathcal{A}_X$

To describe the equations we will use the notation of Miller rather than Volz since it is Miller's argument that we will follow. Miller uses **base** instead of ego and **target** instead of alter. Let β be the rate at which infected individuals infect susceptible neighbors and γ be the recovery rate ($I \to R$ transitions). Let S, I, and R be the fraction of susceptible, infected, and removed nodes at time t. Let $g(x)$ be the probability generating function of the degree distribution p_k and let

$$p_I = \frac{M_{S,I}}{M_S} \quad \text{and} \quad p_S = \frac{M_{S,S}}{M_S}. \tag{4.4.1}$$

Last, but most important, we define an **infectious contact** from v to a neighbor u as a contact that would cause the infection of u if it was susceptible. Here v must be infected but u can be S, I, or R. Let $\theta(t)$ is the probability an oriented edge has not transmitted an infectious contact. To compute this pick an edge at random, and then give it an orientation from v (the base) to u (the target). Let $\theta(\infty)$ be the probability of no infectious contact from v to u.

To calculate the size of an epidemic it suffices to calculate the probability that a vertex that is not allowed to infect its neighbors is infected, since a vertex cannot start a chain of infection that leads to its initial infection. The infections along the different edges that point to such a vertex are independent so

$$S(t) = \sum_{k=0}^{\infty} p_k \theta(t)^k = g(\theta(t)). \tag{4.4.2}$$

Volz's (2008) equations are (we have changed his r to β)

$$\begin{aligned} S &= g(\theta) \\ d\theta/dt &= -\beta p_I \theta && (4.4.3) \\ dI/dt &= \beta p_I \theta g'(\theta) - \gamma I && (4.4.4) \end{aligned}$$

where p_I and p_S satisfy

$$dp_I/dt = \beta p_S p_I \theta \frac{g''(\theta)}{g'(\theta)} - \beta p_I(1 - p_I) - \gamma p_I, \tag{4.4.5}$$

$$dp_S/dt = \beta p_S p_I \left(1 - \theta \frac{g''(\theta)}{g'(\theta)}\right). \tag{4.4.6}$$

We will not give Volz's argument for his equations, but will instead describe the reasoning that led Miller (2011) to his single equation,

$$\theta'(t) = -\beta\theta + \gamma(1 - \theta) + \beta \frac{g'(\theta)}{g'(1)}. \tag{4.4.7}$$

To find the values of S, I, and R from $\theta(t)$ we note that

$$\begin{aligned} S &= g(\theta) \\ I &= 1 - R - S \\ R' &= \gamma I. \end{aligned}$$

Proof (Derivation of (4.4.7).*)* Let ϕ be the probability the base is infected but has not transmitted infection to the target.

$$\theta'(t) = -\beta\phi(t). \tag{4.4.8}$$

An edge no longer satisfies the definition of ϕ when the infection crosses the edge or when the base node recovers. The rate at which neighbors become infectious matches the rate at which neighbors stop being infectious, so

$$\phi'(t) = -(\beta + \gamma)\phi(t) - \sigma'(t), \tag{4.4.9}$$

where $\sigma(t)$ is the probability that a neighbor of a randomly chosen vertex is susceptible. (Miller calls this h but we will have many h's in the next section.) Neighbors of a randomly chosen vertex have the size-biased degree distribution. That is, if $\mu = \sum_j jp_j$ then

$$\sigma(t) = \frac{\sum_{k=1}^{\infty} kp_k \theta(t)^{k-1}}{\mu} = \frac{g'(\theta(t))}{g'(1)}. \tag{4.4.10}$$

Differentiating the last equation, we see that neighbors become infected at rate

$$-\sigma'(t) = -\frac{g''(\theta(t))\theta'(t)}{g'(1)} = \frac{\beta\phi(t)g''(\theta(t))}{g'(1)} \tag{4.4.11}$$

by (4.4.8). Using (4.4.9) and (4.4.11) we have

$$\phi'(t) = \left[-\beta - \gamma + \frac{\beta g''(\theta(t))}{g'(1)}\right]\phi(t). \tag{4.4.12}$$

So using (4.4.8) we have

$$\phi'(t) = \left[1 - \frac{\gamma}{\beta} + \frac{g''(\theta(t))}{g'(1)}\right]\theta'(t).$$

Integrating gives

$$\phi(t) = \theta - \frac{\gamma}{\beta}(1-\theta) - \frac{g'(\theta)}{g'(1)}.$$

Using (4.4.8) gives the result in (4.4.7). □

Final epidemic size. Letting $t \to \infty$ in (4.4.2) we have $S(\infty) = g(\theta_\infty)$. Let

$$\tau = \frac{\beta}{\beta + \gamma}$$

be the probability that a randomly chosen neighbor of u, call it v, has an infectious contact with u given that v has become infected. If σ_∞ is the probability the neighbor is never infected then the probability of an infectious contact is $\tau(1 - \sigma_\infty)$. Thus the probability of not transmitting the infection along the chosen edge is

$$\theta_\infty = 1 - \tau + \tau\sigma_\infty.$$

(4.4.10) tells us that $\sigma(t) = g'(\theta(t))/g'(1) = g_1(\theta(t))$, where $g_1(z) = \frac{g'(z)}{g'(1)}$ is the generating function of the size-biased distribution. So

$$\theta_\infty = 1 - \tau + \tau g_1(\theta_\infty). \tag{4.4.13}$$

The fixed point equation can be solved to determine θ_∞ in the same way that we compute the extinction probability of a branching process.

Example 4.4.1 (Random regular graphs) When all vertices have degree k the generating function of the size-biased degree distribution has $g_1(z) = z^{k-1}$. If $k = 3$ then (4.4.13) is a quadratic equation.

$$\theta = 1 - \tau + \tau\theta^2,$$

or $0 = \tau\theta^2 - \theta + (1-\tau) = (\tau\theta - (1-\tau))(\theta - 1)$. $\theta = 1$ is the trivial solution. $\tau_c = 1/2$ since the mean of the size-biased distribution is 2. For $1/2 \le \tau \le 1$ we have

$$\theta_\infty = \frac{1-\tau}{\tau}.$$

When $\tau = 1/2$, $\theta_\infty = 1$ so $S(\infty) = g(\theta_\infty) = 1$. As $\beta \to \infty$, $\tau = \beta/(\beta+\gamma) \to 1$. When $\tau = 1$, $\phi_\infty = 0$ and $S(\infty) = g(\theta_\infty) = 1$ (Fig. 4.3).

If all vertices have degree $k = 4$ then the equation of θ_∞ is

$$0 = \tau\theta^3 - \theta + (1-\tau) = (\theta - 1)(\tau\theta^2 + \tau\theta - (1-\tau)).$$

$\tau_c = 1/3$ since the mean of the size-biased distribution is 3. For $1/3 \le \tau \le 1$ we have

$$\theta = \frac{-\tau + \sqrt{\tau^2 + 4\tau(1-\tau)}}{2\tau} = \frac{-1 + \sqrt{1 + 4(1-\tau)/\tau}}{2}.$$

Using the first formula, when $\tau = 1/3$,

$$\theta = \frac{-1/3 + \sqrt{1/9 + 8/9}}{2/3} = 1,$$

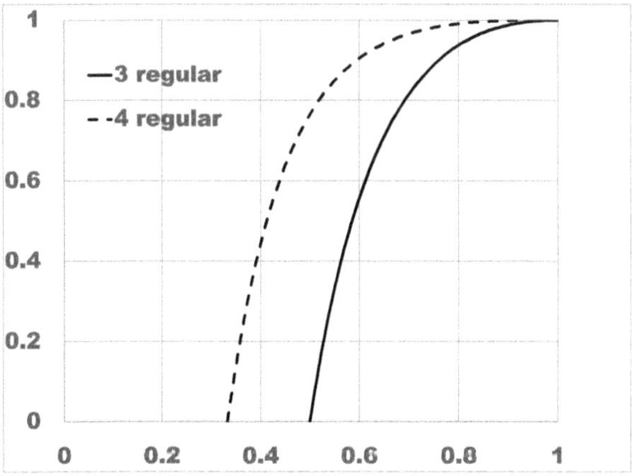

Fig. 4.3 Epidemic size (as fraction of total population) versus $\tau = \beta/(\beta+\gamma)$ for random k-regular graphs. The critical value $\tau_c = 1/(k-1)$

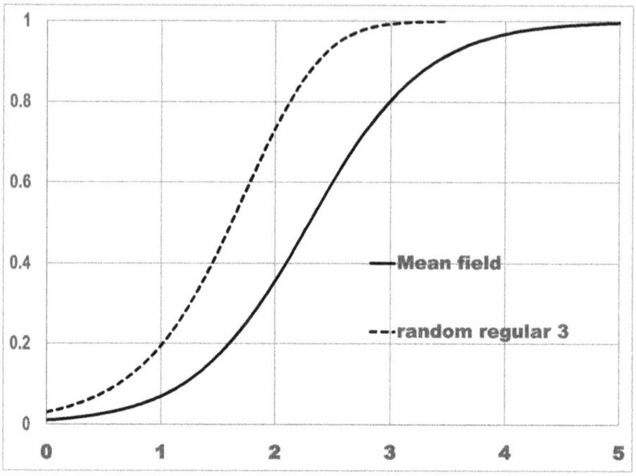

Fig. 4.4 Fraction of infected versus t in SI epidemics: homogeneously mixing versus random regular 3

while when $\tau = 1$, $\theta = (-1 + \sqrt{1})/2 = 0$ (Fig. 4.4).

Solving the equations for SI epidemics. In the homogeneously mixing case when $\gamma = 0$, (4.1.4) becomes

$$\frac{ds}{dt} = -\beta si \quad \text{and} \quad \frac{di}{dt} = \beta si.$$

The equation for i is a special case of the **logistic differential equation**

$$y'(t) = ry(1 - y/K), \tag{4.4.14}$$

where r is the exponential growth rate and K is the carrying capacity. The general solution is

$$y(t) = \frac{Ky(0)e^{rt}}{K + y(0)(e^{rt} - 1)}. \tag{4.4.15}$$

When $K = 1$, $r = \beta$, and $y_\epsilon(0) = \epsilon$ this becomes

$$i_\epsilon^{MF}(t) = \frac{\epsilon e^{\beta t}}{1 - \epsilon + \epsilon e^{\beta t}},$$

where the superscript MF is for mean-field

When $\gamma = 0$, equation (4.4.7) becomes

$$\theta'(t) = \beta(g_1(\theta(t)) - \theta(t)).$$

In the random 3-regular case $g_1(\theta) = \theta^2$, so the equation becomes

$$\theta'(t) = -\beta\theta(1 - \theta) \quad \text{or} \quad [1 - \theta(t)]' = \beta\theta(1 - \theta).$$

Taking $y(t) = 1 - \theta(t)$ this is a special case of the logistic differential equation in (4.4.14). When $K = 1$, $r = \beta$, and $y_\epsilon(0) = \epsilon$ this becomes

$$y_\epsilon(t) = \frac{\epsilon e^{\beta t}}{1 - \epsilon + \epsilon e^{\beta t}},$$

so the solution θ_ϵ with $\theta_\epsilon(0) = 1 - \epsilon$ is

$$\theta_\epsilon(t) = 1 - y_\epsilon(t) = \frac{1 - \epsilon}{1 - \epsilon + \epsilon e^{\beta t}}.$$

We have $s(t) = g(\theta(t)) = \theta^3(t)$ so

$$i_\epsilon^{3R}(t) = 1 - \theta_\epsilon^3(t) = 1 - \left(\frac{1 - \epsilon}{1 - \epsilon + \epsilon e^{\beta t}}\right)^3,$$

where the superscript $3R$ is for three regular.

4.5 Rigorous Derivations of the Equations

Bohman and Picollelli (2012) and Barbour and Reinert (2013) proved results under the assumption that the degree distribution is bounded. Decreusfond et al. (2012) proved a result under the assumption that the fifth moment of the degree distribution was finite. Here we will follow Janson, Luczak and Windridge (2014), a source we call JLW, who proved the result for random graphs with a fixed degree distribution under minimal assumptions.

For each integer n and sequence of nonnegative integers $(d_i)_1^n$ let G be a simple graph (i.e., with no loops or parallel edges) on n vertices chosen at random from all simple graphs with degree sequence $(d_i)_1^n$. We tacitly assume that there is some such graph, so e.g. $\sum_{i=1}^n d_i$ is even.

Given the graph, the SIR model evolves as a Markov chain in which $S - I$ edges turn to $I - I$ at rate β and infecteds become removed at rate γ. (JLW use ρ but we want to minimize the differences with the papers of Miller and Volz.) We assume that initially there are n_S susceptible, n_I infecteds, and n_R removed individuals, where $n_S + n_I + n_R = n$, and let $n_{S,k}$, $n_{I,k}$ and $n_{R,k}$ be the number of such vertices among those with degree k.

JLW make the following assumptions

(D1) $n_S/n \to \alpha_S > 0$, $n_I/n \to \alpha_I$, and $n_R/n \to \alpha_R$.
(D2) $n_{S,k}/n_S \to p_k$ with $\lambda = \sum_k k p_k \in (0, \infty)$.
(D3) $\sum_k k n_k/n \to \mu$.
(D4) $\sum_k k n_{S,k}/n \to \mu_S, \sum_k k n_{I,k}/n \to \mu_I, \sum_k k n_{R,k}/n \to \mu_R$. ($\mu = \mu_S + \mu_I + \mu_R$).

(D5) $\max\{k : n_{I,k} > 0\} = o(n)$.
(D6) $p_1 > 0$.

In the situation considered by Volz and Miller, $\alpha_S = 1 - \epsilon$, $\alpha_I = \epsilon$, $\alpha_R = 0$.

Remark 4.5.1 Assumptions (D1)–(D3) imply $\sum_k kn_{S,k}/n$ is uniformly summable, i.e., for any $\epsilon > 0$ there is a K so that for large n

$$\sum_{k>K} kn_{S,k}/n < \epsilon. \tag{4.5.1}$$

As in Sect. 2.2, it is convenient to work with multigraphs which may have loops or parallel edges. If we condition the multigraph to be simple then the result is uniform on the set of simple graphs. For their proof to work they need the probability that the multigraph is simple to be bounded away from 0, which holds if, see Janson (2009),

(G1) $\sum_k k^2 n_k = O(n)$, which implies $\sum_k k^2 p_k < \infty$.

To study the epidemic on the multigraph JLW reveal the edges of the graph at the same time as the epidemic spreads. They call a half-edge **free** if it has not yet been paired to another half-edge. The construction starts with d_i half-edges attached to vertex i and all half edges free. Half-edges are called susceptible, infective, or removed according to the state of the vertex they are attached to it. The dynamics are as follows

- Each free infective half-edge chooses a free half-edge at rate β at random from all of the free half-edges.
- Together the two half-edges form an edge and are removed from the pool of half-edges.
- If the chosen half-edge is susceptible it becomes infective.
- Infective half edges recover at rate γ.

We stop the process when there are no more infective half-edges, at which point the epidemic stops spreading. Some infective vertices may remain but they will recover without infecting other vertices. Susceptible or removed half-edges may also remain but they can paired at random to complete the construction of the graph.

Let S_t, I_t, and R_t be the number of such vertices at time t. S_t is decreasing in t and R_t is increasing so their limits exist. This implies $I_t = n - S_t - R_t$ converges to a limit, which must be 0. Let $X_{S,t}$, $X_{I,t}$ and $X_{R,t}$ be the number of half-edges with these states at time t. JLW show that under suitable scaling S_t, I_t, R_t, $X_{S,t}$, $X_{I,t}$ and $X_{R,t}$ converge to deterministic limits. The limiting functions will be written in terms of the θ_t, which gives the probability that an initial susceptible half-edge has not been paired with a (necessarily infective) half-edge at time t. If the initial number of infecteds are $o(n)$ then θ will start at 1 and decrease,

JLW define the following notation. (We have set $\mu_R = 0$ for simplicity.)

$$\begin{aligned}
\nu_S(\theta) &= \alpha_S \sum_k p_k \theta^k = \alpha_S g(\theta) \\
h_S(\theta) &= \alpha_S \sum_k k p_k \theta^k = \alpha_S \theta g'(\theta) \\
h_X(\theta) &= \theta^2 \\
h_R(\theta) &= \frac{\mu\gamma}{\beta}\theta(1-\theta) \\
h_I(\theta) &= h_X(\theta) - h_S(\theta) - h_R(\theta) \\
&= \theta^2 - \alpha_S \theta g'(\theta) - \frac{\mu\gamma}{\beta}\theta(1-\theta).
\end{aligned} \tag{4.5.2}$$

To connect with the work of Miller and Volz they let

$$p_I(\theta) = h_I(\theta)/h_X(\theta),$$
$$p_S(\theta) = h_S(\theta)/h_X(\theta).$$

The first results in JLW concerns the situation in which the initial fraction of infecteds $\mu_I > 0$. They have two theorems for this situation: Theorem 2.6 for the multigraph and Theorem 2.7 for the simple graph. In both cases (D1)–(D6) are assumed. In the second they also assume (G1). The conclusions of the two theorems are almost identical and are given in the next result.

Theorem 4.5.2 *(a) There is a unique θ_∞ with $h_I(\theta_\infty) = 0$,*

$$h_I > 0 \text{ on } (\theta_\infty, 1] \text{ and } < 0 \text{ on } (0, \theta_\infty).$$

(b) There is a unique $\theta_t : [0,\infty) \to (\theta_\infty, 1]$ so that $\theta_t \downarrow \theta_\infty$ as $t \to \infty$

$$\frac{d}{dt}\theta_t = -\beta\theta_t p_I(\theta_t). \tag{4.5.3}$$

(c) Let $\hat{I}_t$ be the solution to

$$\frac{d}{dt}\hat{I}_t = -\frac{\beta h_I(\theta_t) h_S(\theta_t)}{h_X(\theta_t)} - \gamma \hat{I}_t. \tag{4.5.4}$$

If we let $\hat{R}_t = 1 - \nu_S(\theta_t) - \hat{I}_t$ then

$$S_t/n \to \nu_S(\theta_t), \qquad I_t/n \to \hat{I}_t, \qquad R_t/n \to \hat{R}_t.$$

$X_{a,t}/n \to h_a(\theta_t)$ for $a = S, I, R$. Hence $X_t/n \to h_X(\theta_t)$ and $S_\infty/n \to \nu_S(\theta_\infty)$.

(4.5.3) is the same as (4.4.3). Using $p_I = h_I(\theta)/h_X(\theta)$ and $h_S(\theta) = \theta g'(\theta)$ we see that (4.5.4) is the same as (4.4.4). To get (4.4.6) we note that

$$p_S(\theta) = \frac{h_S(\theta)}{h_X(\theta)} = \frac{g'(\theta)}{\theta},$$

so differentiating and using (4.4.3) gives

$$\begin{aligned}\frac{d}{dt}p_S(\theta) &= \frac{g''(\theta)\theta - g'(\theta)}{\theta^2} \cdot \frac{d\theta}{dt} \\ &= \frac{g'(\theta)}{\theta} \cdot \left(\frac{g''(\theta)}{g'(\theta)} - \frac{1}{\theta}\right) \cdot (-\beta p_I) = \beta p_I p_S \cdot \left(1 - \theta\frac{g''(\theta)}{g'(\theta)}\right).\end{aligned}$$

Similar calculations, left as an exercise for the reader, show that dp_I/dt is given by (4.4.5). See the proof of (2.3) in JLW.

They also have results in their Theorems 2.9 and 2.10 for the multigraph and for the simple graph when the initial fraction of infecteds $\to 0$. This brings into play the possibility of the infection dying out, so the statements get more complicated. They apply the last two results to the problem of optimal vaccination strategies considered in Britton, Janson, and Martin-Löf (2007).

4.5.1 Main Ideas of the Proof

The most important idea is to speed up the process so that each free susceptible half-edge gets infected at rate 1. Specifically if there are $x_t \geq 1$ free infective half-edges and x free half-edges of any type we multiply the transition rates for such a state by

$$\frac{x-1}{\beta x_t}.$$

JLW add a $'$ to the variables to indicate the time change and use τ and σ for the time variables in the sped-up process, for example $X'_{I,\tau}$ is the number of free infective half-edges in this time-changed process. Let $\tau^* = \inf\{\tau : X'_{I,\tau} = 0\}$.

Lemma 4.5.3 *(Lemma 3.1 in JLW.) Suppose that (D1)–(D5) hold and let* $\tau_1 > 0$. *Then uniformly over* $[0, \tau_1 \wedge \tau^*]$

$$S'_\tau/n \to v_S(e^{-t}), \qquad X'_{S,\tau}/n \to h_S(e^{-t}),$$

$X'_\tau/n \to h_X(e^{-t})$, $X'_{R,\tau}/n \to h_R(e^{-t})$, *and* $X'_{I,\tau}/n \to h_I(e^{-t})$.

Proof Let $S'_\tau(k)$ be the number of susceptible vertices with k half-edges at time τ, which we assume $< \tau^*$. The only jumps in $S'_\tau(k)$ are by -1 and occur when an infective half-edge pairs and chooses one of the susceptible vertices with degree k. Taking into account the rate at which infective half-edges pair with other half-edges, the time change, and the conditional probability of choosing a susceptible half-edge of degree k

$$dS'_\tau(k) = -\beta X'_{I,\tau}\left(\frac{X'_t - 1}{\beta X'_{I,t}}\right)\left(\frac{kS'_\tau(k)}{X'_t - 1}\right)d\tau + dM_{S,\tau(k)}$$
$$= -kS'_\tau(k)d\tau + dM_{S,\tau}(k), \tag{4.5.5}$$

where $M_{S,\tau}(k)$ is a martingale with $M_{S,\tau}(k) = 0$. The result we have just used is often referred to as

Dynkin's formula. If Y_t is a Markov process with infinitesimal generator L then

$$f(X_t) - f(X_0) - \int_0^t Lf(Y_s)\,ds \quad \text{is a martingale.}$$

See Proposition 1.7 in the book by Ethier and Kurtz (1986).

The relationship written informally in (4.5.3) can be written formally as an integral equation

$$S'_\tau(k) - S'_0(k) - k\int_0^\tau S'_\sigma(k)d\sigma + M_{S,\tau}(k).$$

Using Gronwall's inequality (see Appendix 1 of Revuz and Yor (1999))

$$\sup_{\tau\le\tau_1} |S'_\tau(k) - n_{S,k}e^{-k\tau}| \le e^{k\tau_1} \sup_{\tau\le\tau_1} |M_{S,\tau}(k)|.$$

Computing the quadratic variation of the martingale and using the L^2 maximal inequality for martingales, we conclude that the last term in the previous equation is $o(n)$, so we have

$$\sup_{\tau\le\tau_1} |S'_\tau(k)/n - \alpha_S p_k e^{-k\tau}| \to 0.$$

Using (4.5.1) we can sum the last conclusions over k to get

$$\sup_{\tau\le\tau_1} |S'_\tau/n - \nu_S(e^{-\tau})| \to 0$$
$$\sup_{\tau\le\tau_1} |X'_{S,\tau}/n - h_s(e^{-\tau})| \to 0.$$

To derive the next equation we start with

$$dX'_\tau = -2(X'_\tau - 1)\,d\tau + dM_{X,\tau}.$$

Arguing as above gives

$$\sup_{\tau\le\tau_1} |X'_\tau - X'_0 e^{-2\tau}| \to 0.$$

For the fourth equation we begin with

$$dX'_{R,\tau} = [(X'_{R,\tau} - 1) + \gamma\beta^{-1}(X'_\tau - 1)]\,d\tau + dM_{R,\tau}.$$

Differentiating gives

$$\frac{d}{d\sigma} h_R(e^{-\sigma}) = -h_r(e^{-\sigma}) + \gamma\beta^{-1} h_X(e^{-\sigma}),$$

so we arrive at

$$\sup_{\tau \le \tau_1} |X'_{R,\tau} - h_R(-\tau)| \to 0.$$

The final result follows from

$$X'_{I,\tau} = X'_\tau - X'_{I,\tau} - X'_{R,\tau},$$

which completes the discussion of Lemma 4.5.3. □

To return to the original time scale and obtain Theorem 4.5.2 we let

$$A_\tau = \int_0^\tau \frac{1}{\beta}\left(\frac{X'_\sigma - 1}{X'_{1,\sigma}}\right) d\sigma,$$

where we regard the bracketed term as $1/2$ if $X'_{1,\sigma} = 0$, i.e., $\sigma \ge \tau^*$, since until then the term is $\ge 1/2$. A_τ is strictly increasing and continuous. Let $\tau(t)$ be its strictly increasing continuous inverse defined by $A_{\tau(t)} = t$. As demonstrated in Appendix A.1 of JLW the original process can be recovered by

$$S_t = S'_{\tau(t)}, \qquad I_t = I'_{\tau(t)}, \qquad R_t = R'_{\tau(t)}.$$

4.6 Household and Dorm Models

The assumption of a homogeneously mixing population is reasonable for a number of diseases, but in some situations heterogeneities can dramatically change the rate of spread and the severity of the epidemic. A simple example is a community of people who live in separate houses: it is natural to assume that transmission between individuals within a household occurs at a higher rate than between households, while all of the households interact equally.

It would be natural to allow the households to be different sizes, but we will follow Ball, Mollison, and Scalia-Tomba (1997) and consider the simplest situation (i) we let the population of size N be split into m households of size n and (ii) we consider the case of fixed infection time.

Let p_H (H is for household) be the probability an individual i infects individual j, who is another member of the same house.

Let p_G (G is for global) be the probability of directly infecting individual j, who is not in the same house as individual i.

To construct the epidemic we will build a random graph where edges indicate the infections that will occur.

Step 1: Connect each pair of individuals in the same house with probability p_H.

Step 2: Make long distance connections between individuals in different households with probability p_G.

To do computations we need some notation. Consider the infection process restricted to one house, and started with one infected individual. Let π_k be the probability that k individuals are infected. Let $G_H(z) = \sum_{k=0}^{\infty} \pi_k z^k$.

The number of long distance infections caused by one person is $Y_1 \approx \text{Poisson}(Np_G)$ if N is large. Rather than writing G_G, the generating function of Y_1 is denoted by $G_\nu(z) = \exp(Np_G(z-1))$. From steps 1 and 2, we see that

$$R_0 = \mu N p_G.$$

To compute the generating function, we recall that if $X, Y \geq 0$ are independent with generating functions ϕ_X and ϕ_Y, then $X + Y$ has generating function $\phi_X \phi_Y$. This implies that the generating function of the number of individuals infected by a cluster of size k is $\exp(Np_G(z-1))^k$ and hence that the number of individuals infected by the cluster of a randomly chosen individual is

$$\sum_{k=0}^{\infty} \pi_k \exp(Np_G(z-1))^k = G_H(G_\nu(z)).$$

The probability $\rho = 1 - \zeta$ that the process dies out is the fixed point < 1 of the generating function of the offspring distribution:

$$1 - \zeta = G_H(G_\nu(1-\zeta)). \tag{4.6.1}$$

Our epidemic is essentially a percolation process, so the size of a large epidemic is the same as the size of the giant component in percolation, which is the same in our homogeneous graph to the probability of a large epidemic.

Dorm Model. Due to the covid epidemic, the DOMath 2020 summer program was held virtually with students in Long Island, NY, Birmingham, Alabama, Cary, NC, and Taiwan. A big question on everyone's mind was: what would happen when students came back to campus in the Fall of 2020? That motivated us (Borowiak et al. (2021)) to ask: To what extent would the spread of COVID-19 on the Duke campus be reduced if all students had single rooms? To address this question we used variants of the household model.

Single Rooms. The case of single rooms is essentially the household model, except that (i) p_H will instead be written as p_D, since we are concerned with a dorm rather than a household, and G_H will instead be written as G_D^1, with the superscript 1 indicating single rooms. (ii) when we are thinking about households n is typically < 10 but in the case of college dorms, n is large, e.g., 120 is the size of a typical freshman dorm. Before the pandemic, about 6000 students lived on campus, which means 50 dorms.

We first compute R_0. The total number of individuals infected in one dorm as a result of one infected is the total progeny of a branching process. Since n is large, the branching process will have an offspring distribution that is approximately Poisson(λ_1), where $\lambda_1 = np_D$. We assume $np_D < 1$ so that the mean of the total progeny is

$$\mu_1 = 1 + \lambda_1 + \lambda_1^2 + \cdots = \frac{1}{1 - \lambda_1}. \tag{4.6.2}$$

This assumption $np_D < 1$ seems reasonable because if the epidemic is supercritical in each dorm then there is no hope of containing the spread.

Each individual infected in the dorm infects an average of np_G individuals outside the dorm so the basic reproduction number is

$$R_0 = \mu_1 N p_G = \frac{Np_G}{1 - np_D}. \tag{4.6.3}$$

Now, we solve for G_D^1. Breaking things down according to the number of individuals directly infected by the first individual, the generating function of the size of the epidemic within the dorm caused by one infected is

$$\begin{aligned} G_D^1(z) &= z \sum_{k=0}^{\infty} e^{-\lambda_1} \frac{\lambda_1^k}{k!} G_D^1(z)^k \\ &= z \exp(\lambda_1 (G_D^1(z) - 1)) = z G_b^1(G_D^1(z)), \end{aligned} \tag{4.6.4}$$

where $G_b^1(z) = \exp(np_D(z - 1))$, the generating function of the spread of the infection within the dorm. We solve (4.6.4) by iteration. We start with $h_0(z) = z$, which corresponds to 1 individual in generation 0, and then iterate

$$h_\ell(z) = z \exp(\lambda_1 (h_{\ell-1}(z) - 1)). \tag{4.6.5}$$

$h_\ell(z)$ is the generating function of the number of individuals infected in the first ℓ generations of the process, so $h_\ell(z) \downarrow h(z)$, a solution of the recursion.

In the case of single rooms, an explicit solution for G_D^1 can be found by taking the derivative of equation (4.6.4) to get the differential equation

$$G_D^{1'}(z) = G_D^1(z)/z + \lambda_1 z G_D^1(z) G_D^{1'}(z),$$

with the boundary condition $G_D^1(1) = 1$. It is well known that the solution to the differential equation is

$$G_D^1(z) = -W(zW^{-1}(-\lambda_1))/\lambda_1, \tag{4.6.6}$$

where W is the **Lambert W function**, which is defined by $W^{-1}(z) = ze^z$ (Fig. 4.5). By (4.6.1), the probability of a large epidemic is the largest solution of

$$1 - \zeta = G_D^1(\exp(-Np_G\zeta)). \tag{4.6.7}$$

This can be further simplified. Plugging in G_D^1 yields

$$\begin{aligned} 1 - \zeta &= -W(e^{-Np_G\zeta}(-\lambda_1 e^{-\lambda_1}))/\lambda_1 \\ &= -W(e^{-(Np_G+\lambda_1)\zeta}e^{\lambda_1(\zeta-1)}(-\lambda_1))/\lambda_1. \end{aligned}$$

Multiplying through by $-\lambda_1$, then applying W^{-1} to both sides gives

$$\lambda_1(\zeta - 1)e^{\lambda_1(\zeta-1)} = e^{-(Np_G+\lambda_1)\zeta}e^{\lambda_1(\zeta-1)}(-\lambda_1).$$

Recalling that $\lambda_1 = np_D$ and simplifying, we obtain

$$(1 - \zeta) = e^{-(Np_G+np_D)\zeta},$$

which is the survival probability for a branching process with a Poisson($R_0 = Np_G + np_D$) distribution.

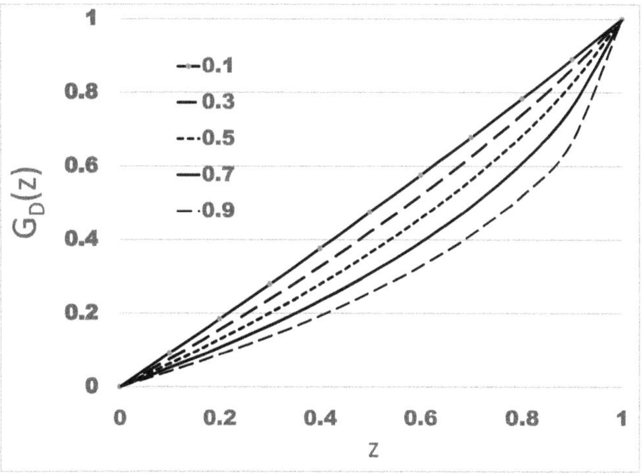

Fig. 4.5 Generating function $G_D^1(z)$ for $np_D = .1, .3, .5, .7, .9$. Solved by iteration with Equation (4.6.6)

Double Rooms. Suppose now that we have m_1 dorms with n_1 double rooms, so that $N = 2m_1n_1$. For the sake of comparison, we let the number of people in a dorm be the same, regardless of whether they have only single rooms or only double rooms, i.e., $n = 2n_1$. Let p_L be the probability of infecting your roommate, and p_D be the probability of infecting an individual within your dorm. To find the analogue of G_b^1 in this scenario, we think of a generation as having a first step where the ancestor splits into two with probability p_L, stays 1 with probability $1 - p_L$ and then a second step in which long range connections are formed within the dorm. The offspring distribution is now a mixture of two Poissons:

$$p_L\text{Poisson}(2 \cdot 2n_1 p_D) + (1 - p_L)\text{Poisson}(2n_1 p_D).$$

The mean number of infections in one generation, λ_2, will be the sum of the means of the two Poisson distributions so $\lambda_2 = (1 + p_L)2n_1 p_D$ (Fig. 4.6).

Imitating the calculations for single rooms, if $\lambda_2 < 1$, the mean of the total progeny of the branching process is

$$\mu_2 = 1 + \lambda_2 + \lambda_2^2 + \cdots = \frac{1}{1 - \lambda_2}, \tag{4.6.8}$$

and the basic reproduction number is

$$R_0 = \mu_2 N p_G = \frac{N p_G}{1 - (1 + p_L)2n_1 p_D}. \tag{4.6.9}$$

To compute G_D^2, the analogue of G_D^1 for the double room scenario, we note that the generating function of the number of infections in one generation is

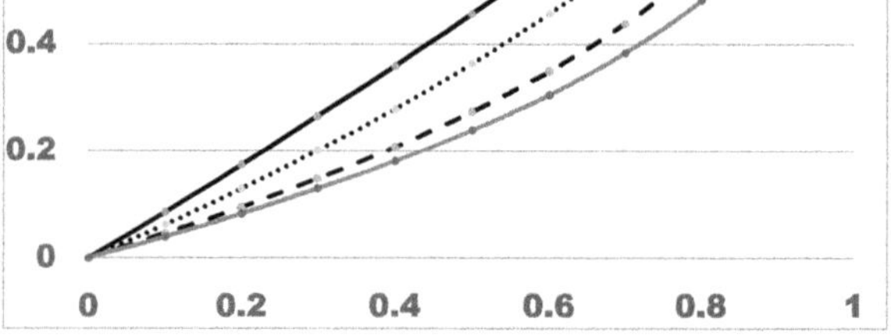

Fig. 4.6 Generating function $G_D^2(z)$ for $2n_1 p_D = .1, .3, .5, .5882$ while $p_L = .7$. Solved by iteration. When $p_L = .7$ and $\lambda_2 > .5882$, there is a large epidemic within the dorm

$$G_b^2(z) = p_L \exp(2n_1 p_D(z-1))^2 + (1-p_L)\exp(2n_1 p_D(z-1)),$$

so G_D^2 satisfies

$$G_D^2(z) = zG_b^2(G_D^2(z)). \tag{4.6.10}$$

G_D^2 can then be solved using iteration.

Having computed G_D^2 we can compute the probability of a large epidemic as in the previous case by numerically solving

$$1-\zeta = G_D^2(\exp(-Np_G\zeta)). \tag{4.6.11}$$

A comparison of the epidemic probabilities ζ for double-room and single-room dorms is given in Fig. 4.7. Note that in the double-room situation when $2n_1 p_D = 0.7$ or 0.9, there is a positive probability of a large epidemic within a single dorm, so the curves are positive when $Np_G = 0$. When $2n_1 p_D = 0.3$ or 0.5, p_G needs to exceed a positive threshold for there to be an epidemic in a university with double-room dorms, but those thresholds are higher when there are only single-room dorm.

4.7 Epidemics on $\mathbb{Z}^2$

Mollison (1977, 1978) was one of the first to study spatial models for epidemics. With Kuulasmaa (1985) he used the theory he developed to study fox rabies in Europe. The spatial epidemic model independently arose in the physics literature as a description of a forest fire (infected = on fire, recovered = burnt). McKay and Jan (1984) is one example of a physics paper, but there were many others.

To construct the epidemic model on a graph we have i.i.d. positive random variables T_x for the vertices with a distribution F that give the duration of the infection at x. Associated with each oriented edge (x, y) is an independent exponential(λ) random variable $I_{x,y}$, which is the amount of time after the infection of x until x tries to infect y. To identify the individuals infected in the epidemic we draw an oriented

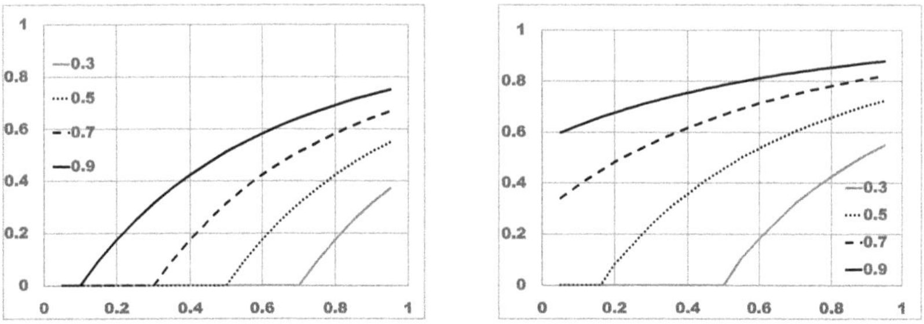

Fig. 4.7 Epidemic size in the single room (left panel) and double room (right panel) scenarios of the household model when $np_D = 2n_1 p_D = 0.3, 0.5, 0.7, 0.9$ and Np_G varies

edge from x to y if (and only if) $T_x > I_{x,y}$ and let C_0 be the set of sites that can be reached from 0 by a path of edges in the graph.

4.7.1 Probability of a Large SIR Epidemic

We set $\lambda = \alpha/2d$, where α is total rate at which a site sends out germs to be sent to a randomly chosen neighbor. If x and y are neighbors, $x \sim y$, then the probability x tries to infect y

$$p_d = 1 - \int_0^\infty e^{-\alpha s/4}\, dF(s).$$

In ordinary bond percolation there is one (unoriented) edge connecting x to y. Here, we have two oriented edges connecting each pair of neighbors, but Frisch and Hammersley (1963) have shown that the critical value remains the same and that for any two sets of sites U and V the probability there is a path from U to V is the same in the two models. This result is part of a circle of ideas that led to the "clutter percolation" which encapsulates a number of comparisons between percolation processes, see McDiarmid (1980).

Kuulasmaa (1982), who called these structures **locally dependent random graphs**, has derived a number of properties of the model. If $A \subset \{z : |z| = 1\}$ let $\phi(A)$ be the probability that all edges $(0, z)$ are closed for $z \in A$. Let B be a collection of paths in $\mathbb{Z}^2$ and let $\mathcal{B}$ be the event that some path is open.

Lemma 4.7.1 *Suppose we have graphs G_1 and G_2 with a common set of vertices. If $\phi_1(A) \geq \phi_2(A)$ for all $A \subset \{z : |z| = 1\}$ then $P_1(\mathcal{B}) \leq P_2(\mathcal{B})$.*
Proof See Kuulasmaa (1982) pages 749–750. □

Since $\phi_{bond}(A) = (1 - p)^{|A|}$ and $\phi_{site}(A) = (1 - p)$. It follows that for any model

$$P_{site}(\mathcal{B}) \leq P(\mathcal{B}) \leq P_{bond}(\mathcal{B}).$$

Let $\alpha_c(F)$ be the smallest value of α for which $P(C_0 = \infty) > 0$, and let

$$p_c(F) = 1 - \int_0^\infty \exp(-\alpha_c(F)s/4)\, dF(s),$$

then we have

$$p_c^{bond} \leq p_c(F) \leq p_c^{site}.$$

In two dimensions the lower bound is 1/2 from Kesten's rigorous result for bond percolation, see Kesten (1982), while the upper bound is 0.5927 if we use the consensus of a large number of studies in the physics literature.

4.7.2 Shape Theorem

Cox and Durrett (1988) studied the model on $\mathbb{Z}^2$. Formulating the model in forest fire terms, where $S =$ a tree, $I =$ a tree on fire, and $R =$ a burnt tree.

1. A burning tree emits sparks at rate α.
2. A spark emitted at x goes to one of the four nearest neighbors chosen at random.
3. If a spark hits a live tree it catches fire immediately and begins to emit sparks.
4. It burns for an exponentially distributed amount of time then burns out. The main result of Cox and Durrett (1988) is a **shape theorem**.

Suppose that initially the origin is infected while all the other sites are susceptible, let $\zeta_t = \{x : \eta_t(x) = 0\}$ be the set of removed individuals, and let $\xi_t = \{x : \eta_t(x) = i\}$ be the set of infected individuals.

Theorem 4.7.2 *Consider the epidemic model on* $\mathbb{Z}^2$ *and assume that* $\int_0^\infty s^2\, dF(s) < \infty$ *and* $\alpha > \alpha_c(F)$. *Then there is a convex set* D *so that for any* $\epsilon > 0$ *there is a* $t_0(\epsilon)$ *so that*

(i) $P(C_0 \cap (1-\epsilon) \subset \zeta_t \subset t(1+\epsilon)D\ t \ge t_0(\epsilon)) = 1.$
(ii) $P(\xi_t \subset t(1+\epsilon)D - t(1-\epsilon)D\ t \ge t_0(\epsilon)) = 1.$

In words, the epidemic spreads linearly and has an asymptotic shape. (i) says that inside $(1-\epsilon)$ times the growing ball all the individuals that will be ever be infected are in the removed state, while (ii) says that all of the active infections are near the boundary of the growing region. Since the shape is convex and has the same rotational and reflection symmetries as $\mathbb{Z}^2$ it will look roughly like a disk.

History lesson. Proving shape theorems for interacting particle systems was a popular enterprise when Cox and Durrett (1988) was written. The earliest result was due to Richardson (1973) who proved a general result which can be used to prove a shape theorem for Eden's (1961) growth model, which can be succinctly defined as first passage percolation on $\mathbb{Z}^2$ with exponential(1) passage times. Now this is more commonly done by using Liggett's version of Kingman's subadditive ergodic theorem, see Sect. 6.5 in PTE5.

Cox and Durrett (1981) proved a shape theorem for first passage percolation on $\mathbb{Z}^d$ with necessary and sufficient conditions. Bramson and Griffeath (1980,1981) proved a shape result for the **biased voter model** on $\mathbb{Z}^d$, also known as the **Williams-Bjerknes tumor growth model** (Williams and Bjerknes (1972)). In this system vertices in state 1 change to 0 at a rate equal to the fraction of neighbors in state 0, vertices in state 0 change to 1 at rate $\lambda > 1$ times the fraction of neighbors in state 1. They showed that starting from a single 1 at the origin then if the 1's do not die out the region that they occupy grows linearly and has an asymptotic shape.

Durrett and Griffeath (1982) proved a shape theorem for the d-dimensional contact process (see Chapter 5 for the definition) under the assumption that the birth rate $\lambda > \lambda_c(\mathbb{Z})$, the critical value for the model on $\mathbb{Z}$. They showed that if we start from a single 1 at the origin and the process does not die out then with high probability inside $(1 -$

$\epsilon)tD$ the state of the process can be coupled to the upper invariant measure, while outside $(1+\epsilon)tD$ there are no particles. The proof under the natural assumption $\lambda > \lambda_c(\mathbb{Z}^d)$ had to wait for Bezuidenhout and Grimmett (1990) to provide an important tool for studying the multi-dimensional contact process.

Main ideas of the proof. To begin we prove a fact that was stated earlier:

Lemma 4.7.3 *The cluster containing the origin in the graph C_0 is the set of sites that will ever become infected if initially the origin is infected and all other sites are susceptible. Furthermore a site $y \in C_0$ first becomes infected at the first passage time from 0 to y.*

Proof Clearly if y become infected at some time then $y \in C_0$ since y was infected by a neighbor, who was infected by a neighbor ... who was infected by 0. We argue the other inclusion by induction on the length of the shortest path. If the path has length 1 this is clear because when 0 tries to infect y either (a) y is not infected or (b) it has already been infected. In either case the result is true.

If $y \in C_0$ and $0, y_1, \ldots, y_n = y$ is the shortest path then we apply the last argument with 0 replaced by y_{n-1}. The induction hypothesis implies $y_{n-1} \in C_0$. When y_{n-1} tries to infect y_n then either (a) or (b) holds and in either case the result is true. To get the second conclusion, note that if the path achieves the first passage time then alternative (b) is impossible. □

The proof follows the approach of Cox and Durrett (1981). To carry this out, Sect. 2 of their paper extends results from ordinary percolation to the new model with bidirected edges. The proofs are old-school in that they use the machinery of sponge crossings developed by Kesten (1982). The next step, taken in Sect. 3, is to prove the existence of radial limits for the first passage times. Intuitively this means that on the set where the infection does not die out $\lim_{n\to\infty} \tau(0, nx)/n$ exists. To deal with the fact that nx may not be in C_0 we find a ring of sites $\Delta(x)$ around x that are infected and we define a passage time $\hat{t}(x, y)$ from $\tilde{\Delta}(x)$ to $\tilde{\Delta}(y)$ where $\tilde{\Delta}(x)$ is $\Delta(x)$ enlarged to include all the open edges inside the ring.

Once it has been shown radial limits exist, then the machinery that was developed to prove the other results listed above takes over to prove the result. An early lemma in the proof of these results (see e.g., Bramson and Griffeath (1981)) is a lemma that shows the existence of a linearly growing solid ball that grows at some small rate $\delta > 0$. Using a finite set of radial limits that are sufficiently dense in the limiting shape D and the small ball result proves that all sites in C_0 inside $(1-\epsilon)tD$ are infected. To prove there are no infecteds outside $(1+\epsilon)tD$, we use the small ball result to show that if there are, then one of the radial limits in a finite set inside $2tD$ is going too fast.

Open problem. Use the version of the shape theorem for first passage percolation proved in Kesten (1986) to extend the argument to d dimensions.

4.7.3 SIRS Epidemics

Given the recent covid epidemic, most readers should be familiar with the fact that individuals resistant to the disease can again become susceptible, in that case because the virus mutates and evades the immunity to previous strains. On a slower time scale, the regrowth of the susceptible population in the case of measles occurs when immune individuals die and are replaced by new individuals susceptible to the disease.

Our main purpose in this section is to state a result of Durrett and Neuhauser (1991). They formulated their result using the forest fire interpretation. 0 = live tree, 1 = on fire, 2 = burnt. They add a fifth transition to the four described earlier. 5. Burnt trees return to life at rate β.

At first glance, the spontaneous reappearance of trees may not seem reasonable. This can be justified biologically by the fact that in real forests, trees regrow from seeds in the ground. However, the real motivation for making this assumption is that if we assume new trees appear by births from existing trees the next result is a difficult open problem.

Theorem 4.7.4 *Suppose that* $\alpha > \alpha_c(0)$*, the critical value for the epidemic with no regrowth. Then for all* $\beta > 0$ *there is a stationary distribution that puts no mass on the all healthy state.*

The proof of the result uses a block construction. A number of examples are described in Durrett's 1993 St. Flour notes but a detailed description is beyond the scope of this book. The key fact, which requires considerable ingenuity to prove, is that if we have a large patch of woods (an $L \times L$ box) then the fire will burn for a long time. To do this they show if that if the fire has dwindled to only $L^{1/2}$ burning trees then it will with high probability return to size $\delta L^{0.7}$ before hitting 0. The challenge is to prove this happens no matter how the burning, live, and burnt trees are distributed.

References

Ball F, Mollison D, Scalia-Tomba G (1997) Epidemics with two levels of mixing. Ann Appl Probab 7:46–89

Barbour AD, Mollison D (1990) Epidemics and random graphs. In: Stochastic processes in epidemic theory, Springer Lecture Notes in Biomathematics, pp 86–89

Barbour AD, Reinert G (2013) Approximating the epidemic curve. Electron J Probab 18:paper 54

Bezuidenhout C, Grimmett G (1990) The critical contact process dies out. Ann Probab 18:1462–1482

Bohman T, Picollelli M (2012) SIR epidemics on random graphs with fixed degree distributions. Rand Struct Alg 41:179–214

Borowiak M, Ning F, Pei J, Zhao S, Tung H-R, Durrett R (2021) Controlling the spread of COVID-19 on college campuses. Math Biosci Eng 18(1):551–563

Bramson M, Griffeath D (1980) On the Williams-Bjerknes tumor growth model. II Proc Camb Phil Soc 88:339–357

Bramson M, Griffeath D (1981) On the Williams-Bjerknes tumor growth model. I Ann Probab 9:173–185
Britton T, Janson S, Martin-Löf A (2007) Graphs with specified degree distributions, simple epidemics and local vaccination strategies. Adv Appl Prob 39:922–948
Cox JT, Durrett R (1981) Some limit theorems for percolation processes with necessary and sufficient conditions. Ann Probab 30:583–603
Cox JT, Durrett R (1988) Limit theorems for the spread of epidemics and forest fires. Stoch Proc Appl 30:171–193
Decreusfond L, Dhersin J-S, Moyal P, Tran VC (2012) Large graph limit for an SIR process in random network with heterogeneous connectivity. Ann Appl Probab 22:541–575
Durrett R, Griffeath D (1982) Contact processes in several dimensions. Probab Theory Rel Fields 59:535–552
Durrett R, Neuhauser C (1991) Epidemics with recovery in $d = 2$. Ann Appl Prob 1(1991):189–206
Eden M (1961) A two-dimensional growth process. In: Proceedings of fourth Berkeley symposium on mathematics, statistics, and probability, vol 4. University of California Press, Berkeley, pp 223–239
Ethier SN, Kurtz TG (1986) Markov processes: characterization and convergence. Wiley Series in Probability and Statistics
Frisch HL, Hammersley JM (1963) Percolation processes and related topics. J SIAM 11:894–918
Grassberger P (1983) On the critical behavior of the general epidemic process and dynamical percolation. Math Biosci 63:157–172
Jacquez JA (1987) A note on chain-binomial models of epidemic spread: what is wrong with the Reed-Frost formulation? Math Biosci 87:73–82
Janson S (2009) The probability that a multigraph is simple. Comb Probab Comput 18:205–225
Janson S, Luczak M, Windridge P (2014) Law of large numbers for the SIR epidemic on a random graph with given degrees. Rand Struct Alg 45:726–763
Kesten H (1982) Percolation theory for mathematicians. Birkhäuser, Boston
Kesten H (1986) École d'été de probabilités de Saint-Flour, XIV-1984 Springer Lecture Notes in Math 1180. Aspects of first passage percolation
Kuulasmaa K (1982) The spatial general epidemic and locally dependent random graphs. J Appl Prob 19:745–758
Kuulasmaa K (1985) The spatial general epidemic and locally dependent random graphs. J Appl Prob 22:548–562
Martin-Löf A (1986) Symmetric sampling procedures, general epidemic processes and their threshold limit theorems. J Appl Prob 23:265–282
McDiarmid C (1980) Clutter percolation and random graphs. Math Program Study 13:17–25
McKay G, Jan N (1984) Forest fires as critical phenomena. J Phys A 17:L757–L760
Miller JC (2011) A note on a paper by Erik Volz: SIR dynamics in random networks. J Math Biol 62:349–358
Mollison D (1977) Spatial contact models ecological and epidemic spread. J R Stat Soc B 39:283–326
Mollison D (1978) Markovian contact processes. Adv Appl Prob 10:85–108
Newman MEJ (2002) Spread of epidemic disease on networks. Phys Rev E 66, paper 016128
Revuz D, Yor M (1999) Continuous martingales and Brownian motion. Grundlehren der mathematischen Wissenschaften, vol 293, 3rd edn. Springer
Richardson D (1973) Random growth in a tessellation. Proc Camb Phil Soc 74:515–528
Volz E (2008) SIR dynamics in random networks with heterogeneous connectivity. J Math Biol 56:293–310
Williams T, Bjerknes R (1972) Stochastic model for abnormal clone spread through epithelial basal layer. Nature 236:19–21

Chapter 5
Contact Process

Harris (1974) introduced the contact process on $\mathbb{Z}^d$. Generalizing his definition to a graph G, the state at time t is a set $\xi_t \subset G$ of sites occupied by particles. Particles die at rate 1. Each occupied site x gives birth onto each neighbor y at rate λ. If y is vacant then it becomes occupied. If y is occupied then no change occurs. Let ξ_t^x be the contact process starting from one particle at x and let

$$\lambda_c = \inf\{\lambda : P(\xi_t^x \neq \emptyset \text{ for all } t) > 0\}.$$

This value is independent of x on connected graphs. Interest focuses on estimating the value of λ_c and understanding the qualitative behavior of the model when $\lambda < \lambda_c$ (subcritical phase) and $\lambda > \lambda_c$ (supercritical). One is also interested in the behavior of critical systems ($\lambda = \lambda_c$) but that is a very difficult problem.

In Sect. 5.1 we construct the contact process from a **graphical representation** and introduce basic concepts from the theory of interacting particle systems that will be useful for its study. The reader can consult Liggett's (1999) book for more information.

In Sect. 5.2 we calculate the predictions of **mean-field theory**, which is a typical first step for physicists (and probabilists) in determining the properties of a stochastic spatial model. These calculations can be done in two ways: (i) compute the evolution under the assumption that the states of sites are always independent, or (ii) consider the system on the complete graph. Suppose we have a vertex transitive graph (it looks the same from every point) in which all sites have degree m and we let $\beta = \lambda m$. Using either viewpoint we conclude that $\beta_c = 1$, the process dies out exponentially fast when $\beta < 1$, and converges to an equilibrium with particle density $(\beta - 1)/\beta$ when $\beta > 1$.

The second viewpoint enables us to conclude that if we start the contact process on the complete graph from all n sites occupied and $\beta > 1$ then the system survives for time $\exp(c(\beta)n)$. In Sect. 5.3 we consider the contact process on bounded degree graphs such as the random regular graph. From our study of the geometry of random

R. Durrett, *Dynamics on Graphs*, Probability Theory and Stochastic Modelling 110,
https://doi.org/10.1007/978-3-032-21908-4_5

graphs we know that this graph is locally tree-like. So to prepare for the study of these examples we recall that Pemantle (1992) studied the contact process on the tree $\mathbb{T}^d$ in which each vertex has degree $d+1$ and found that in this situation if $d \geq 2$ (i.e., degree ≥ 3) then the contact process has two critical values:

$$\lambda_1 = \inf\{\lambda : P_0(\xi_t \neq \emptyset \text{ for all } t) > 0\}$$
$$\lambda_2 = \inf\{\lambda : \liminf_{t\to\infty} P_0(0 \in \xi_t) > 0\}.$$

For many graphs with bounded degree it can be proved that if there are n vertices and we start from all sites occupied then there are constants that depend on λ so that if 2 is another randomly chosen vertex

the process survives for a time $\leq c \log n$ when $\lambda < \lambda_1$,

the process survives for time $\geq \exp(cn)$ when $\lambda > \lambda_1$.

People with some experience with the contact process on trees may wonder why the critical value here is λ_1 instead of λ_2. For an explanation, see the discussion after Theorem 5.3.3.

In Sect. 5.4 we investigate the contact process on an Erdős-Rényi graph. The degree distribution is unbounded, but has a very thin tail. Using the existence of long paths proved in Section 1.5 and comparing with the contact process on an interval we can prove survival for time $\exp(cn)$ for $\lambda > \lambda_c(\mathbb{Z})$. Also in Sect. 5.4 we describe a very clever proof of Cator and Don (2021) that lower bounds the survival time of the contact process on an Erdős-Rényi graph by comparing the number of occupied sites with a birth and death chain.

In Sect. 5.5 we begin the study of the contact process on graphs with a power law degree distribution, $p_k \sim Ck^{-\alpha}$. Using **degree based mean-field theory** physicists calculated that

- If $\alpha \leq 3$ then $\lambda_c = 0$.
- If $3 < \alpha < 4$, then $\lambda_c > 0$ but the critical exponent $\beta > 1$.
- If $\alpha > 4$ then $\lambda_c > 0$ and $\rho(\lambda) \sim C(\lambda - \lambda_c)$ so the critical exponent $\beta = 1$.

In contrast, Chatterjee and Durrett (2009) proved

Theorem 5.5.1 *Consider a graph G_n with n vertices generated by the configuration model with $P(d_i = k) \sim Ck^{-\alpha}$ where $\alpha > 3$ and $P(d_i \leq 2) = 0$. Let $\xi_t^1, t \geq 0$ denote the contact process on G_n starting from all sites occupied. Then for any $\lambda > 0$ there is a positive constant $p(\lambda) > 0$ so that for any $\delta > 0$*

$$\inf_{t \leq \exp(n^{1-\delta})} P\left(|\xi_t^1|/n \leq p(\lambda)\right) \to 1 \qquad \text{as } n \to \infty.$$

Mountford, Valesin and Yao (2013) extended the results to cover the powers $2 < \alpha \leq 3$, and computed the correct values of the critical exponent β. Later Mountford, Mourrat, Valesin and Yao (2016) improved the survival time estimate to $\geq e^{cn}$.

In Sect. 5.6 and 5.7 we will explain an improved proof of Theorem 5.5.1 due to Huang and Durrett (2020b). In the first section we give results for the survival time on the star graph that are the key to the proof, since they give lower bounds of how long the infection persists on vertices of large degree. In the second section we apply these results to show $\lambda_2 = 0$ when the degree distribution is subexponential. The method of proof is due to Pemantle (1992), who, in his remarkable paper which initiated the study of the contact process on trees, proved that $\lambda_2 = 0$ for degree distributions that are **stretched exponential** $p_k \sim C \exp(-ck^\gamma)$.

The goal of Huang and Durrett's (2020b) paper was to show that if the degree distribution is exponentially bounded, i.e., $\sum_k p_k e^{\theta k} < \infty$ for some $\theta > 0$ then $\lambda_1 > 0$. In Sect. 5.8 we describe the solution of this problem due to Bhamidi, Nam, Nguyen, and Sly in 2021. In hindsight the proof is not very difficult. They add a new vertex ρ^+ above the root of the tree ρ that is always infected, and they do not allow recovery at the root until all other vertices have recovered. This modification of the graph enables one to write recursive equations, but you have to be smart (or brave) enough to realize this does not give away too much.

The paper of BNNS proves a lot, but there are still significant open problems. For example Theorem 5.8.5 shows that (i) for $\lambda < \lambda_i$ the process on a finite configuration model graph with exponentially bounded degrees and n vertices survives for time $n^{1+o(1)}$ and (ii) when $\lambda > \lambda_{ii}$ the survival time is $e^{\Theta(n)}$. One would like to prove that $\lambda_i = \lambda_{ii} = \lambda_c$, and there is the second question: does the subcritical survival result give the right answer?

The chapter concludes in Sect. 5.9 with a look at the threshold-θ contact process, a discrete time system in which sites that have at least θ occupied neighbors at time n will be occupied with probability p at time $n + 1$. Chatterjee and Durrett (2013) showed that if $\theta \geq 2$ and we consider the system on a random r-regular graph with $r \geq \theta + 2$ then the phase transition is discontinuous. That is, as p is reduced from 1, the equilibrium density drops from a positive level at p_c to 0 for $p < p_c$. The system has some other unusual properties that suggest that when we leave the world of additive processes (e.g. the contact process and voter model) with their nice dual processes, strange things can happen.

5.1 Basic Properties

A rich theory has been developed for the contact process on $\mathbb{Z}^d$ and on regular trees. See Liggett's (1999) book for a summary of much of what is known in these settings. In this section we will review some of the definitions and results that are the most important for our work. The first item on our agenda is a special construction of the process from a **graphical representation** that is built using independent Poisson processes.

- For each site there is an independent rate 1 Poisson process T_n^x, $n \geq 1$. At the arrivals of this process if there is a particle at x, it will die. To facilitate later definitions we write a dot at x at the times T_n^x. In early versions of the construction people wrote a δ for death, instead of or in addition to our •.

- For each oriented edge (x, y) we have a rate λ Poisson process $T_n^{x,y}$. At these times we draw an arrow from x to y to indicate that if x is occupied there will be a birth from x to y. If y is vacant it will become occupied. If y is occupied there is no change (Fig. 5.1).

Let ξ_t^A denote the contact process starting from A occupied at time 0. A point $y \in \xi_t^A$ if for some $x \in A$ there is path from $(x, 0)$ to (y, t) that goes up the graphical representation without passing through •s and crosses edges in the direction of their orientation. A nice feature of the graphical representation is that it allows us to construct all of the ξ_t^A on the same space in such a way that

$$\text{if } A \subset B \text{ then } \xi_t^A \subset \xi_t^B. \tag{5.1.1}$$

When a set-valued Markov process has this property, it is called **attractive**. The term originated from the Ising model in which each site was in state 1 (spin up) or -1 (spin down). In this case attractive meant that the spins had a tendency to align. An important consequence of a process being attractive is

Theorem 5.1.1 *If we let ξ_t^1 be the system starting from all sites occupied then ξ_t^1 converges in distribution to a limit ξ_∞^1, which is a stationary distribution.*

The reader can find the proofs of these and other assertions in Liggett's (1999) book. Due to the monotonicity property in (5.1.1), ξ_∞^1 is the largest possible stationary distribution. Thus, if $\xi_\infty^1 = \delta_\emptyset$ then there are no nontrivial stationary distributions.

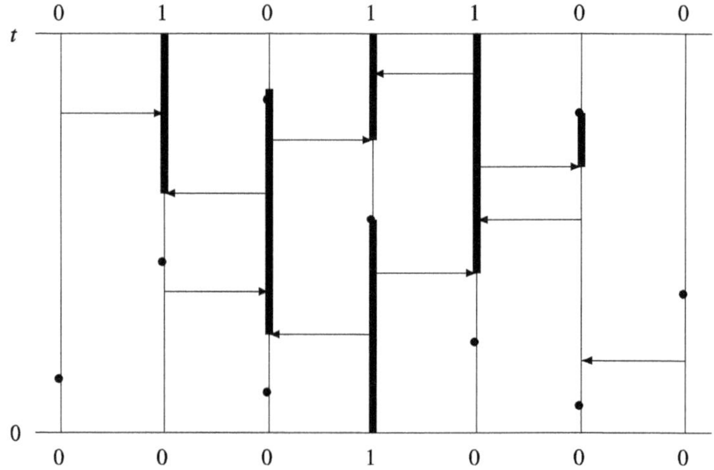

Fig. 5.1 Graphical representation for the contact process. We think of fluid flowing up the structure and across arrows in the direction of their orientation, but being stopped by dots. Thick lines indicate occupied sites

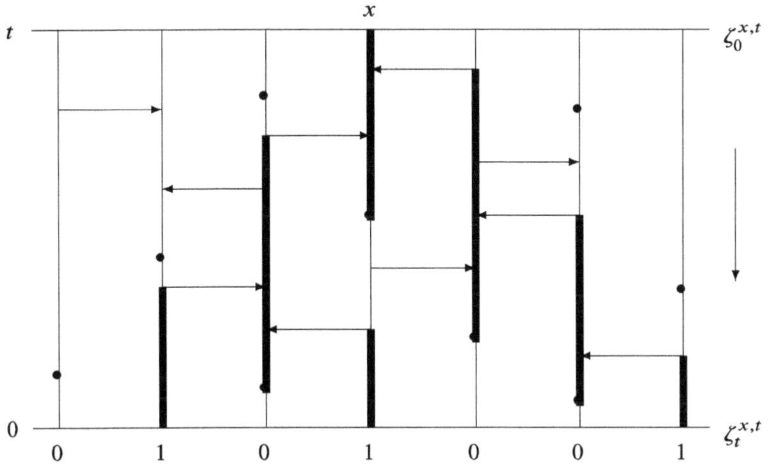

Fig. 5.2 Contact process duality. We think of fluid flowing down the structure and across arrows in the direction opposite their orientation, but being stopped by dots

A second important consequence of the construction is that it allows us to define for each x and time t a **dual process** $\zeta_s^{x,t}$, $s \le t$, that works backwards in time to answer the question "Is the site x occupied at time t?" The dual process can be constructed by a variant of the rule used for the contact process: $y \in \zeta_s^{x,t}$ if there is a path from (x, t) to $(y, t - s)$ that goes down the graphical representation without passing through •s and crosses edges in the direction opposite their orientation (Fig. 5.2).

We extend the definition of the dual to an initial set B by setting

$$\zeta_s^{B,t} = \cup_{x \in B} \zeta_s^{x,t}.$$

A little thought shows that

$$\{\xi_t^A \cap B = \emptyset\} = \{A \cap \zeta_t^{B,t} = \emptyset\}. \tag{5.1.2}$$

In words there is a path from some $(x, 0)$ with $x \in A$ to some (y, t) with $y \in B$ if and only if there is a path down from some (y, t) with $y \in B$ to some $(x, 0)$ with $x \in A$.

The almost sure equality in (5.1.2) is convenient for establishing the equation but it is useful to rewrite the equality without the superscript t. To do this we note that if $x \in \mathbb{Z}^d$ and $t < t'$ then the joint distribution of the $\zeta_s^{x,t}$, $s \le t$ with $x \in \mathbb{Z}^d$ is the same as that of $\zeta_s^{x,t'}$, with $x \in \mathbb{Z}^d$ when $s \le t$, so using the Kolmogorov extension theorem there is a family of processes ζ_s^x whose joint distributions agree with $\zeta_s^{x,t}$ on $s \le t$. Defining $\zeta B_t = \cup_{x \in B} \zeta_t^x$ then

$$P(\xi_t^A \cap B \ne \emptyset) = P(A \cap \zeta_t^B \ne \emptyset). \tag{5.1.3}$$

With future generalizations in mind we will ignore for the moment that contact process on a graph G is self dual. Taking $A = G$ and $B = \{x\}$ the duality equation becomes

$$P(x \in \xi_t^1) = P(\zeta_t^x \neq \emptyset).$$

Since the empty set is an absorbing state for the dual

$$\lim_{t \to \infty} P(x \in \xi_t^1) = P(\zeta_t^x \neq \emptyset \text{ for all } t).$$

In words, the density of particles in ξ_∞^1 is the probability that the dual process lives forever starting from one site. Replacing x by a finite set B

$$\lim_{t \to \infty} P(\xi_t^1 \cap B \neq \emptyset) = P(\zeta_t^B \neq \emptyset \text{ for all } t).$$

The fact that this holds for all B implies that ξ_t^1 converges in distribution.

A consequence of the construction from a graphical representation is that the interaction is **additive**

$$\xi_t^A \cup \xi_t^B = \xi_t^{A \cup B}.$$

Harris (1978) identified the processes with this property and showed that they had dual processes satisfying (5.1.3). The study of this class of processes was greatly advanced by the work of Griffeath (1979). In Chapter 7 we will encounter another member of the family of additive processes, the voter model.

5.2 Mean-Field Theory

A typical first step in the investigation of an interacting particle system is to examine its properties when we get rid of the spatial correlations that make it difficult to analyze. To motivate the simplifications we will use, we begin with an exact equation.

$$\frac{d}{dt} P(x \in \xi_t) = -P(x \in \xi_t) + \lambda \sum_{y \sim x} P(x \notin \xi_t, y \in \xi_t), \tag{5.2.1}$$

where $y \sim x$ means y is a neighbor of x. Let $\mathcal{N}_x = \{y : y \sim x\}$. Suppose we are on a **vertex transitive graph**. This implies (i) that the number of neighbors $|\mathcal{N}_x|$ does not depend on x and (ii) if we start from all sites occupied then at any time t the probabilities $P(y \in \xi_t)$ do not depend on y, so we can call it $u(t)$. Using the fact that $P(x \notin \xi_t, y \in \xi_t) \leq P(y \in \xi_t)$, (5.2.1) becomes

$$\frac{du}{dt} \leq (\beta - 1)u(t), \tag{5.2.2}$$

where $\beta = \lambda|\mathcal{N}_x|$. If $\beta < 1$ and $u(0) > 0$ then the last equation implies that $u(t) \to 0$. Using Theorem 5.1.1 now, we conclude that on any vertex transitive graph the critical total birth rate for the existence of a nontrivial stationary distribution $\beta_c \geq 1$.

We get more information if we have an equality rather than an inequality. The first and simplest thing to do is to pretend that the states of sites are independent and to compute the behavior of $u(t) = P(x \in \xi_t)$. When we use our independence assumption (5.2.1) becomes

$$\begin{aligned}\frac{du}{dt} &= -u(t) + \beta u(t)(1-u(t)) \\ &= u(t)(\beta - 1 - \beta u(t)) \equiv g(u(t)),\end{aligned} \tag{5.2.3}$$

where $\equiv$ indicates that the second equality defines the function g.

Theorem 5.2.1 *Let $\beta_0 = (\beta-1)/\beta$. If $u(0) \in (0,1]$ and $\beta > 1$ then in (5.2.3) $u(t) \to \beta_0$ as $t \to \infty$. In words, the mean-field contact process has a nontrivial equilibrium frequency $\beta_0 > 0$ in addition to the trivial fixed point at 0.*

Proof When $\beta > 1$ it is easy to see that $g(u) > 0$ for $u < \beta_0$ and $g(u) < 0$ when $u > \beta_0$. From this the result follows easily. The monotonicity of $u(t)$ implies that the limit exists, and a simple argument shows that it can only converge to a point v where $u(v) = 0$. □

The contact process on the complete graph provides a second approach to mean-field theory, which is a little more satisfying than just pretending sites are independent. The complete graph $\mathbb{K}_n$ on n vertices has vertex set $\{1, \ldots, n\}$ and all pairs of vertices are connected by an edge. On the complete graph it is convenient to reformulate the definition of the contact process as: an occupied site produces new particles at rate β and sends them to a site chosen at random from the graph (including x itself). The number of occupied sites, X_t, is a birth and death chain with birth and death rates

$$b_m = \beta \cdot m \cdot \left(1 - \frac{m}{n}\right) \quad \text{and} \quad d_m = m, \tag{5.2.4}$$

where m is the number of occupied sites.

The first step in analyzing a birth and death chain is to find a function $\phi(x)$, often called the **natural scale**, so that $\phi(X_t)$ is a martingale, i.e., $(d/dt)E\phi(X_t) = 0$. In terms of the rates

$$0 = b_m \cdot (\phi(m+1) - \phi(m)) + d_m \cdot (\phi(m-1) - \phi(m)). \tag{5.2.5}$$

Rearranging and using (5.2.4) gives

$$(\phi(m+1) - \phi(m)) = \frac{d_m}{b_m}(\phi(m) - \phi(m-1)) = \frac{1}{\beta(1-m/n)}(\phi(m) - \phi(m-1)).$$

Setting $\phi(0) = 0$ and $\phi(1) - \phi(0) = 1$

$$\phi(m+1) - \phi(m)) = \prod_{k=1}^{m} \frac{1}{\beta(1-k/n)} \approx (1/\beta)^m, \tag{5.2.6}$$

when $m \ll n$. Summing we have (again for $m \ll n$.) (Fig. 5.3)

$$\phi(m) \approx \sum_{k=0}^{m-1} \phi(k+1) - \phi(k) = \sum_{k=0}^{m-1} (1/\beta)^k = \frac{1-(1/\beta)^m}{1-1/\beta}. \tag{5.2.7}$$

From (5.2.7) we see that

$$\text{if } m \to \infty \text{ and } m/n \to 0 \text{ then } \phi(m) \to \beta/(\beta-1). \tag{5.2.8}$$

Eventually we will prove that if $\beta > 1$ the equilibrium frequency of 1's on the complete graph is $(\beta-1)/\beta$, in agreement with the ODE calculation. To do this the following monotonicity result, which follows from the formula for the rates in (5.2.4), is useful.

Lemma 5.2.2 *If $\beta > 1$ we have $b_m \geq d_m$ when $\beta(1-m/n) \geq 1$ or $m \leq m_0 = \lfloor n(\beta-1)/\beta \rfloor$. We have $b_m < d_m$ when $m > m_0$.*

Using this result we can extend (5.2.8)

Lemma 5.2.3 *As $n \to \infty$, $\phi(m_0) \to \beta/(\beta-1)$.*

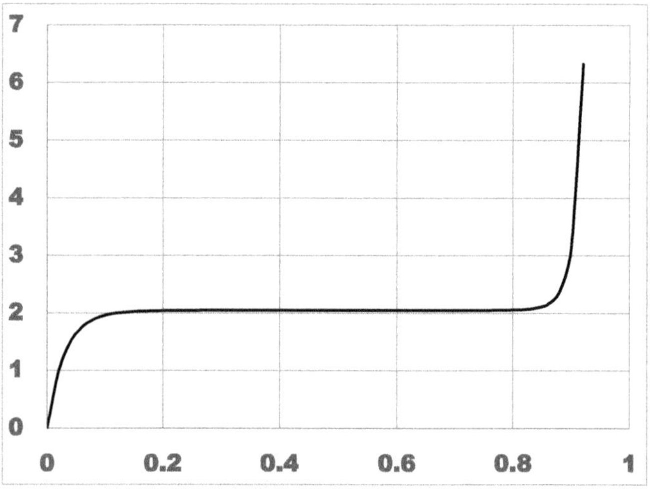

Fig. 5.3 Natural scale $\phi(x)$ when $\beta = 2$ and $n = 50$, and hence the equilibrium density $\beta_0 = (\beta-1)/\beta = 1/2$

Proof Lemma 5.2.2 implies that the slope $\phi(m+1) - \phi(m)$ is decreasing when $m \le m_0$. Let $m_1 = C_1 \log(n)$ where $C_1 = -2/\log(1/\beta)$. Since $m_1/n \to 0$, we have $\phi(m_1) \to (\beta - 1)/\beta$. Using (5.2.6) $\phi(m_1+1) - \phi(m_1) \approx n^{-2}$. Using the fact that $\phi(m+1) - \phi(m)$ is decreasing for $m_1 \le m \le m_0$, and there are only $O(n)$ points in this range, it follows that $\phi(m_0) - \phi(m_1) \to 0$, which completes the proof. □

Let $T_k = \min\{t : X_t = k\}$. Since $\phi(X_t)$ is a martingale, $\phi(0) = 0$, and $\phi(1) = 1$

$$1 = \phi(1) = \phi(m_0) P_1(T_{m_0} < T_0) \approx \frac{\beta}{\beta - 1} P_1(T_{m_0} < T_0),$$

by 5.2.3, so as $n \to \infty$

$$P_1(T_{m_0} < T_0) \to \frac{\beta - 1}{\beta}.$$

On any finite graph, the contact process dies out but as we will see in Sect. 5.3 that on $\mathbb{K}_n$ and many other graphs the contact has a **quasi-stationary distribution** or **metastable state** with density $(\beta - 1)/\beta$.

Prolonged persistence. Our next goal is to show that when $\beta > 1$ then the contact process on the complete graph with n vertices survives for time $\approx \exp(C(\beta)n)$. This is the best we can hope for since the probability that in one unit of time no births occur and all particles die is

$$\ge [(1 - e^{-\beta})e^{-1}]^n. \tag{5.2.9}$$

We will eventually show that $C(\beta) = \log \beta - (\beta - 1)/\beta$ provides a sharp estimate on the exponential survival time. To do this we will use two methods to obtain matching upper and lower bounds.

Method 1: Bounding the extinction time using hitting times. The first step is to construct a martingale for the mean-field contact process in continuous time X_t, which has jump rates given in (5.2.4). In order for $\psi(X_t)$ to be a martingale while $X_t \in (0, m_0)$ we must have

$$0 = b_m(\psi(m+1) - \psi(m)) + d_m(\psi(m-1) - \psi(m)) \quad \text{for } 0 < m < m_0.$$

If $\psi(m_0) = 0$ and $\psi(m) = \sum_{k=m}^{m_0-1} s_k$ for $0 < m < m_0$ then $\psi(X_t)$ will be a martingale if

$$0 = -b_m s_m + d_m s_{m-1}.$$

Thus we want

$$s_{m-1} = s_m b_m / d_m = s_m \beta(1 - m/n). \tag{5.2.10}$$

To get the recursion started we set $s(m_0) = 1$.

If we let $L = m_0$ and $h_L = P_{L-1}(T_0 < T_L)$ then using the fact that $\psi(X_t)$ is a martingale and $\psi(L) = 0$ we have $\psi(L-1) = h_L \psi(0)$ and hence

$$h_L = \psi(L-1)/\psi(0) \le s_{L-1}/s_0. \tag{5.2.11}$$

Using the recursion (5.2.10)

$$\log(s_0/s_{L-1}) = \sum_{k=1}^{L-1} \log(s_{k-1}/s_k) = \sum_{k=1}^{L-1} \log(\beta(1-k/n)).$$

Note that by the definition of $L = m_0$ the terms in the final sum > 0. Multiplying by n/n and turning the sum into an approximation of an integral the RHS is

$$\approx n \int_0^{\beta_0} \log\beta + \log(1-x)\,dx = \beta_0 \log\beta + \int_0^{\beta_0} \log(1-x)\,dx, \tag{5.2.12}$$

where $\beta_0 = (\beta-1)/\beta$. The antiderivative of $\log(1-x)$ is $(x-1)\log(1-x) - x$ so the above is

$$\begin{aligned} B &\equiv \beta_0 \log\beta + (\beta_0 - 1)\log(1-\beta_0) - \beta_0 \\ &= \frac{\beta-1}{\beta}\log\beta - \frac{1}{\beta}\log(1/\beta) - \frac{\beta-1}{\beta} = \log(\beta) - \frac{\beta-1}{\beta} \end{aligned}$$

and using (5.2.11) we have $h_L \le \exp(-nB)$. Since each excursion away from L takes at least $1/L$ unit of time on the average we have

Lemma 5.2.4 *If $\beta > 1$ and $\epsilon > 0$ then the extinction time for the contact on the complete graph with n vertices takes time $\ge (1/L)\exp(n(B-\epsilon))$ with high probability.*

Method 2: Hitting times for birth and death processes. Our second approach is to compute the expected hitting time directly using methods that hold for many birth and death chains. Let $T_{b,a}$ = time to go from a to b. The subscripts are backwards from how I would have chosen them, but it is easier to live with the notation than to rewrite all of the equations in Sect. 6.7 in Allen (2003). Our first identity is obvious from the definitions:

Lemma 5.2.5 *If $a > b$ then $T_{b,a} = T_{a-1,a} + T_{a-2,a-1} + \cdots + T_{b,b+1}$.*

Let μ_i be the death rate at i (rate for jumps $i \to i-1$) and λ_i be the birth rate at i (rate for jumps $i \to i+1$). It is easy to see that

$$ET_{i-1,i} = \frac{\mu_i}{\lambda_i + \mu_i} \cdot \frac{1}{\lambda_i + \mu_i} + \frac{\lambda_i}{\lambda_i + \mu_i} \cdot \left(\frac{1}{\lambda_i + \mu_i} + ET_{i,i+1} + ET_{i-1,i} \right).$$

To explain this equality note that the chain stays at i for a time with mean $1/(\lambda_i + \mu_i)$. If it jumps to $i-1$ (with probability $\mu_i/(\mu_i + \lambda_i)$) then we have gotten to $i-1$. If not, then to get to $i-1$ we have to go from $i+1$ to i and from i to $i-1$. Rearranging gives (Fig. 5.4)

$$ET_{i-1.i}\left(1-\frac{\lambda_i}{\lambda_i+\mu_i}\right)=\frac{1}{\lambda_i+\mu_i}+\frac{\lambda_i}{\lambda_i+\mu_i}\cdot ET_{i,i+1},$$

which gives us our recursion:

$$ET_{i-1.i}=\frac{1}{\mu_i}+\frac{\lambda_i}{\mu_i}\cdot ET_{i,i+1}. \tag{5.2.13}$$

Suppose now that we want to compute $ET_{0,n}$ for a birth and death process on $\{0,1,\ldots,n\}$. The first jump can only be from n to $n-1$, so using the recursion (5.2.13) for later terms, we have

$$\begin{aligned}
ET_{n-1,n} &= \frac{1}{\mu_n},\\
ET_{n-2,n-1} &= \frac{1}{\mu_{n-1}}+\frac{\lambda_{n-1}}{\mu_{n-1}}\cdot\frac{1}{\mu_n},\\
ET_{n-3,n-2} &= \frac{1}{\mu_{n-2}}+\frac{\lambda_{n-2}}{\mu_{n-2}}\cdot\left(\frac{1}{\mu_{n-1}}+\frac{\lambda_{n-1}}{\mu_{n-1}}\cdot\frac{1}{\mu_n}\right)\\
&= \frac{1}{\mu_{n-2}}\left(1+\frac{\lambda_{n-2}}{\mu_{n-1}}+\frac{\lambda_{n-2}\lambda_{n-1}}{\mu_{n-1}\mu_n}\right),\\
ET_{n-4,n-3} &= \frac{1}{\mu_{n-3}}\left(1+\frac{\lambda_{n-3}}{\mu_{n-2}}+\frac{\lambda_{n-3}\lambda_{n-2}}{\mu_{n-2}\mu_{n-1}}+\frac{\lambda_{n-3}\lambda_{n-2}\lambda_{n-1}}{\mu_{n-2}\mu_{n-1}\mu_n}\right).
\end{aligned}$$

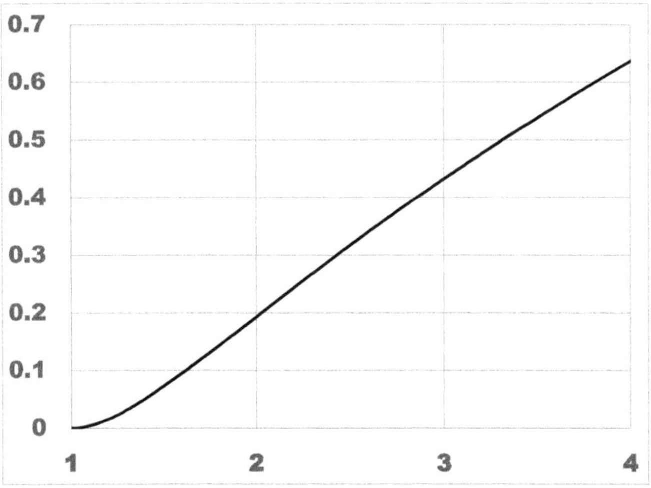

Fig. 5.4 Plot of $B=\beta\log\beta-(\beta-1)/\beta$ versus β

Generalizing from these formulas, we see that

$$ET_{n-k-1,n-k} = \frac{1}{\mu_{n-k}}\left(1 + \sum_{j=n-k+1}^{n} \prod_{i=n-k+1}^{j} \frac{\lambda_{i-1}}{\mu_i}\right)$$
$$= \frac{1}{\mu_{n-k}}\left(\sum_{j=n-k}^{n} \prod_{i=n-k+1}^{j} \frac{\lambda_{i-1}}{\mu_i}\right),$$

where in the last step we have absorbed the 1 into the sum by using the convention that the empty product $\prod_{i=n-k+1}^{n-k} = 1$. Using Lemma 5.2.5 we have

$$ET_{0,n} = \sum_{k=0}^{n-1} ET_{n-k-1,n-k}.$$

When we combine the last two expressions we get a sum of roughly $n^2/2$ products of λ_{i-1}/μ_i from $i = n-k+1$ to $j \in [n-k+1, n]$. When $\beta > 1$, if $\lambda_k = \beta(1 - k/n)$ and $\mu_k = k$, the largest of these products is the one with $i = 1$ and $j = n\beta_0$,

$$\pi = \prod_{i=1}^{n\beta_0} \beta(1 - i/n). \tag{5.2.14}$$

If we take logs

$$B = \log \pi = \beta_0 \log(\beta) + \frac{1}{n}\sum_{i=1}^{\beta_0 n} \log(1 - i/n).$$

Using a calculation from the end of the section on Method 1, see (5.2.12),

$$B = \log \beta - (\beta - 1)/\beta.$$

The sum which gives $T_{0,n}$ is bigger than the term in (5.2.14) and smaller than n^2 times it, so we have

Lemma 5.2.6 *If n is large then the extinction time for the contact on the complete graph takes time $\le Cn^2 \exp(nB)$ with high probability.*

This result and Lemma 5.2.4 provide complementary upper and lower bounds on the survival time starting from all sites occupied at time 0, so we can conclude that if $\beta > 1$,

$$\frac{1}{n} \log \tau_n \to B = \log \beta - (\beta - 1)/\beta. \tag{5.2.15}$$

5.3 Bounded Degree Graphs

Pemantle (1992) was the first to study the contact process on the tree $\mathbb{T}^d$ in which each vertex has degree $d + 1$. Here, and in what follows, we assume $d \geq 2$ since $\mathbb{T}^1 = \mathbb{Z}$. Let 0 be the root of the tree and let P_0 be the probability measure for the process starting from only the root occupied. Pemantle found that the contact process on $\mathbb{T}^d$ has two critical values:

$$\lambda_1 = \inf\{\lambda : P_0(\xi_t \neq \emptyset \text{ for all } t) > 0\},$$
$$\lambda_2 = \inf\{\lambda : \liminf_{t\to\infty} P_0(0 \in \xi_t) > 0\}.$$

In words, the contact process **survives** when $\lambda > \lambda_1$ but **survives locally** when $\lambda > \lambda_2$, i.e., with positive probability 0 is occupied infinitely many times, though the definition we have chosen is stronger than that.

By deriving bounds on the critical values, he showed that $\lambda_1 < \lambda_2$ when $d \geq 3$. Liggett (1996) settled the case $d = 2$ by showing

$$\lambda_1 < 0.605 < 0.609 < \lambda_2.$$

At about the same time, Stacey (1996) gave a proof that $\lambda_1 < \lambda_2$ that did not rely on numerical bounds on the critical value.

Open problem. Stacey's argument is simple and elegant but relies heavily on the fact that the graph is a homogeneous tree. It would be nice to generalize it to Galton-Watson trees, but the randomness of the graph seems to pose a substantial problem. A simpler problem is to consider the Big World of Durrett and Jung (2007), shown in Figure 5.5. Algebraically it is the free product $\mathbb{Z}^d * \{0, 1\}$. Geometrically, the Big World is the covering space of the Bollobás-Chung (1988) small world, which was defined much earlier than the more highly cited small world of Watts and Strogatz (1998). In the BC small world one starts with a circle with an even number of vertices, pairs them at random and connects each pair by an edge. The Big World gives the limit as $n \to \infty$ of the view point of a bug walking on the small world. Moving through a long range edge brings it to a new copy of $\mathbb{Z}$.

Contact process on $\{-n, \ldots, n\}$ with edges connecting nearest neighbors. Suppose it starts from all sites occupied and let $\tau_n = \inf\{t : \xi_t = \emptyset\}$. Combining results of Durrett and Liu (1988) and Durrett and Schonmann (1988) gives the following results

Theorem 5.3.1 *Let λ_c be the critical value for the nearest neighbor contact process on $\mathbb{Z}$.*

(i) If $\lambda < \lambda_c$ then there is a constant $\gamma_1(\lambda)$ so that

$$\tau_n / \log n \to \gamma_1(\lambda) \quad \textit{in probability.}$$

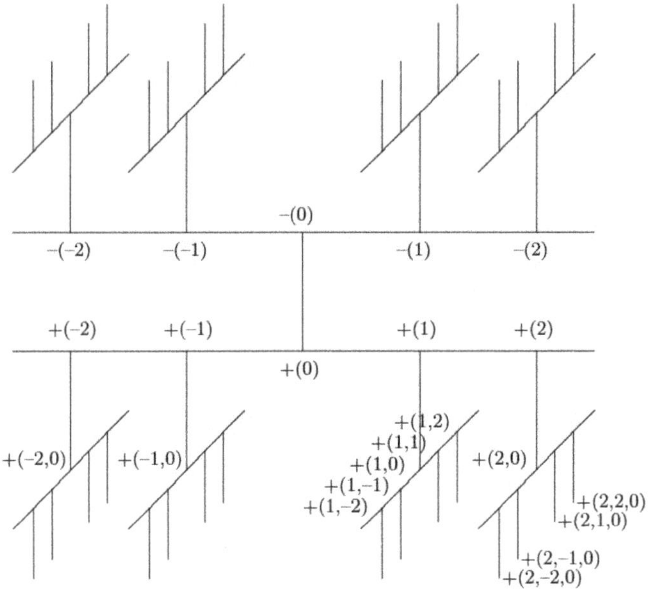

Fig. 5.5 The Big World graph

(ii) If $\lambda > \lambda_c$ then there is a constant $\gamma_2(\lambda)$ so that

$$(\log \tau_n)/n \to \gamma_2(\lambda) \quad \textit{in probability.}$$

(iii) When $\lambda > \lambda_c$ there is "metastability":

$$\tau_n / E\tau_n \Rightarrow \textit{exponential}(1).$$

Here $\Rightarrow$ means convergence in distribution. Intuitively, when $\lambda > \lambda_c$ the process on the interval stays exponentially long in a state that looks like the stationary distribution for the process on $\mathbb{Z}$ restricted to the interval $[-n, n]$. The lack of memory property of the survival time suggests that the death of the process comes suddenly and without warning, but we know of no result that makes this precise.

Open Problem 5.3.2 *Consider first the much simpler problem of the contact process on the complete graph with n vertices and $\beta > 1$. Let $mn(\beta - 1)/\beta$. To get an idea of what the contact looks like when it is dying, start the contact process from $m - 1$ and condition it to hit 0 before m. If $h(x) = P_x(T_0 < T_m)$ where T_k is the time of the first visit to k then the conditioned process has transition probability*

$$\bar{p}_t(x, y) = p_t(x, y)h(y)/h(x).$$

It is "routine" to compute h but it is not so easy to get a good enough formula to understand the behavior of the conditioned process.

Results on $\mathbb{Z}^d$ with $d > 1$ had to wait for the work of Bezuidenhout and Grimmett (1990), who showed that in $d > 1$ the contact process dies out at the critical value. To do this they introduced a **block construction** that can be used to study the supercritical process. See Sect. 2 of part I of Liggett. Mountford proved the metastability result in 1993 and that $(\log \tau_n)/n^d \to \gamma(\lambda)$ in 1999.

Stacey (2001) studied the contact process on a tree truncated at height ℓ, $\mathbb{T}^d_\ell$. To be precise, the root has degree d, vertices at distance $0 < k < \ell$ from the root have degree $d + 1$, while those at distance ℓ have degree 1. Cranston, Mountford, Mourrat, and Valesin (2014) improved Stacey's result to establish that the time to extinction starting from all sites occupied τ^d_ℓ satisfies

Theorem 5.3.3 *(a) For any $0 < \lambda < \lambda_2(\mathbb{T}^d)$ there is a $c \in (0, \infty)$ so that as $\ell \to \infty$*

$$\tau^d_\ell / \log |\mathbb{T}^d_\ell| \to c \quad \textit{in probability.}$$

(b) For any $\lambda_2(\mathbb{T}^d) < \lambda < \infty$ there is a $c \in (0, \infty)$ so that as $\ell \to \infty$

$$\log(\tau^d_\ell)/|\mathbb{T}^d_\ell| \to c \quad \textit{in probability.}$$

In case (b), $\tau^d_\ell / E\tau^d_\ell$ converges to a mean one exponential.

When a tree is truncated at a finite distance, a positive fraction of the sites are on the boundary. A more natural finite version of a tree is a **random regular graph** in which all vertices have degree $d + 1$. In this case there is no boundary and the graph has the same distribution viewed from any point. If there are n vertices, the graph looks like $\mathbb{T}^d$ in neighborhoods of a point that have $\le n^{1/2-\epsilon}$ vertices. See Theorem 1.2.3. Mourrat and Valesin (2016) have shown for a random regular graph that the time to extinction starting from all sites occupied τ_n satisfies:

Theorem 5.3.4 *(a) For any $0 < \lambda < \lambda_1(\mathbb{T}^d)$ there is a $C < \infty$ so that as $n \to \infty$*

$$P(\tau_n < C \log n) \to 1.$$

(b) For any $\lambda_1(\mathbb{T}^d) < \lambda < \infty$ there is a $c > 0$ so that as $n \to \infty$

$$P(\tau_n > e^{cn}) \to 1.$$

Notice that the threshold in this theorem comes at λ_1, while the one in Stacey's result comes at λ_2. The difference in the two results is that when $\lambda \in (\lambda_1, \lambda_2)$ on the infinite tree the origin is in the middle of a linearly growing vacant region. On the truncated tree the system dies out when the vacant region is large enough, or more poetically the particles like lemmings run over the edge of cliff. However, on the random regular graph the occupied sites will later return to the origin. Durrett and Jung (2007) investigated the qualitative differences between $\lambda \in (\lambda_1, \lambda_2)$ and $\lambda > \lambda_2$ on the small world graph.

Lalley and Su (2017) also proved Theorem 5.3.4. In addition they established an interesting "cutoff" result about convergence to equilibrium. Their Theorem 1.1 shows that if we start from only 1 occupied then there are constants c_λ and $g_n(\epsilon) \to 0$ so that

$P(2 \in \xi_t)$	at times
$\le g_n(\epsilon)$	$\le (1-\epsilon)c_\lambda \log n$
$\ge \rho^2(1 - g_n(\epsilon))$	$\ge (1+\epsilon)c_\lambda \log n$

where ρ is the survival probability, which by duality is the fraction of sites that are occupied.

Mourrat and Valesin (2016) extended their result on random regular graphs given in Theorem 5.3.4 to graphs with bounded degree. The next result is their Theorem 1.3.

Theorem 5.3.5 *Let $\mathcal{G}_{n,d}$ be the set of connected graphs with n vertices and degree bounded by $d+1$.*

(a) For any $0 \in (0, \lambda_c(\mathbb{T}^d))$ there is a constant $C < \infty$ so that

$$\lim_{n\to\infty} \inf_{G \in \mathcal{G}_{n,d}} P(\tau_G < C \log n) = 1.$$

(b) For any $\lambda \in (\lambda_c(\mathbb{Z}), \infty)$ there is a constant $c > 0$ so that

$$\lim_{n\to\infty} \inf_{G \in \mathcal{G}_{n,d}} P(\tau_G > e^{cn}) = 1.$$

Note that the constants are uniform over $G \in \mathcal{G}_{n,d}$. The form of the result in (b) suggests the method of proof: they find a large path in the graph and compare with the one dimensional contact process.

Schapira and Valesin (2015) extended the results of Mountford, Mourrat, Valesin and Yao (2016) in two ways. They proved without any restriction in the graph G that if the infection rate is larger than the critical value of the one-dimensional contact process, $\lambda_c(\mathbb{Z})$, then the extinction time τ_G starting from all sites occupied grows faster than $\exp(|G|/(\log|G|)^\kappa)$ for any $\kappa > 1$. Also for general graphs they showed that the extinction time divided by its expectation converges to the exponential distribution with mean 1. Even though this is a section titled bounded degrees, they do not make that assumption. To state their results formally using the numbers in their paper.

Theorem 5.3.6 *(Theorem 1.2) For any $\lambda > \lambda_c(\mathbb{Z})$ there is a constant c_ϵ such that for any connected graph G that has at least 2 vertices*

$$E_\lambda[\tau_G] \geq (\log |G|)^{1+\epsilon} \left(\frac{c_\epsilon |G|}{|G|^{1+\epsilon}} \right).$$

Clearly there is no hope when there is only one vertex since in that case it is not possible to give birth.

Theorem 5.3.7 *(Theorem 1.3.) For any $\lambda > \lambda_c(\mathbb{Z})$ and any sequence of graphs G_n with $|G_n| \to \infty$*

$$\frac{\tau_{G_n}}{E\tau_{G_n}} \Rightarrow \textit{exponential}(1).$$

There are a large number of details involved in carrying out these proofs, so we will content ourselves to describe the main new ideas

Lemma 5.3.8 *(Proposition 2.7.) There is a $c_{comp} > 0$ so that for any $n \geq 2$ and any tree G with n vertices*

$$P\left(\xi_t^A \neq \emptyset, \xi_t^A \neq \xi_t^1\right) \leq \exp\left(-c_{comp} \left\lfloor \frac{t}{n(\log n)^3} \right\rfloor\right),$$

for all $t > 0$ and $A \neq \emptyset$.

Once this result is established Theorem 1.3 follows fairly easily, since the proof that the event on the left has small probability is a key step in showing that limits of $\tau_{G_n}/E\tau_{G_n}$ have the lack of memory property, and hence must be exponential(1). This idea already appeared as Lemma A.1 in MMVY (2016). Proposition 2.7 is an immediate consequence of

Lemma 5.3.9 *(Lemma 2.8.) There is a $c_1 < 1$ so that for all G with $n \geq 2$ vertices and $t \geq n(\log n)^3$*

$$P\left(\xi_t^A \neq \emptyset, \xi_t^A \neq \xi_t^1\right) < c_1.$$

Proposition 2.7 leads to the second main ingredient.

Lemma 5.3.10 *(Proposition 2.9.) There exists a constant c_{split} so that for any tree G containing N connected and disjoint subtrees $G_1, \ldots, G_N$*

$$E[\tau_G] \geq \frac{c_{split}}{(|G|^3)^{N+1}} \prod_{i=1}^{N} E[\tau_{G_i}].$$

This result is interesting in its own right. The authors use it to give a simple proof of part (b) of Theorem 5.3.5 without thc assumption of bounded degree. For more information about the proof the reader will have to consult Schapira and Valesin (2015). In their (2018) paper they use these methods to analyze the behavior of eight different sequences of graphs G_n.

5.4 Erdős-Rényi Graphs

In the last section we considered graphs with bounded degree distributions, e.g., trees and random regular graphs. In next several sections we will consider random graphs with degree distributions that are power-law, subexponential, and have exponentially bounded tails. The Erdős-Rényi graphs we consider here have unbounded degrees but the tails decay faster than exponential. As in the previous section, we will be concerned with showing that if the contact process is sufficiently supercritical, then starting from all sites occupied the process survives for time $\exp(cn)$, which as remarked earlier, see (5.2.9), is the best possible result for graphs with n vertices and $\leq Bn$ edges.

Proving survival for time $\exp(cn)$ in the Erdős-Rényi case is easy thanks to Theorem 1.5.3, which shows that if the mean degree is $\lambda > 1$ then there is a path with length $\geq c_\lambda n$. Comparing the contact process on the path with the contact process on an interval, see Theorem 5.3.1, it follows that if we start with all sites occupied then survival occurs for time $\geq \exp(cn)$.

A new approach due to Cator and Don (2021) relies on uniformly bounding the infection rate from below for all sets with a fixed number of vertices k. Once the minimum infection rate for sets of size k has been estimated and it is $\geq \rho k$ for a range of values, a simple comparison with a birth and death chain will show exponential growth of the infection time.

To illustrate this method we consider the Erdős-Rényi(n, p) graph with $np = \sigma$. Their method will clearly fail if $\sigma < 1$, but as they explain it will also fail if the fraction of sites in the giant component is $< 1/2$, because in that case there will be sets of size $n/2$ that have no neighbors (take the complement of the giant component). The probability ρ that a vertex does not belong to the giant component satisfies $\rho = \exp(-\sigma(1-\rho))$ so recalling that the generating function $\phi(z) = \exp(\sigma(1-z)$ is convex and has $\phi'(1) = \sigma$ we see that $\rho > 1/2$ if $1/2 < \exp(-\sigma/2)$, i.e., $\sigma < 2\log 2 = 1.3862$. For technical reasons that will become clear later they have to choose $\sigma > 4\log 2$.

Large deviations for the number of connections to a set. Suppose we have a graph with n vertices, i.e., $|V| = n$, and a set $S \subset V$ of size k. The number of potential connections from S to $V - S$ is $k(n-k)$ and each is independently present with probability σ/n so the number of links from one fixed set S, $L_S = \text{binomial}(k(n-k), \sigma/n)$. The number of sets of vertices size k is $\binom{n}{k}$ so it is a simple exercise in large deviations to find a lower bound $M_{n,k}$ on the minimum number of connections from S to $V - S$ for a set S of size k.

Fix n and $k = \gamma n$, $p = \sigma/n$. L_S is binomial$(k(n-k), p)$ so $EL_S = k(n-k)\sigma/n$ and

$$E\exp(-\theta L_S) = \left(1 - \frac{\sigma}{n} + \frac{\sigma}{n}e^{-\theta}\right)^{k(n-k)} = \exp\left(-(\sigma/n)(1-e^{-\theta})k(n-k)\right).$$

Thus if $\theta > 0$, Markov's inequality implies

$$\exp(-\theta\rho EL_S)P(L_S \le \rho E(L_S)) \le \exp(-(\sigma(1-e^{-\theta})k(n-k)/n).$$

Filling in the formula for EL_S and rearranging

$$P(L_S \le \rho E(L_S)) \le \exp\left(-\frac{\sigma}{n}k(n-k)\cdot(-\theta\rho + 1 - e^{-\theta})\right).$$

The next step is to optimize this bound over θ. Differentiating gives

$$(-\theta\rho + 1 - e^{-\theta})' = -\rho + e^{-\theta},$$

so we let $\theta_0 = -\log\rho$. In this case

$$(-\theta_0\rho + 1 - e^{-\theta_0}) = \rho\log\rho + (1-\rho) \equiv G(\rho).$$

If we let $\gamma = k/n$ then we have

$$P(L_S \le \rho E(L_S)) \le \exp(-\sigma G(\rho)\gamma(1-\gamma)). \tag{5.4.1}$$

This is (3.12) in Cator and Don (2021), which they prove by invoking the "Chernoff-Hoeffding inequality," and cite Hoeffding (1963).

Before we move on note that $G(1) = 0$, which is natural since it gives us $P(L_S \le EL_S)$, which by the central limit theorem has limit 1/2.

$$\text{As } \rho \to 0,\ G(\rho) \to 1. \tag{5.4.2}$$

At first it may be surprising that $G(\rho) \not\to \infty$ as $\rho \to 0$ but then you remember than if X has a Poisson(bn) distribution then $P(X = 0) = e^{-bn}$ (Fig. 5.6).

Later we will want to know that $G(\rho) > (1-\sqrt{\rho})^2$ for $0 < \rho < 1$. We content ourselves with a proof by picture.

Counting the number of sets of size k, $\gamma = k/n$.

Lemma 5.4.1 *If $H(\gamma) = -\gamma\log(\gamma) - (1-\gamma)\log(1-\gamma)$ then for large n*

$$\binom{n}{\gamma n} \le e^{nH(\gamma)}. \tag{5.4.3}$$

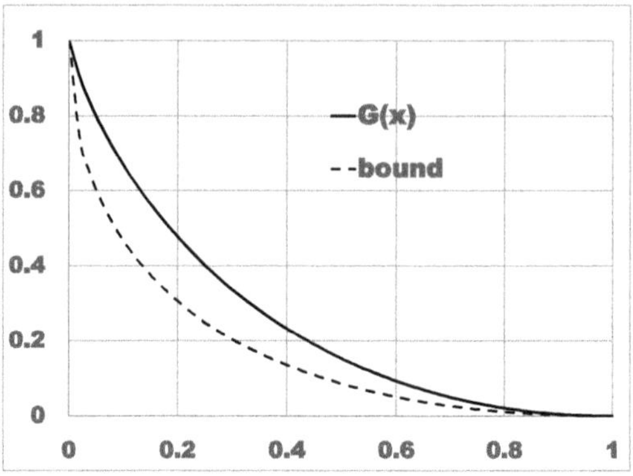

Fig. 5.6 Picture proof of $G(\rho) > (1 - \sqrt{\rho})^2$

Proof To prove this, we use Stirling's formula $m! \sim m^m e^{-m}\sqrt{2\pi m}$ to conclude that (the e^{-m} terms in Stirling's formula cancel)

$$\begin{aligned}\binom{n}{\gamma n} &= \frac{n!}{(\gamma n)!((1-\gamma)n)!}\\ &\sim \frac{n^n}{(\gamma n)^{\gamma n}((1-\gamma)n)^{(1-\gamma)n}} \cdot \frac{\sqrt{2\pi n}}{\sqrt{2\pi\gamma n}\sqrt{2\pi(1-\gamma)n}}\\ &= \frac{1}{\gamma^{\gamma n}(1-\gamma)^{(1-\gamma)n}} \cdot \frac{1}{\sqrt{2\pi\gamma(1-\gamma)n}}.\end{aligned}$$

The last term is < 1 for large n, which completes the proof. □

Combining (5.4.1) and (5.4.3) we have

$$P\{\exists S : |S| = k, L_S \le \rho E L_S\} \le e^{nH(\gamma)} \cdot C e^{-\sigma G(\rho)\gamma(1-\gamma)}.$$

The exponent is negative if

$$G(\rho) \cdot \sigma\gamma(1-\gamma) > H(\gamma). \tag{5.4.4}$$

The functions $\sigma\gamma(1-\gamma)$ and $H(\gamma)$ are symmetric about 1/2. It is easy to see that $\gamma(1-\gamma)$ is maximized at 1/2, where the value is 1/4. A little calculus shows that $H(\gamma)$ is maximized at 1/2, where the value is $\log 2$. If we suppose that $\sigma > 4\log 2$ then $\gamma(1-\gamma) > H(\gamma)$ when $\gamma = 1/2$ so using (5.4.2) it is possible to pick $\rho > 1$ so that (5.4.4) holds at $\lambda = 1/2$. If we consider sets of size $k = n/2$ and know that $L_S \ge \rho E L_S$ then the

birth rate $b_k = \lambda\rho(\sigma/n)k(n-k) = \lambda\rho\sigma n/4$,

death rate $d_k = k = n/2$.

Note that once we are able to choose $\rho > 0$ then we can get $b_k > d_k$ by choosing λ large.

In our simple calculation once we know we can choose $\rho > 0$ at 1/2 then we can widen this to $(1/2 - \delta, 1/2 + \delta)$ for some $\delta > 0$ and we can get the exponential survival result by coupling the contact process to a birth and death chain and using calculations at the end of Sect. 5.2. Cator and Don (2021), see their Lemma 3.2, have a better way of generating an interval of values with $\rho > 0$.

Lemma 5.4.2 *Fix $\sigma > 4\log 2$ and consider an Erdős-Rényi graph with n vertices and edge probabilities σ/n. Choose*

$$\gamma_\sigma = \frac{1}{2} - \sqrt{\frac{1}{4} - \frac{\log 2}{\sigma}} \in (0, 1/2) \quad \textit{and} \quad \alpha_\sigma = \frac{2\log(1 - 2\sqrt{(\log 2)/\sigma}}{\log(1/4 - (\log 2)/\sigma)} \in (0, 2).$$

For $\gamma_\sigma < \gamma < 1 - \gamma_\sigma$ let

$$\rho(\gamma) = (\gamma(1-\gamma) - (\log 2)/\sigma)^{\alpha_\sigma} \in (0, 1) \quad \textit{and} \quad M_{n,k} = \rho(k/n) \cdot \sigma k(n-k).$$

Then with high probability $L_S \geq M_{n,k}$ for all $\gamma_\sigma n < k < (1-\gamma_\sigma)n$ and all $S \in \mathcal{S}_k$.

Proof A little calculus shows that for $\gamma_\sigma < (1 - \gamma_\sigma)$

$$\alpha_\sigma \geq \frac{2\log\left(1 - 2\sqrt{\frac{H(\gamma)}{\sigma\gamma(1-\gamma)}}\right)}{\log(\gamma/(1-\gamma) - (\log 2)/\sigma} \in (0, 2). \tag{5.4.5}$$

Since $G(\rho) > (1 - \sqrt{\rho})^2$ we have

$$\begin{aligned} \sigma G(\rho)\gamma(1-\gamma) &> \sigma\left(1 - \sqrt{(\gamma(1-\gamma) - (\log 2)/\sigma)^{\alpha_\sigma}}\right)^2 \gamma(1-\gamma) \\ &\geq \sigma\left(1 - \left(1 - \sqrt{\frac{H(\gamma)}{\sigma\gamma(1-\gamma)}}\right)\right)^2 \gamma(1-\gamma) = H(\gamma), \end{aligned}$$

where the second inequality follows from (5.4.5). It follows that

$$\sup_{\gamma_\sigma < \gamma < 1-\gamma_\sigma)} H(\gamma) - \sigma G(\rho)\gamma(1-\gamma) < 0,$$

which gives the desired result. □

Using this result with results from their Sect. 2 on birth and death processes, they are able to prove (see their Theorem 3.2)

Theorem 5.4.3 *Consider an Erdős-Rényi graph G_n with n vertices and edge probabilities σ/n where $\sigma > 4\log 2$. There are functions $\lambda_0(\sigma) = (1 + o(1))/\sigma$ and*

$\epsilon(\sigma) = o(1)$ so that if $\lambda > \lambda_0(\sigma)$ then T_n, the survival time of contact process starting from all sites occupied, satisfies

$$\frac{1}{n} \log ET_n \geq (1 - \epsilon(\sigma) \log(\lambda\sigma) + \frac{1 - \epsilon(\sigma)}{\lambda\sigma} - 1,$$

with high probability.

5.5 Power-Law Random Graphs

Pastor-Satorras and Vespignani, who we will abbreviate PSV (2001a), have made an extensive study of the contact process on "scale-free" random networks using mean-field methods. For this and many other related results see the survey paper by Pastor-Satorras, Castellano, van Meighem, and Vespignani (2015).

5.5.1 *Mean-Field Theory*

To be precise we will use what is called **degree-based mean-field theory**. The results here are from PSV (2001b) although we will write the arguments differently. Let $\rho_k(t)$ denote the fraction of vertices of degree k that are infected at time t, and $\theta(\lambda)$ be the probability that a given link points to an infected vertex. If we make the assumption that there is no correlation between the degree of a site and the state of the vertex pointed to then

$$\frac{d\rho_k(t)}{dt} = -\rho_k(t) + \lambda k[1 - \rho_k(t)]\theta(\lambda).$$

This will turn out to be a bad assumption but it is needed to conclude that the equilibrium frequency ρ_k satisfies

$$0 = -\rho_k + \lambda k[1 - \rho_k]\theta(\lambda). \tag{5.5.1}$$

Solving and writing $\theta(\lambda)$ simply as θ we have

$$\rho_k = \frac{k\lambda\theta}{1 + k\lambda\theta}.$$

Suppose p_k is the degree distribution in the graph. The probability that a given link points to a vertex of degree k is the size-biased degree distribution $q_k = kp_k/\mu$ where $\mu = \sum_j jp_j$, so we have the following self-consistent equation for θ:

$$\theta = \sum_k q_k \rho_k = \sum_k q_k \frac{k\lambda\theta}{1+k\lambda\theta}. \tag{5.5.2}$$

Once θ is computed we can compute the fraction of occupied sites from

$$\rho = \sum_k p_k \frac{k\lambda\theta}{1+k\lambda\theta}. \tag{5.5.3}$$

To reduce the problem to computing θ we begin by noting that in all of the examples we will consider $\mu = \sum_k kpk < \infty$. In examples the critical value λ_c may be 0 or positive, but in all cases $\theta(\lambda) \downarrow 0$ as $\lambda \to \lambda_c$. Using the dominated convergence theorem

$$\rho(\lambda) = \lambda\theta \sum_k kp_k \frac{1}{1+k\lambda\theta} \sim \lambda\theta(\lambda)\mu. \tag{5.5.4}$$

Analysis of (5.5.2), which we are about to describe in some detail, suggests the following conjectures about the contact process on power law graph with degree distribution $p_k \sim Ck^{-\alpha}$. Here β is the critical exponent for the equilibrium density $\rho(\lambda) \approx (\lambda - \lambda_c)^\beta$ as $\lambda \downarrow \lambda_c$.

- If $\alpha \le 3$ then $\lambda_c = 0$.
- If $3 < \alpha < 4$, $\lambda_c > 0$ but the critical exponent $\beta > 1$.
- If $\alpha > 4$ then $\lambda_c > 0$ and $\rho(\lambda) \sim C(\lambda - \lambda_c)^1$ so $\beta = 1$.

In Section 1.9 we defined β and other exponents associated with the phase transition in Erdős-Rényi graphs. In Section 2.7 we computed the exponent β associated with the percolation phase transition in power-law graphs generated by the configuration model. We invite the reader to compare the results here in Figure 5.7 to those in the table after Theorem 2.7.1.

The results depend on the power α in the degree distribution so our discussion is organized by its value. For reasons that will become clear in the discussion we have abandoned our usual γ and will denote the power law by α. We begin in the middle of the range of values.

$\alpha = 3$

Since we are following in the footsteps of physicists, we will use the continuous approximation $p(x) = 2/x^3$ for $x \ge 1$, and enjoy the fact that it simplifies computations. The size-biased distribution has $q(x) = 1/x^2$ for $x \ge 1$ and (5.5.2) becomes

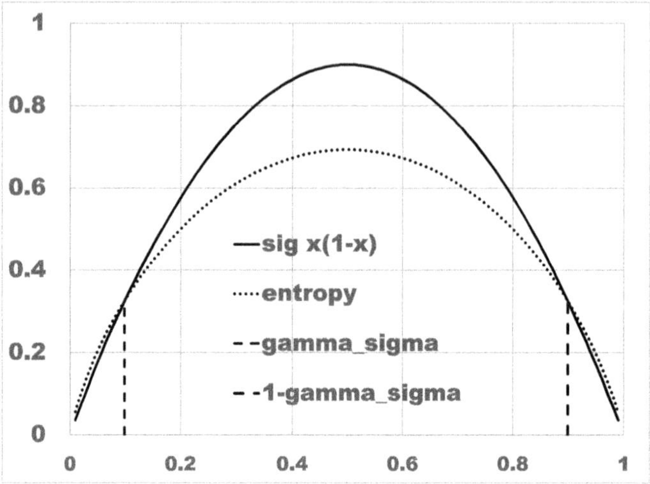

Fig. 5.7 Comparison of $0.9\gamma(1-\gamma)$ and the entropy $H(\gamma)$

$$\theta = \int_1^\infty \frac{1}{x} \cdot \frac{\lambda\theta}{1+\lambda\theta x}\, dx = \int_1^\infty \frac{\lambda\theta}{x} + \frac{\lambda\theta}{x}\left(\frac{1}{1+\lambda\theta x} - 1\right) dx$$
$$= \int_1^\infty \frac{\lambda\theta}{x} - \frac{(\lambda\theta)^2}{1+\lambda\theta x}\, dx.$$

The two parts of the last integrand are not integrable separately, but if we replace the upper limit of ∞ by M the integral is

$$\lambda\theta \log M - \lambda\theta\{\log(1+\lambda\theta M) - \log(1+\lambda\theta)\}$$
$$= -\lambda\theta \log(\lambda\theta + 1/M) + \lambda\theta \log(1+\lambda\theta),$$

so letting $M \to \infty$ the integral is

$$-\lambda\theta\big[(\log(\lambda\theta) - \log(1+\lambda\theta)\big] = \lambda\theta \log\left(1 + \frac{1}{\lambda\theta}\right).$$

The equation we want to solve is $1 = \lambda \log(1 + 1/\lambda\theta)$. Dividing by λ and exponentiating

$$e^{1/\lambda} = 1 + \frac{1}{\lambda\theta}.$$

Solving for θ now we have

$$\theta(\lambda) = \frac{1}{\lambda(e^{1/\lambda} - 1)} = (1/\lambda)e^{-1/\lambda}(1 - e^{-1/\lambda})^{-1}. \tag{5.5.5}$$

Using (5.5.4) and dropping constants

$$\rho(\lambda) \sim \lambda\theta(\lambda) \approx e^{-1/\lambda}.$$

Notice that $\rho(\lambda) \to 0$ exponentially fast. This agrees with the result for percolation on graphs with $p_k \sim Ck^{-3}$ in Section 2.7 but as we will see later in Figure 5.8 it is not accurate for the contact process.

$2 < \alpha < 3$

In this and future examples, we will let the degree distribution be $p(x) = (1 + \gamma)x^{-2-\gamma}$ for $x \geq 1$, so the size-biased distribution is $q(x) = \gamma x^{-1-\gamma}$. In the new notation $2 < \alpha < 3$ is $0 < \gamma < 1$ and (5.5.2) becomes

$$1 = \int_1^\infty \frac{\gamma}{x^\gamma} \cdot \frac{\lambda}{1 + \lambda\theta x}\, dx \equiv F(\lambda, \theta).$$

$\theta \to F(\lambda, \theta)$, a decreasing function of θ that is ∞ when $\theta = 0$ and $\to 0$ when $\theta \to \infty$, so we know there is a unique solution. Changing variables $x = u/\lambda\theta, dx = du/(\lambda\theta)$ we have

$$1 = \lambda^\gamma \theta^{\gamma-1} \int_{\lambda\theta}^\infty \gamma u^{-\gamma} \frac{1}{1+u}\, du.$$

Since $\gamma < 1$ the integral on the right has a limit c_γ as $\lambda\theta \to 0$. Rearranging we have

$$\theta \sim C\lambda^{\gamma/(1-\gamma)}. \tag{5.5.6}$$

Using (5.5.4) the fraction of occupied sites

$$\rho(\lambda) \sim C'\lambda^{1/(1-\gamma)}. \tag{5.5.7}$$

$\alpha > 3$

In the new notation this case is $\gamma > 1$ and (5.5.2) becomes

$$1 = \int_1^\infty \frac{\gamma}{x^\gamma} \cdot \frac{\lambda}{1 + \lambda\theta x}\, dx. \tag{5.5.8}$$

However, now the integral converges when $\theta = 0$, so for a solution to exist we must have

$$F(\lambda, 0) = \int_1^\infty \frac{\lambda\gamma}{x^\gamma}\,dx > 1 \quad \text{or} \quad \lambda > \lambda_c = 1\Big/\int_1^\infty \frac{\gamma}{x^\gamma}\,dx = \frac{\gamma-1}{\gamma}.$$

For fixed $\lambda > \lambda_c$ we want to solve $F(\lambda, \theta(\lambda)) = 1$. If $\lambda > \lambda_c$, $F(\lambda, 0) = \lambda/\lambda_c > 1$. To find the point where $F(\lambda, \theta)$ crosses 1 we begin by noting that

$$\frac{\partial F}{\partial \theta} = -\int_1^\infty \frac{\gamma}{x^\gamma}\frac{\lambda^2 x}{(1+\lambda\theta x)^2}\,dx. \tag{5.5.9}$$

When $\gamma \le 2$, $\partial F/\partial\theta \to \infty$ as $\theta \to 0$ so we will begin with the case $\gamma > 2$ where

$$\frac{\partial F}{\partial \theta}(0) = -b_{\gamma,\lambda},$$

so we have

$$1 - \frac{\lambda}{\lambda_c} = F(\lambda, \theta(\lambda)) - F(\lambda, 0) \sim -b_{\gamma,\lambda_c}\theta(\lambda),$$

and it follows that

$$\theta(\lambda) \sim \frac{\lambda - \lambda_c}{\lambda_c b_{\gamma,\lambda_c}}. \tag{5.5.10}$$

Using (5.5.4) we conclude

$$\rho(\lambda) \sim C(\lambda - \lambda_c) \quad \text{as } \lambda \to \lambda_c. \tag{5.5.11}$$

Turning now to $\gamma < 2$, changing variables $y/\theta = x$ (5.5.9) becomes

$$-\int_\theta^\infty \frac{\gamma\theta^\gamma}{y^\gamma}\frac{\lambda^2 y/\theta}{(1+\lambda y)^2}\frac{dy}{\theta} \sim -\theta^{\gamma-2}\int_0^\infty \frac{\gamma}{y^{\gamma-1}}\frac{\lambda^2}{(1+\lambda y)^2}\,dy.$$

Writing $c_{\gamma,\lambda}$ for the integral (which is finite) and integrating

$$1 - \frac{\lambda}{\lambda_c} = F(\lambda, \theta(\lambda)) - F(\lambda, 0) \sim -c_{\gamma,\lambda}\theta^{\gamma-1}/(\gamma-1).$$

Rearranging

$$\theta(\lambda) \sim C(\lambda - \lambda_c)^{1/(\gamma-1)}.$$

Using (5.5.4) we conclude that for $3 < \alpha < 4$

$$\rho(\lambda) \sim C(\lambda - \lambda_c) \quad \text{as } \lambda \to \lambda_c. \tag{5.5.12}$$

5.5.2 Rigorous Results

The first result about the long time survival of the contact process was proved by Berger, Borgs, Chayes, and Saberi in 2005. They considered the preferential attachment model which has a power law degree distribution with $\alpha = 3$, so when they proved that $\lambda_c = 0$ they confirmed the physicists' prediction. Chatterjee and Durrett (2009) showed that $\lambda_c > 0$ is not correct when $\alpha > 3$.

Theorem 5.5.1 *Consider a graph G_n with n vertices generated by the configuration model with $P(d_i = k) \sim Ck^{-\alpha}$ with $\alpha > 3$ and $P(d_i \le 2) = 0$. Let $\xi_t^1, t \ge 0$ denote the contact process on G_n starting from all sites occupied. Then for any $\lambda > 0$ there is a positive constant $p(\lambda) > 0$ so that for any $\delta > 0$*

$$\inf_{t \le \exp(n^{1-\delta}} P\left(|\xi_t^1|/n \le p(\lambda)\right) \to 1 \qquad \text{as } n \to \infty.$$

Sects. 5.6 and 5.7 are devoted to the proof of this result.

Mountford, Valesin, and Yao (2013) extended the results of Chatterjee and Durrett to include $2 < \alpha \le 3$ and proved upper and lower bounds that had the same dependence on λ but different constants, showing that

$$\rho(\lambda) \sim \begin{cases} \lambda^{1/(3-\alpha)} & 2 < a \le 5/2, \\ \lambda^{2\alpha-3} \log^{2-\alpha}(1/\lambda) & 5/2 < \alpha \le 3, \\ \lambda^{2\alpha-3} \log^{4-2\alpha}(1/\lambda) & 3 < \alpha. \end{cases} \qquad (5.5.13)$$

The result for $2 < \alpha \le 5/2$ agrees with the mean-field calculations quoted above but that formula is claimed to hold for $2 < \alpha < 3$. Figure 5.8 gives a visual comparison of the mean-field and rigorous results for critical exponents. For more about why the change occurs at 5/2 see the 2013 paper cited above. Three years later, Mountford, Mourrat, Valesin, and Yao (2016) showed that for all $\lambda > 0$, there exists some $c > 0$ so that the survival time $\ge e^{cn}$ with high probability.

5.6 Results for the Star Graph

Let G_k be a star graph with center 0 and leaves $1, 2, \ldots, k$ and let ξ_t be set of vertices infected in the contact process at time t. Write the state ξ_t as (i, j) where i is the number of infected leaves and $j = 1$ if the center is infected and $j = 0$ otherwise. We write $P_{i,j}$ for the law of the process starting from (i, j).

Here, following the approach in Chatterjee and Durrett (2009), we will reduce to a discrete time one-dimensional chain by only looking at times when $j = 1$. When the state is $(i, 0)$ with $i > 0$, the next event will occur after exponential time with mean $1/(i\lambda + i)$. The probability that it will be the reinfection of the center is $\lambda/(\lambda + 1)$. The probability it will be the recovery of a leaf is $1/(\lambda + 1)$. Thus, the number of

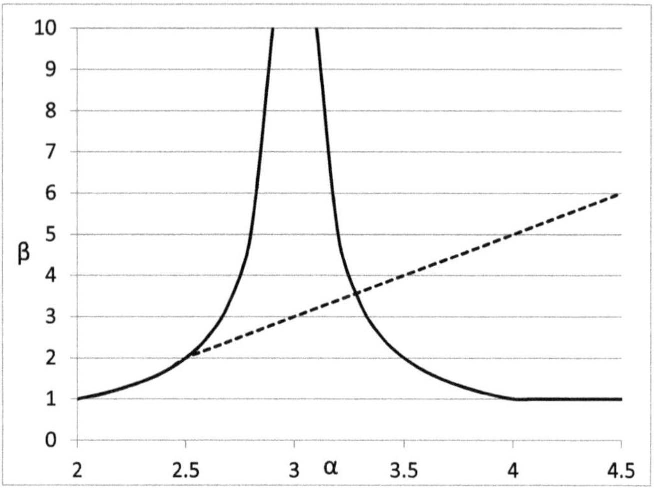

Fig. 5.8 Mean field critical exponents (solid line) versus rigorous results (dashed line) given in (5.5.13) as α varies from 2 to 4.5

leaf infections N that will be lost while the center is healthy has a shifted geometric distribution with success probability $\lambda/(\lambda+1)$, i.e.,

$$P(N=j)=\left(\frac{1}{\lambda+1}\right)^j\cdot\frac{\lambda}{\lambda+1}\quad\text{for } j\geq 0.$$

Note that since this version of the geometric counts the number of failures before the first success

$$EN=\frac{\lambda+1}{\lambda}-1=\frac{1}{\lambda}.$$

The next step is to modify the chain so that the infection rate is 0 when the number of infected leaves is at least

$$L=pk\quad\text{where}\quad p=\lambda/(1+2\lambda). \tag{5.6.1}$$

To motivate the truncation note that it guarantees that the number of infected leaves in the modified chain is always $\leq pk$ and the number of uninfected leaves is $\geq (1-p)k$. Thus if we look at the embedded discrete time process for the contact process on the star and only look at times when the center is infected, the process dominates Y_n where

$$\begin{array}{lr} \text{jump} & \text{with prob} \\ Y_n \to Y_n - 1 & pk/D, \\ Y_n \to \min\{Y_n + 1, pk\} & \lambda(1-p)k/D, \\ Y_n \to Y_n - N & 1/D. \end{array}$$

Here N is independent of Y_n and the denominator

$$D = pk + \lambda(1-p)k + 1 \le k + \lambda k + 1 \le (2+\lambda)k.$$

The fact that Y_n has a reflecting barrier at pk will simplify computations. We will use the process to lower bound survival times.

Lemma 5.6.1 *Let $L = pk$ where $p = \lambda/(1+2\lambda)$. Let $e^\theta = 1/(1+\lambda/2)$. If k is large enough $e^{\theta Y_n}$ is a supermartingale while $Y_n \in (0, pk)$.*

Proof We begin by noting that

$$E(\exp(\theta Y_{n+1}) - \exp(\theta Y_n)|Y_n = y) = e^{\theta y}(e^\theta - 1)\lambda(1-p)k/D \tag{5.6.2}$$
$$+ e^{\theta y}(e^{-\theta} - 1)pk/D + \frac{e^{\theta y}}{D}\left[\sum_{j=0}^\infty \left(\frac{e^{-\theta}}{1+\lambda}\right)^j \left(\frac{\lambda}{1+\lambda}\right) - 1\right].$$

The term in square brackets is

$$\frac{1}{1 - e^{-\theta}/(1+\lambda)} \cdot \frac{\lambda}{1+\lambda} - 1 = \frac{\lambda}{1+\lambda - e^{-\theta}} - 1 = \frac{e^{-\theta} - 1}{1+\lambda - e^{-\theta}} \ge 0.$$

Note that this implies we must take $e^{-\theta} < 1 + \lambda$.

The first two terms are

$$\frac{e^{\theta y}k}{D}\left((e^\theta - 1)\lambda(1-p) + (e^{-\theta} - 1)p\right),$$

so we begin by solving

$$(e^\theta - 1)\lambda(1-p) + (e^{-\theta} - 1)p = 0.$$

Rearranging and setting $x = e^\theta$ we want

$$x^2\lambda(1-p) - [\lambda(1-p) + p]x + p = 0.$$

Factoring we have

$$(\lambda(1-p)x - p)(x-1) = 0.$$

Since $p = \lambda/(1+2\lambda)$ the smaller root is

$$\frac{p}{\lambda(1-p)} = \frac{\lambda/(1+2\lambda)}{\lambda(1+\lambda)/(1+2\lambda)} = \frac{1}{1+\lambda}.$$

We let $e^\theta = 1/(1+\lambda/2) \in (1/(1+\lambda), 1)$ so that there is a $\delta > 0$ with

$$e^\theta \lambda(1-p) + e^{-\theta} p = [\lambda(1-p) + p] - \delta,$$

and hence

$$(e^\theta - 1)\lambda(1-p)k + (e^{-\theta} - 1)pk + \frac{e^{-\theta} - 1}{1+\lambda - e^{-\theta}} = -\delta k + \frac{e^{-\theta} - 1}{1+\lambda - e^{-\theta}}.$$

From this we see that if k is large enough $e^{\theta Y_n}$ is a supermartingale while $Y_n \in (0, pk)$. □

Let $T_\ell^- = \inf\{n : Y_n \le \ell\}$ and let $T_m^+ = \inf\{n : Y_n \ge m\}$.

Lemma 5.6.2 *Let $a, b \in (0, L)$. If $b < a$ then*

$$P_a(T_b^- < T_L^+) \le (1+\lambda/2)^{b-a}.$$

Proof To estimate the hitting probability let $\phi(x) = \exp(\theta x)$ where we take $e^\theta = 1/(1+\lambda/2)$ and note that if $\tau = T_b^- \wedge T_L^+$ then $\phi(Y(t \wedge \tau))$ is a supermartingale. Let $q = P_a(T_b^- < T_L^+)$. Using the optional stopping theorem we have

$$q\phi(Y_b^-) + (1-q)\phi(Y_L^+) \le \phi(a).$$

It is possible that $Y_b^- < b$. Note that since $\theta < 0$, we have $\phi(x) \ge \phi(b)$ for $x \le b$. Hence,

$$q\phi(b) + (1-q)\phi(L) \le \phi(a).$$

Dropping the second term on the left, $q \le \phi(a)/\phi(b) = (1+\lambda/2)^{b-a}$, which completes the proof. □

Lemma 5.6.3 *If $R_L = \inf\{n > T_{L-1}^- : Y_n = L\}$ and $b \in [0, L)$ then*

$$P_L(T_b^- < R_L) \le (2+\lambda)(1+\lambda/2)^{b-L}.$$

Remark. Here, and in later lemmas, the computation of explicit constants is somewhat annoying. However, when we consider asymptotics for critical values, λ will go to 0, so we will need to know how the constants depend on λ.

Proof To compute the left-hand side we break things down according to the first jump. The definition of R_L allows us to ignore the attempted upward jumps that do nothing. Recall that $L = pk$. The jump is to $L - 1$ with probability $pk/(pk + 1)$ and to $L - j$ with probability $\frac{\lambda}{(1+\lambda)^{j+1}} \cdot \frac{1}{1+pk}$. In the first case the probability of returning to L before going below b is

$$\leq (1 + \lambda/2)^{b-(L-1)} = (1 + \lambda/2) \cdot (1 + \lambda/2)^{b-L}.$$

In the second we have to sum over the possible values of $L - j$. Using Lemma 5.6.2

$$\begin{aligned}
&\leq (1 + \lambda/2)^{b-L} \sum_{j=1}^{\infty} \frac{\lambda}{(1 + \lambda)^{j+1}} (1 + \lambda/2)^j + \frac{\lambda}{1 + \lambda} P_L(T_b^- < R_L) \\
&\leq (1 + \lambda/2)^{b-L} \frac{\lambda}{\lambda + 1} \cdot \sum_{j=0}^{\infty} \left(\frac{1 + \lambda/2}{1 + \lambda} \right)^j + \frac{\lambda}{1 + \lambda} P_L(T_b^- < R_L) \\
&= 2(1 + \lambda/2)^{b-L} + \frac{\lambda}{1 + \lambda} P_L(T_b^- < R_L).
\end{aligned}$$

Noting that $\max\{2, 1 + \lambda/2\} \leq 2(1 + \lambda/2) - \delta$ for some small $\delta < \lambda$, we have the following relation,

$$P_L(T_b^- < R_L) \leq \frac{\lambda}{(1 + \lambda)(1 + pk)} P_L(T_b^- < R_L) + (2 + \lambda - \delta)(1 + \lambda/2)^{b-L}.$$

Hence for k sufficiently large, we have

$$P_L(T_b^- < R_L) \leq (2 + \lambda)(1 + \lambda/2)^{b-L}. \qquad (\square)$$

Lemma 5.6.4 *Let $b = \epsilon L$ and $S = \frac{1}{(2+\lambda)2k}(1 + \lambda/2)^{L(1-2\epsilon)}$*

$$P_{L,1}\left(\inf_{t \leq S} \xi_t \leq b \right) \leq (3 + \lambda)(1 + \lambda/2)^{-L\epsilon}.$$

Proof Let $M = (1 + \lambda/2)^{L(1-2\epsilon)}$. By Lemma 5.6.3 the probability that the chain fails to return M times to L before going below ϵL is

$$\leq (2 + \lambda)(1 + \lambda/2)^{-L\epsilon}.$$

Using Chebyshev's inequality on the sum S_M of M exponentials with mean 1 (and hence variance 1),

$$P(S_M < M/2) \leq 4/M.$$

When the number of infected leaves is $\leq L$ maximum jump rate is $D \leq (2 + \lambda)k$, so

$$P\left(\frac{S_M}{(2 + \lambda)k} < \frac{(1 + \lambda/2)^{L(1-2\epsilon)}}{2(2 + \lambda)k} \right) \leq 4(1 + \lambda/2)^{-L(1-2\epsilon)}.$$

Adding up the error probabilities completes the proof. □

Up to this point we have shown that if a star has L infected leaves it will remain infected for a long time. To make this useful, we need estimates about what happens when the star starts with only the center infected.

Lemma 5.6.5 *Let $\lambda > 0$ be fixed and $K = \lambda k^{1/3}$. Then for large k*

$$P_{0,1}(T_K^+ > T_{0,0}) \le 2\lambda k^{-1/3},$$
$$P_{K,1}(T_{0,0} < T_L^+) \le k^{-1/3},$$
$$E_{0,1}(T_L^+ | T_L^+ < T_{0,0}) \le 2/\lambda.$$

Proof Clearly

$$P_{0,1}(T_K^+ < T_{0,0}) \ge \prod_{j=0}^{K-1} \frac{(k-j)\lambda}{1 + (k-j)\lambda + j},$$

so subtracting the last inequality from $1 = \prod_{j=0}^{K-1} 1$ and using Lemma 3.4.3 from PTE5

$$P_{0,1}(T_K^+ > T_{0,0}) \le \sum_{j=0}^{K-1} \frac{1+j}{(k-j)\lambda} \le \frac{\lambda^2 k^{2/3}}{(k - \lambda k^{1/3})\lambda} \le 2\lambda k^{-1/3}.$$

For the second result we use the supermartingale $e^{\theta Y_n}$ from Lemma 5.6.1. If $q = P_{K,1}(T_{0,0} < T_L^+)$, using optional stopping theorem we have

$$q \cdot 1 + (1-q)e^{\theta L} \le e^{\theta K}.$$

Dropping the second term on the left,

$$q \le e^{\theta K} = (1 + \lambda/2)^{-K} \le k^{-1/3}.$$

To bound the time we return to continuous time

jump	at rate
$Y_t \to Y_t - 1$	pk
$Y_t \to \min\{Y_t + 1, pk\}$	$\lambda(1-p)k$
$Y_t \to Y_t - N$	1

Before time $V_L = T_{0,0} \wedge T_L^+$ the drift of Y_t is at least

$$\mu = \lambda(1-p)k - pk - 1/\lambda = \lambda pk - 1/\lambda \tag{5.6.3}$$

so $Y_t - \mu t$ is a submartingale. Stopping this martingale at the bounded stopping time $V_L \wedge t$

$$EY(V_L \wedge t) - \mu E(V_L \wedge t) \ge EY_0 \ge 0.$$

Since $EY(V_L \wedge t) \leq L$, it follows that

$$E(V_L \wedge t) \leq \frac{L}{\mu} = \frac{pk}{\lambda pk - 1/\lambda},$$

where $p = \lambda/(1 + 2\lambda)$, so if λ is fixed and k is large

$$E(V_L \wedge t) \leq 2/\lambda,$$

which completes the proof. □

Combining Lemmas 5.6.4 and 5.6.5 we have the following

Lemma 5.6.6 *Let A_t denote the number of infected leaves at time t and take S as in Lemma 5.6.4. Define $G = \{\inf_{k^{2/3} \leq t \leq S} |A_t| \geq \epsilon L\}$. If $\lambda > 0$ is fixed*

$$P_{0,1}(G) \geq 1 - C_\lambda k^{-1/3}, \tag{5.6.4}$$

for some constant C_λ.

When G occurs, we say **the star at 0 is good.**

5.7 Subexponential Degree Distributions

Given an offspring distribution p_k, we construct a Galton-Watson tree as follows. Starting with the root, each individual has k children with probability p_k. Pemantle (1992) has shown in his Theorem 3.2 that

Theorem 5.7.1 *There are constants c_2 and c_3 so that if $\mu > 1$ is the mean of the offspring distribution, then for any $k > 1$, if we let $r_k = \max\{2, c_2 \log(1/kp_k)/\mu\}$,*

$$\lambda_2 < c_3 \sqrt{r_k \log r_k \log(k)/k}. \tag{5.7.1}$$

If the offspring distribution is a stretched exponential $p_k = c_\gamma \exp(-k^\gamma)$ with $\gamma < 1$ then $\log(1/kp_k) \sim k^\gamma$ and hence $\lambda_2 = 0$.

Huang and Durrett (2020b) extended the last result to subexponential distributions, which satisfy

$$\limsup_{k \to \infty} (1/k) \log p_k = 0.$$

Theorem 5.7.2 *If the offspring distribution p_k for a Galton-Watson tree is subexponential and has mean $\mu > 1$ then $\lambda_2 = 0$.*

This result was proved as a consequence of their study of the case in which degrees have a geometric distribution, $p_k = (1-p)^{k-1} p$ for $k \geq 1$. The goal of their paper was to prove that $\lambda_1 > 0$. The solution of that problem will be described in the next section. The upper bound on λ_1 is easy.

Theorem 5.7.3 *Under the assumptions of Theorem 5.7.2, $\lambda_1 \le p/(1-p)$.*

Proof Modify the contact process so that births from a site can only occur on sites further from the root. Each vertex x will be occupied at most once. If x is occupied then it will give birth with probability $\lambda/(\lambda+1)$ onto each neighbor y. The birth events are not independent but that is not important. If we let Z_n be the number of sites at distance n that are ever occupied, Z_n is a branching process in which the offspring distribution has mean $\lambda/((\lambda+1)\cdot p)$ which is > 1 if $\lambda > p/(1-p)$. □

When $p_k = (1-p)^{k-1}p$, $\log(1/kp_k) \sim c_p k$, so (5.7.1) gives a finite upper bound on λ_2. However, the resulting bound is much worse than the following:

Theorem 5.7.4 *If $p_k = 2^{-k}$ for $k \ge 1$, then $\lambda_2 \le 2.5$.*

The proof works for a general geometric $p_k = (1-p)^{k-1}p$, $k \ge 1$. Huang and Durrett (2020b) could not get a nice formula for the upper bound as a function of p but the upper bounds can easily be computed numerically and graphed. These upper bounds are only interesting for small p. A Galton-Watson tree with $p_0 = 0$ and $p_1 < 1$ contains a copy of $\mathbb{Z}$ (start with a vertex with two children and follow their descendants) so using the bound on $\lambda_c(\mathbb{Z})$ proved in Liggett (1995) we conclude $\lambda_2 \le 2$ for all $0 < p < 1$ (Fig. 5.9).

Proof for $p_n = 2^{-n}, n \ge 1$. Our proof follows the outline of the proof of Theorem 3.2 in Pemantle (1992), see pages 2109–2110. We can suppose without loss of generality that the root has degree k. Otherwise examine the children of the root until we find one with degree k and apply the argument to the children of this vertex. There are two steps in the proof.

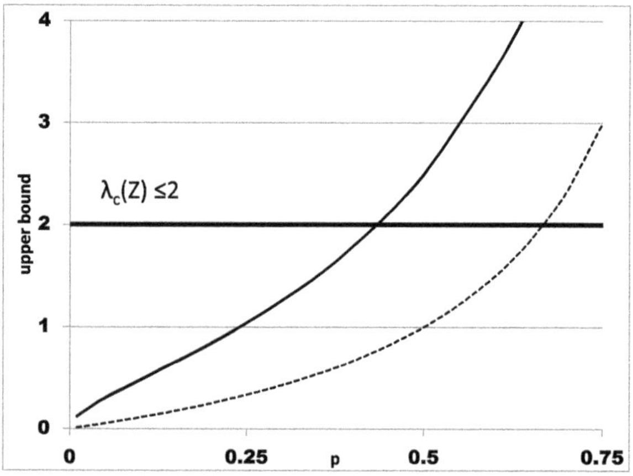

Fig. 5.9 Upper bounds on λ_1 (dotted line) from Theorem 5.7.4 and on λ_2 (solid line) from (5.7.4) as a function of p for the geometric degree distribution. The horizontal line is the bound that comes from using the existence of a copy of $\mathbb{Z}$ in the graph

1. Push the infection to vertices at a distance $r = k$ that have degree k.
2. Bring the infection back to the root at time t.

To push the infection in either direction we use the following results.

Lemma 5.7.5 *Let $v_0, v_1, \ldots, v_r$ be a path in a graph. Suppose that v_0 is infected at time 0 and that it is good in the sense of Lemma 5.6.6. Then there is a $\gamma > 0$ so that the probability that v_r will become infected by time $2r$ is*

$$\geq \left(\frac{\lambda}{\lambda+1}\right)^r (1 - \exp(-\gamma r)).$$

If $\epsilon > 0$ and we let $\hat{\lambda} = (1-\epsilon)\lambda/(\lambda+1)$ then for large r this probability is $\geq \hat{\lambda}^r$.

Proof The probability that v_{i-1} infects v_i before it is cured is $\lambda/(1+\lambda)$. When this transfer of infection occurs the amount of time is t_i, which is an independent exponential with rate $1+\lambda$. By large deviations for the exponential distribution $P(t_1 + \cdots + t_r > 2r) \leq e^{-\gamma r}$ for some $\gamma > 0$. □

Lemma 5.7.6 *Run the contact process on a graph consisting of a star with k leaves, to which there has been added a single chain $v_1, \ldots, v_r$ of length r where v_1 is a neighbor of 0, the center of the star. Suppose that at time 0 there are L infected leaves. For large r the probability that v_r will not be infected before time $T = m(2r+1)$ is*

$$\leq (1 - \hat{\lambda}^r)^m.$$

Proof Consider a sequence of times $t_i = (2r+1)i$ for $i \geq 1$. The center 0 may not be infected at time t_i but since the star at 0 is good the number of infected neighbors is $\geq \epsilon L$ and it will with high probability be infected by time $t_i + 1$. By Lemma 5.7.5 the probability v_r is successfully infected in $[t_i, t_{i+1})$ is $\geq \hat{\lambda}^r$ when 0 is good, even if we condition on the events up to time t_i. The desired result follows. □

To use the two lemmas to prove the main result we need the next weird but wonderful result, which is Lemma 2.4 from Pemantle (1992). Let $\varphi(x) = \sum_{n=0}^{\infty} p_n x^n$ be the generating function of the degree distribution. We will apply Lemma 5.7.7 to

$$f(t) = P(0 \in \xi_t^0) \geq p_k P(0 \in \xi_t^0 |\, 0 \text{ has at least } k \text{ children}).$$

Lemma 5.7.7 *Let H be any nondecreasing function on the nonnegative reals with $H(x) \geq x$ when $x \in [0, x_0]$. If f satisfies (i) $\inf_{0 \leq t \leq L} f(t) > 0$ and (ii) $f(t) \geq H(\inf_{0 \leq s \leq t-L} f(s))$ for $t \geq L$ some $L > 0$ then* $\liminf_{t\to\infty} f(t) > 0$.

Proof For any t_0 and $\epsilon > 0$, (ii) implies that there is a decreasing sequence t_i with $t_{i+1} \leq t_i - L$ and $t_k < L$ for some k

$$f(t_i) \geq H(f(t_{i+1})) - \epsilon 2^{-i}.$$

If $f(t_i) < x_0$ for all $1 \le i \le k$ then

$$f(t_i) \ge f(t_{i+1}) - \epsilon 2^{-i}$$

and summing gives $f(t_0) > f(t_k) - \epsilon$ which gives the desired result. Suppose now that j is the smallest index with $f(t_j) > x_0$. If $j = 0$ we have $f(t_0) > x_0$. If $j = 1$ we have $f(t_0) \ge H(x_0)$. If $j \ge 2$ we have

$$f(t_0) \ge f(t_{j-1}) - \epsilon \ge H(x_0) - \epsilon$$

so in all cases we get the desired conclusion. □

Step 1. The mean of the offspring distribution is 2. Let Z_r be the number of vertices at distance r from 0 and let $v_r^1, \ldots, v_r^J$ be the subset of those that have exactly k children, where J is a random variable that represents the number of such vertices.

Since the root has degree k and $p_k = 2^{-k}$ if we set $r = k$

$$EJ \ge k\mu^{r-1} p_k = k/2,$$

where $\mu = 2$ is the mean offspring number.

If we condition on the value of $W = Z_r/(k\mu^{r-1})$ and let $\bar{J} = (J|W)$ be the conditional distribution of J given W then

$$\bar{J} = \text{binomial}(k2^{r-1}W, 2^{-k}).$$

Let M be the random number of vertices among $v_r^1, \ldots, v_r^J$ that are infected before time

$$S = \frac{1}{2k(2+\lambda)}(1+\lambda/2)^{L(1-2\epsilon)}.$$

By Lemma 5.7.6 the probability a given vertex will not become infected by time S is

$$p_{noi} \le (1-\hat{\lambda}^k)^m \quad \text{where} \quad \hat{\lambda} = (1-\epsilon)\frac{\lambda}{\lambda+1} \quad \text{and}$$

$$m = \frac{S}{2k+1} = \frac{(1+\lambda/2)^{L(1-2\epsilon)}}{2k(2k+1)(2+\lambda)} \quad \text{with} \quad L = \frac{\lambda k}{1+2\lambda}.$$

Combining the definitions and using $(1-x) \le e^{-x}$ we have

$$p_{noi} \le \exp\left(-\frac{\Gamma^k}{2k(2k+1)(2+\lambda)}\right) \quad \text{where} \quad \Gamma = \hat{\lambda}(1+\lambda/2)^{(1-2\epsilon)\lambda/(1+2\lambda)}.$$

When $\lambda = 2.5$

$$\frac{\lambda}{\lambda+1}(1+\lambda/2)^{\lambda/(1+2\lambda)} = 1.0014 > 1, \qquad (5.7.2)$$

so $\Gamma > 1$ when ϵ is small and $p_{noi} \to 0$ as $k \to \infty$. From this we see that if $\delta > 0$ then for large k

$$EM \geq (1-\delta)EJ.$$

The remark after Lemma 5.7.6 implies that if we condition on the value of W and let $\bar{M} = (M|W)$ then

$$\bar{M} \geq \text{binomial}(k2^{r-1}W, 2^{-k}(1-\delta)).$$

To prepare for the following two generalizations of the result for geometric(1/2) offspring distribution we ask the reader to verify that in Step 2, all we use is the fact that (5.7.2) implies the bounds on EM and $\bar{M}$.

Step 2. Let $H_1(t) = P(v_r^i \in \xi_{t-S}$ for some $1 \leq i \leq J)$ and

$$H_2(t) = P(0 \in \xi_t | v_r^i \in \xi_{t-S} \text{ for some } 1 \leq i \leq J),$$

so that $f(t) \geq H_1(t)H_2(t)$. Fix $t > 2S$ and let

$$\chi(t) = \inf\{f(s) : s \leq t - S\}.$$

Since t is fixed, we simplify the notation and write $\chi(t)$ as χ.

Ignore all but the first infection of each v_r^i by its parent. Any of these will evolve independently from the time $s < S$ it is first infected, and will be infected at time $t - S$ with probability at least χ. Thus given M the number of infected sites at time $t - S$ will dominate $N = \text{binomial}(M, \chi)$. If we let $\bar{N} = \text{binomial}(\bar{M}, \chi)$ and let $\delta > 0$, then by Lemma 2.3 in Pemantle (1992) we see that there exists a $\varepsilon > 0$ such that

$$P(\bar{N} \geq 1) \geq (1-\delta)\chi EM \wedge \varepsilon.$$

Therefore $H_1(t) \geq (1-\delta)\chi EM \wedge \varepsilon$ when $t > 2S$.

Finally, if some v_r^i is infected at time $t - S$ then the probability of finding 0 infected at time t is bounded below by $\rho_1\rho_2$ where

- ρ_1 is the probability that the contact process starting with only v_r^i infected at time $t - S$ infects 0 at some time s with $t - S \leq s \leq t$. By Lemmas 5.6.5, 5.6.6, and 5.7.6, $\rho_1 \geq 1 - \delta$.
- ρ_2 is the probability 0 is infected at time t given the infection of 0 at such a time s. For any $\epsilon > 0$, by Lemma 5.7.6 the probability that 0 has not been infected by time $S/2$ is less than ϵ when k is sufficiently large. By Lemma 5.6.6, with probability $\geq 1 - (2 + 2\lambda)k^{-1/3}$ there should be at least ϵL infected leaves at time $t - \epsilon$. Hence 0 is infected at t with probability at least $(1 - e^{-\lambda\epsilon^2 L})e^{-\epsilon}$, where the second term guarantees that the root is infected at time t. Choosing ϵ sufficiently small and k sufficiently large gives $\rho_2 \geq 1 - \delta$.

Thus

$$f(t) \geq \begin{cases} \chi(t)EM(1-\delta)^3 \wedge \varepsilon & t > 2S, \\ \inf_{0 \leq s \leq 2S} f(s) & S \leq t \leq 2S. \end{cases}$$

We can take $\varepsilon < \inf_{0 \leq s \leq 2S} f(s)$ so that $f(t) \geq \chi(t)EM(1-\delta)^3 \wedge \varepsilon$ for all $t \geq S$. The result now follows from Lemma 5.7.7 with $L = S$ and $H(x) = (1-\delta)^3(EM)x \wedge \varepsilon$.

Proof for $p_n = (1-p)^{n-1}p$. It is now straightforward to replace 1/2 by p. We only have to pick k and r so that we can prove the analogue of (5.7.2). The mean of the offspring distribution is $1/p$. Let Z_r be the number of vertices at distance r from 0 and let $v_r^1, \ldots, v_r^J$ be those that have exactly k children. Since the root has degree k and $p_k = (1-p)^{k-1}p$

$$EJ \geq k(1/p)^{r-1}(1-p)^{k-1}p. \tag{5.7.3}$$

In this case we want to pick r so that $(1/p)^r(1-p)^k \approx 1$. Hence EJ can be large when k is large. Ignoring the fact that r and k must be integers this means

$$r/k = \log(1-p)/\log p.$$

Let M be the random number of vertices among $v_r^1, \ldots, v_r^J$ that are infected before time S. By Lemma 5.7.6 the probability a given vertex will not become infected is

$$\leq (1-\hat{\lambda}^r)^{\lceil S/(2r+1) \rceil} \leq \exp\left(-\frac{-\Gamma^k}{2k(2r+1)(2+\lambda)}\right)$$

where $\Gamma = \hat{\lambda}^{r/k}(1+\lambda/2)^{(1-2\epsilon)\lambda/(1+2\lambda)}$. That is, if we choose λ such that

$$\left(\frac{\lambda}{\lambda+1}\right)^{r/k} \cdot (1+\lambda/2)^{\lambda/(1+2\lambda)} > 1, \tag{5.7.4}$$

then we have $\Gamma > 1$ for large k. By the same reasoning as before this choice of λ gives an upper bound on λ_2.

If we want to graph the bound as a function of p it is better to work backwards. Given λ the second factor is > 1 so we can easily find the value of r/k that makes this 1. Having done this we can easily compute the value of p for which λ gives the upper bound on λ_2.

Proof for subexponential distributions. We suppose that the mean of the offspring distribution is $\mu > 1$. If p_k is subexponential, i.e.,

$$\limsup_{k \to \infty} (1/k) \log p_k = 0,$$

then for any δ there is a k with $p_k \geq (1-\delta)^k$. It follows from the same reasoning as in (5.7.3) that we can take r such that

$$\frac{r}{k} = -\frac{\log(1-\delta)}{\log \mu}.$$

Given any $\lambda > 0$, (5.7.4) will hold if δ is small enough, which implies local survival of the process. Therefore $\lambda_2 = 0$.

5.8 Exponentially Bounded Degrees

In the decade after my 2009 work with Shirshendu Chatterjee on the contact process on power law graphs, one of my favorite open problems was proving that the critical value for the contact process on a Galton-Watson tree is positive if it has an **exponentially bounded degree distribution** D, i.e., $Ee^{\theta D} < \infty$ for some $\theta > 0$. I was so excited by this question that when I gave a talk at the Northeast Probability Seminar held in November 2018 I followed in Erdős's footsteps and offered \$1000 for a proof.

Danny Nam and Oanh Nguyen, two students of Allan Sly at Princeton, were in the audience and it was not long until in joint work with their advisor and Shankar Bhamidi of UNC the problem was solved. It is somewhat surprising that the proof is not very complicated. Ménard and Singh (2016) studied random geometric graphs in d-dimensions, graphs that are built by connecting points in a spatial Poisson process that are within distance R. The proof that $\lambda_c > 0$ for these graphs, which was done by studying "cumulative merging on a weighted graph," was arduous (and an impressive achievement). In contrast, the BNNS proof is short and sweet. As the paper explains, there are two important new ideas:

(i) They modified the process by adding a new vertex ρ^+ above the root ρ that is always infected. Recoveries at the root are not allowed unless all of the descendants of the root are healthy. One might worry that this assumption gives away too much, but the approach has the advantage that while the root is infected then the subtrees below its descendants are independent and this greatly facilitates recursion.
(ii) The second, somewhat more technical, idea is to prove that the probability that the infection goes deeper than depth h decays exponentially fast in h. See their Theorem 3.4 (Theorem 5.8.3 below). To see why this is useful the reader will need to wait to see it used in the proof.

Turning to the details, let D be a distribution on the positive integers with an exponential tail, as defined above. Suppose that $ED > 1$ so that the Galton-Watson tree with this offspring distribution $\mathbb{GW}(D)$ has positive probability of surviving forever.

Theorem 5.8.1 *(Theorem 1) Suppose the degree distribution D has an exponential tail. Then there is a λ_0 so that for $\lambda < \lambda_0$ the contact process starting from a single infection at the root dies out with probability 1.*

5.8.1 Expected Survival Time

Let L be a large positive integer, let $\mathcal{T}_L$ denote the depth L Galton-Watson tree and let $\mathbb{CP}^\lambda(\mathcal{T}_L; 1_\rho)$ be the contact process on $\mathcal{T}_L$ in which initially there is one infected at ρ and the other sites are healthy. The first of two main steps in the proof of Theorem 5.8.1 is to show:

Theorem 5.8.2 *(Theorem 3.1) Let R_L be the first time that $\mathbb{CP}^\lambda(\mathcal{T}_L; 1_\rho)$ reaches $\mathbf{0}$. There are constants C and $\lambda_0 > 0$ so that for any $\lambda \le \lambda_0$, $ER_L \le C$ for all integers L.*

A recursive equation. As mentioned in (i) we add a new vertex ρ^+ as a parent of the root of the tree ρ and set ρ^+ to be permanently infected. Since ρ^+ is always in state 1, we define the state space of the process to be $\{0, 1\}^{\mathcal{T}_L}$. Let X_t be the contact process modified to have permanent infection at ρ^+ and let $\tilde{X}_t$ be X_t further modified to not allow recovery at ρ until none of its descendants are infected. Let S_L be the time for X_t to reach the all healthy state $\mathbf{0}$ and let $\tilde{S}_L$ be the time for $\tilde{X}_t$ to reach the all healthy state. The first event in $\tilde{X}_t$ is either

A. ρ recovers.
B. ρ infects one of its children v_i.

In case A, $\tilde{S}_L$ is the exponential(1) time it took for ρ to recover. If the degree of the root ρ is D the probability of event A is $1/(1 + \lambda D)$. In case B, the evolution of the contact process on the subtrees $\mathcal{T}_{v_i}$ are independent of each other until the time $\tilde{S}_L^\otimes$ at which all of them are completely healthy. Let $X_t^\otimes$ be the contact process on the union of the subtrees, which has initial state 1_{v_i}, run until the time $\tilde{S}_L^\otimes$ they are all completely healthy (Fig. 5.10).

At time $\tilde{S}_L^\otimes$ the process $\tilde{X}_t$ is again in state 1_ρ. After a geometric number of returns to this state, we will finally achieve outcome A so

$$E(\tilde{S}_L|\mathcal{T}_L) = \sum_{k=0}^{\infty} \left(\frac{\lambda D}{1+\lambda D}\right)^k \frac{1}{1+\lambda D} \cdot \left[(k+1) \cdot \frac{1}{1+\lambda D} + kE\left(\tilde{S}^\otimes \middle| \mathcal{T}_{v_i}\right)\right], \tag{5.8.1}$$

where $|\mathcal{T}_{v_i})$ is short for conditioning on all of the subtrees of the v_i for $1 \le i \le D$. The mean of the geometric(p) distribution $\sum_{k=0}^{\infty}(1-p)^k p(k+1) = 1/p$ so the above is

$$= (1 + \lambda D) \cdot \frac{1}{1 + \lambda D} + (1 + \lambda D - 1) E(\tilde{S}^\otimes | D),$$

which implies

$$E(\tilde{S}_L|D) = 1 + \lambda D E(\tilde{S}^{\otimes}|D). \tag{5.8.2}$$

Estimating $E(\tilde{S}^{\otimes}|D)$

We do this by relating this expected value to the stationary distribution of the root-added contact process, which, conditional on $\mathcal{T}_L$, is irreducible on a finite state space and hence a unique stationary distribution. Let π^D be the stationary distribution of the product chain $\mathbb{CP}_\rho^{\otimes}(\mathcal{T}_L)$. If we let π_i be the stationary distribution of $\mathbb{CP}_\rho^{\lambda}(\mathcal{T}_i^+)$, where $\mathcal{T}_i^+$ is $\mathcal{T}_{v_i}$ with ρ added as a permanently infected added root, then

$$\pi^D = \otimes_{i=1}^D \pi_i.$$

For any state η on $\mathcal{T}_L - \{\rho\}$, $\pi^D(\eta)$ is proportional to the time that the chain $X_t^{\otimes} \sim \mathbb{CP}_\rho^{\otimes}(\mathcal{T}_L)$ stays at state η. The expected time to stay at $\mathbf{0}$ (conditional on D) is $(\lambda D)^{-1}$. After escaping from $\mathbf{0}$ it spends expected time $E(\tilde{S}_L^{\otimes}|\mathcal{T}_L)$ before returning to $\mathbf{0}$. Therefore

$$\pi^D(\mathbf{0}) = \frac{(\lambda D)^{-1}}{(\lambda D)^{-1} + E(\tilde{S}_L^{\otimes}|\mathcal{T}_L)} = \frac{1}{1 + \lambda D E(\tilde{S}_L^{\otimes}|\mathcal{T}_L)}.$$

Similarly we have

$$\pi_i(\mathbf{0}) = \frac{1}{1 + \lambda E(\tilde{S}_{L-1}|\mathcal{T}_{v_i})}$$

where S_{L-1} is the first time $X_t^i \sim \mathbb{CP}_\rho^{\lambda}(\mathcal{T}_{v_i}^+, 1_{v_i})$ reaches state $\mathbf{0}$. Therefore we obtain

$$1 + \lambda D E(\tilde{S}^{\otimes}|\mathcal{T}_L) = \prod_{i=1}^{D} [1 + \lambda E(\tilde{S}_{L-1}|\mathcal{T}_{v_i})]. \tag{5.8.3}$$

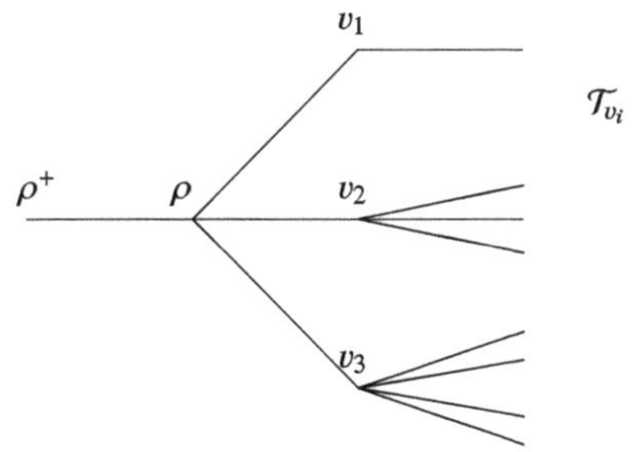

Fig. 5.10 A picture of the notation used in the proof

Since T_{v_i}, $i \geq 1$ are i.i.d. $\mathbb{GW}(D)_{L-1}$ integrating out the randomness of T_{v_i}, $i \geq 1$, we have

$$1 + \lambda D E(\tilde{S}^{\otimes}|D) = (1 + \lambda E(\tilde{S}_{L-1}))^D.$$

Using $1 + x \leq e^x$ this becomes

$$\leq \exp(\lambda E(\tilde{S}_{L-1})).$$

Using this with (5.8.2) we get

$$E(\tilde{S}_L|D) \leq \exp(\lambda E(\tilde{S}_{L-1}) \cdot D)).$$

and it follows that

$$ES_L \leq E\tilde{S}_L \leq E_D[\exp(\lambda E(S_{L-1}|D)]. \tag{5.8.4}$$

To complete the proof now let $c > 0$ be so that $M = E\exp(cD) < \infty$. Define K and λ so that

$$K = e \cdot \max\{\log M, 1\}, \qquad \lambda_0 = c/K.$$

To prove Theorem 5.8.2 by induction note that $ES_0 = 1$ and suppose that $ES_{L-1} \leq e$. If $\lambda < \lambda_0$ we have

$$\gamma \equiv \frac{\lambda ES_{L-1}}{c} \leq \frac{\lambda_0 e}{c} = \frac{e}{K} < 1,$$

so using Jensen's inequality

$$E\exp(\gamma c D) \leq (E\exp(cD))^\gamma = M^\gamma = \exp(\log M \cdot (e/K)) \leq e,$$

which completes the proof of Theorem 5.8.2.

5.8.2 *Exponential Decay of Infection Depth*

We continue to use the notation introduced previously. Recall that $\mathcal{T}_L = \mathbb{GW}(D)_L$ and $\mathcal{T}_L^+$ is the graph obtained by adding a permanently infected vertex ρ^+ above the root. For each state η of the contact process on $\mathcal{T}_L$ define the depth of η to be

$$r(\eta) = \max\{d(\rho^+, v) : \eta(v) = 1\},$$

with $r(\mathbf{0}) = 0$. This definition is natural since ρ^+ will be permanently occupied. Consider the root added process $X_t \sim \mathbb{CP}^{\lambda}_{\rho^+}(\mathcal{T_L}^+, 1_\rho)$ and let S_l be the first time the process reaches $\mathbf{0}$. Let $H = \max\{r(X_t) : t \in [0, S_L]\}$ be the maximum depth reached. The second main step in the proof of Theorem 5.8.1 is

Theorem 5.8.3 *(Theorem 3.4) Let $L > 0$ be an integer. There are constants K, $\lambda_0 > 0$ depending only on the degree distribution D so that for all $\lambda \le \lambda_0$, $h > 0$ and $m > 0$ we have*

$$P(H > h|\mathcal{T}_L) \le 2m(K\lambda)^h,$$

for a collection of values of $\mathcal{T}_L$.with probability $\ge 1 - m^{-1}$.

Delayed Contact Process

Let $\mathcal{S}^+$ be a finite graph rooted at ρ^+. We assume that the graph is finite so that the stationary distributions we define later in this paragraph will exist. Here we will only consider $\mathcal{S} = \mathcal{T}_L^+$, but the general notation saves some typing. Let $\mathcal{S} = \mathcal{S}^+ - \{\rho^+\}$. For any two states $\eta, \zeta \in \{0, 1\}^S$ let $Q_{\eta,\zeta}$ be the transition rate from η to ζ in $\mathbb{CP}^{\lambda}_{\rho^+}(\mathcal{S}^+)$. For a fixed constant $\theta \in (0, 1)$, the delayed contact process, denoted by $\mathbb{DP}^{\lambda,\theta}_{\rho^+}(\mathcal{S}^+, \eta_0)$, is the continuous time Markov process on $\{0, 1\}^S$ with initial state η_0 and transition rate

$$Q^{\theta}_{\eta,\zeta} = \theta^{r(\eta)} Q_{\eta,\zeta}.$$

If π_S and ν^{θ}_S are the stationary distribution for the ordinary and delayed contact processes then it is immediate from the definitions that

$$\nu^{\theta}_{\mathcal{S}}(\eta) = \frac{\theta^{-r(\eta)}\pi_{\mathcal{S}}(\eta)}{\sum_{\zeta}\theta^{-r(\zeta)}\pi_{\mathcal{S}}(\zeta)}. \tag{5.8.5}$$

Lemma 5.8.4 *(Lemma 3.6) Let $L > 0$ be an integer. There are constants K and $\lambda_0 > 0$ so that for all $\lambda \le \lambda_0$ and L*

$$E\left[\nu^{\theta}_{\mathcal{T}_L}(\mathbf{0})^{-1}\right] \le 2.$$

The proof is similar to that of Theorem 5.8.2, but requires more work, so we refer the reader to the paper for details.

Proof of Theorem 5.8.3. To simplify the notation write π_L for $\pi_{\mathcal{T}_L}$ and ν_L for $\nu^{\theta}_{\mathcal{T}_L}$ with $\theta = K\lambda$ where K is the constant from Lemma 5.8.4. Let $A = \{\eta : r(\eta) \ge h\}$ and note that (5.8.5) implies

$$\frac{\pi_L(A)}{\pi_L(\mathbf{0})} \le \theta^h \frac{\nu_L(A)}{\nu_L(\mathbf{0})}.$$

Lemma 5.8.4 and Markov's inequality imply that, for a collection of $\mathcal{T}_L$ with probability $\ge 1 - m^{-1}$, we have $\nu_L(\mathbf{0})^{-1} \le 2m$ and hence

$$\frac{\pi_L(A)}{\pi_L(\mathbf{0})} \leq 2m(K\lambda)^h. \tag{5.8.6}$$

Now if $X_t \sim \mathbb{CP}^{\lambda}_{\rho^+}(\mathcal{T}_L^+)$ hits A before 0 then the time needed to escape from A is at least exponential with mean 1, since escape can only happen when there is exactly one infected at distance h from ρ^+. Thus if $\gamma(h) = |\{t \in [0, S_L] : X_t \in A\}|$, where $|\cdot|$ denotes Lebesgue measure, then

$$E(\gamma(h)|H \geq h, \mathcal{T}_L) \geq 1.$$

Combining this with (5.8.6) and noting that $\gamma(h) = 0$ on $H < h$

$$\begin{aligned} P(H \geq h|\mathcal{T}_L) &\leq E(\gamma(h)|H \geq h, \mathcal{T}_L) \cdot P(H \geq h|\mathcal{T}_L) \\ &\leq E(\gamma(h)|\mathcal{T}_L) \leq \frac{\pi_L(A)}{\pi_L(\mathbf{0})} \leq 2m(K\lambda)^h, \end{aligned}$$

which completes the proof. □

At this point the rest is routine.

Proof of Theorem 5.8.1. Let $\mathcal{T} \sim \mathbb{GW}(D)$, ρ be its root, and let $X_t \sim \mathbb{CP}(\mathcal{T}, 1_\rho)$. Let K, λ_0 be given in Theorem 5.8.3 and let $\lambda \leq \lambda_0$ so that $K\lambda \leq 1$. Let $\delta > 0$ be small and pick h so that $(K\lambda)^h = \delta^2/8$. Let $E(h)$ be the event that the infection inf X_t does not go deeper than h before dying out. Taking $m = 2/\delta$ in Theorem 5.8.3 we see that

$$P(H \geq h|\mathcal{T}_L) \leq (4/\delta)(\delta^2/8) = \delta/2,$$

for a collection of values of $\mathcal{T}_L$ with probability $\geq 1 - \delta/2$ so $P(E(h)) \geq 1 - \delta$.

Let $\mathcal{T}_h$ be $\mathcal{T}$ truncated at depth h and couple the processes $\mathbb{CP}(\mathcal{T}, 1_\rho)$ and $\mathbb{CP}(\mathcal{T}_h, 1_\rho)$ by identifying the recoveries and infections inside $\mathcal{T}_h$. Let R and R_h be the times $\mathbb{CP}(\mathcal{T}, 1_\rho)$ and $\mathbb{CP}(\mathcal{T}_h, 1_\rho)$ reach $\mathbf{0}$. Then Theorem 5.8.2 tells us that

$$E(R|E(h)) = E(R_h|E(h)) \leq \frac{ER_h}{P(E(h))} < \infty.$$

Thus $P(X_t \neq \mathbf{0}$ for all $t \geq 0) \leq \delta$. Since δ is arbitrary the desired result follows. □

5.8.3 Bounds on Survival Time

Theorem 5.8.5 *(Theorem 3) Suppose that G_n is a graph on n vertices generated by the configuration model with degree distribution D with (i) $ED(D-2) > 0$, so there is a giant component, and (ii) $E(e^{cD}) < \infty$ for some $c > 0$. Consider the contact process on G_n starting from all sites infected. There are constants $0 < \lambda_i \leq \lambda_{ii} < \infty$ so that*

(i) For $\lambda < \lambda_i$ the process survives for time at most $n^{1+o(1)}$,

(ii) for $\lambda > \lambda_{ii}$ the process survives for time at least $e^{\Theta(n)}$.

The proof of this result requires hard work, so we will only explain a few ideas from the proof.

Proof of (i). It is enough to prove

Theorem 5.8.6 *(Theorem 4.1) Fix an arbitrary vertex $v \in G$ and let T_v be the time when $\mathbb{CP}^\lambda(G; 1_v)$ reaches the root 0. There is an event H so that $P(G \in H) = 1 - o(1)$ and constants $B, \lambda_0 > 0$ depending on μ so that for all $\lambda \in (0, \lambda_0)$ we have*

$$E(T_V | G \in H) \le B.$$

The result follows from Markov's inequality. Using this approach we have to estimate the probability the survival time $> n^{1+o(1)}$ so that n times the error probability goes to 0.

To prove Theorem 5.8.6, they show that the graph looks locally like a Galton-Watson tree with an augmented distribution defined as follows. Suppose for simplicity that $k_{max} = \max\{k : p_k > 0\}$ and let

$$k_0 = \max\{k : \sum_{j \ge k} \sqrt{p_j} \ge 1/2\}.$$

They define the augmented distribution by

$$\hat{\mu}(j) = \begin{cases} p_j/2Z & \text{if } j \le k_0 \\ \sqrt{p_j}/Z & \text{if } j > k_0 \end{cases}$$

where Z is chosen to make the sum of the $\hat{\mu}(j) = 1$. In addition to changing the tails of the distribution, one cannot assume that the graph is tree-like, but one must allow an extra edge that may form a cycle.

Proof of (ii). To prove prolonged persistence they show that in Sect. 7 the random graph contains a large (α, R)-embedded expander. A subset W_0 of vertices has this property if for every set $A \subset W_0$ with $|A| \le \alpha |W_0|$

$$|N(A, R) \cap W_0| \ge 2\, A|,$$

where $N(A, R)$ is the collection of vertices in G_n within distance at most R of A.

Once this is done they can prove

Lemma 5.8.7 *(Lemma 5.4) There are constants λ_0, C, C', and β so that for all $\lambda \ge \lambda_0$ and integer $a \in (0, \alpha\beta n]$*

$$P(|X^0_{t+C}| \le 5a/4 | X^0_t = a) \le 2\exp(-a/C').$$

This leads easily to the lower bound on the survival times. For more details see the paper.

Further Reading. Huang (2020a) shows that if a contact process on a Galton-Watson tree survives then it dominates a Crump-Mode-Jagers branching process. This implies that the survival probability $p(\lambda)$ is continuous at $\lambda = \lambda_1$

Nam, Nguyen, and Sly (2022) have studied the asymptotics of critical values λ_1 and λ_2 for the contact process on Galton-Watson trees and finite graphs when the mean degree $ED \to \infty$ and degree distribution satisfies a concentration condition for D/ED. Theorem 1 shows that $\lambda_1/ED \to 1$. Theorem 2 shows that the same result holds for the short and long time survival thresholds.

5.9 Threshold-θ Contact Process

Reproduction of particles in the contact process is asexual. An individual at x gives birth to a new individual at a neighboring site y at rate λ. As a consequence, the process is additive: when built on the graphical representation

$$\xi_t^{A \cup B} = \xi_t^A \cup \xi_t^B.$$

A number of particle systems have been defined that have sexual reproduction in the sense that at least two particles are needed to create a new one. Here we will describe results for one non-additive process with nonlinear flip rates that has been studied on random regular graphs: the threshold $\theta \geq 2$ contact process of Chatterjee and Durrett (2013). Along the way we will mention some of the other models. The introduction of CD (2013) has examples that have been studied on $\mathbb{Z}^d$. Before we get lost in the details of what is in CD (2013), we should note that Danny Nam (2019) has proved results for graphs that have more general degree distributions. Since our proof of Proposition 5.9.8 does not work when the degree is not constant his proof required some substantial new ideas.

Let G_n be a random r-regular graph with $r \geq 3$ on n vertices, constructed for example by the algorithm used in Section 2.1 for the configuration model. Let $\mathbb{P}$ denote the distribution of G_n, which is the first of several probability measures we will define. We condition on the event E_n that the graph is **simple**, i.e., it does not contain a self-loop at any vertex, or more than one edge between two vertices. Theorem 2.1.1 implies that $\mathbb{P}(E_n)$ converges to a positive limit c_r as $n \to \infty$, and hence

$$\text{if } \tilde{\mathbb{P}} = \mathbb{P}(\cdot|E_n), \text{ then } \tilde{\mathbb{P}}(\cdot) \leq c_r \mathbb{P}(\cdot) \text{ for some constant } c_r > 0. \tag{5.9.1}$$

Thus conditioning on the event E_n will not have much effect on the distribution of G_n. It is easy to see that the distribution of G_n under $\tilde{\mathbb{P}}$ is uniform over the collection of all r-regular (simple) graphs on the vertex set V_n. (We put simple in parentheses since it is redundant: graphs by definition are simple.) We choose G_n according to the distribution $\tilde{\mathbb{P}}$ on simple graphs, and once chosen the graph remains fixed through time.

Having defined the graph, the next step is to introduce the dynamics on the graph. For the proofs it is crucial that we work in discrete time. We write $x \sim y$ to mean that x is a neighbor of y, and let

$$\mathcal{N}_y = \{x \in V_n : x \sim y\} \tag{5.9.2}$$

be the set of neighbors of y. The distribution $P_{p,\theta}^{G_n}$ of the (discrete time) threshold-θ contact process $\xi_t \subseteq V_n$ with parameters p and θ conditioned on G_n can be described as follows:

$$P_{p,\theta}^{G_n}(x \in \xi_{t+1} \mid |\mathcal{N}_x \cap \xi_t| \geq \theta) = p \text{ and}$$
$$P_{p,\theta}^{G_n}(x \in \xi_{t+1} \mid |\mathcal{N}_x \cap \xi_t| < \theta) = 0,$$

where the decisions for different vertices at time $t+1$ are made independently. Let $\xi_t^A \subseteq V_n$ denote the threshold-θ contact process starting from $\xi_0^A = A$, and let ξ_t^1 denote the special case when $A = V_n$.

Being an attractive process, the threshold-θ contact process on an r-regular tree has a translation-invariant upper invariant measure, ξ_∞^1, that is the limit as $t \to \infty$ for the system starting from all 1's. There are three basic questions for our models.

Q1. Let ξ_t^p be the system starting from product measure with density p, i.e., $\xi_0^p(x)$ are independent and equal 1 with probability p. Does ξ_t^p die out for small p? That is, do we have $P(\xi_t^p(x) = 1) \to 0$ as $t \to \infty$ if $p \leq p_0(\lambda)$?

Q2. Let $\rho(\lambda) = P(\xi_\infty^1(x) = 1)$ and let $\lambda_c = \inf\{\lambda : \rho(\lambda) > 0\}$. Is $\rho(\lambda)$ discontinuous at λ_c? If so, then soft results imply that $P(\xi_\infty^1(x) = 1) > 0$ when $\lambda = \lambda_c$. (As $\lambda \downarrow \lambda_c$ the upper invariant measures ξ_∞^1 decrease to a limit which must be the upper invariant measure at λ_c.)

Q3. Let $\xi_\infty^{0,\beta}$ be the limit as $t \to \infty$ for the system starting from all 0's when sites become occupied spontaneously at rate β along with the original dynamics. Is $\lim_{\beta \to 0} P(\xi_\infty^{0,\beta}(x) = 1) = 0$? If so, we say that **0 is stable under perturbation**, and it follows that there are two nontrivial stationary distributions when $\beta > 0$ is small. (To see this, note that $\xi_\infty^{1,\beta}$ is larger than $\xi_\infty^{1,0}$, so we have two stationary distributions if the density of particles in $\xi_\infty^{0,\beta}$ is less than the density in $\xi_\infty^{1,0}$.

5.9.1 Results

The first step is to prove that the threshold-θ contact process dies out for small values of p and survives for p close to 1. It is easy to see that on any graph in which all vertices have degree r the threshold-θ contact process dies out rapidly if $p < 1/r$, because an occupied site has at most r neighbors that it could cause to be occupied at the next time step suggesting $E^{G_n}_{p,\theta}\xi^1_t \le n(rp)^t$.

Survival from initial density close to 1

Our next result shows that if $\theta \ge 2$, $r \ge \theta + 2$ and p is sufficiently close to 1, then with high probability the fraction of occupied vertices in the threshold-θ contact process on G_n starting with all 1's stays above $1 - \epsilon_1$ for an exponentially long time.

Theorem 5.9.1 *Suppose $\theta \ge 2$ and $r \ge \theta + 2$. There are constants $\epsilon_1, \gamma_1 > 0$, and a good set of graphs $\mathcal{G}_n$ with $\tilde{\mathbb{P}}(G_n \in \mathcal{G}_n) \to 1$ so that if $G_n \in \mathcal{G}_n$ and $p \ge p_1 = 1 - \epsilon_1/(3r - 3\theta)$, then*

$$P^{G_n}_{p,\theta}\left(\inf_{t \le \exp(\gamma_1 n)} \frac{|\xi^1_t|}{n} < 1 - \epsilon_1\right) \le \exp(-\gamma_1 n).$$

Here and in what follows, all constants will depend on the degree r and threshold θ. If they depend on other quantities, that will be indicated.

The reason for the restriction to $r \ge \theta + 2$ comes from Proposition 5.9.8 (with $j = r - \theta + 1$) below. When $r \le \theta + 1$, it is impossible to pick $\eta > 0$ so that $(1 + \eta)/(r - \theta) < 1$. There may be more than algebra standing in the way of constructing a proof. We conjecture that the result is false when $r \le \theta + 1$. To explain our intuition in the special case $\theta = 2$ and $r = 3$, consider a rooted binary tree in which each vertex has two descendants and hence, except for the root, has degree three. If we start with a density u of 1's on level k and no 1's on levels $m < k$, then at the next step the density will be $g(u) = pu^2 < u$ on level $k - 1$. When each vertex has three descendants instead of two, then

$$g(u) = p(3u^2(1 - u) + u^3),$$

which has a nontrivial fixed point for $p \ge 8/9$ (divide by u and solve the quadratic equation).

As the next result shows, there is a close relationship between the threshold-θ contact process ξ_t on a random r-regular graph and the corresponding process ζ_t on the homogeneous r-tree. Following the standard recipe for attractive interacting particle systems, if we start with all sites on the tree occupied, then the sequence $\{\zeta^1_t\}$ of sets of occupied vertices decreases in distribution to a limit ζ^1_∞, which is called the *upper invariant measure*, since it is the stationary distribution with the most 1's. Here and later we denote by $\mathbf{0}$ any fixed vertex of the homogeneous tree. Writing $P_{p,\theta}$ for the distribution of ζ_t with parameters p and θ, the critical value is defined by

$$p_c(\theta) := \sup\{p : P_{p,\theta}(\zeta^1_\infty(\mathbf{0}) = 1) = 0\}.$$

Corollary 5.9.2 *Suppose $\theta \geq 2$, $r \geq \theta + 2$ and that p_1 and ϵ_1 are the constants in Theorem 5.9.1. If $p \geq p_1$, then there is a translation-invariant stationary distribution for the threshold-θ contact process on the homogeneous r-tree in which each vertex is occupied with probability $\geq 1 - \epsilon_1$.*

Fontes and Schonmann (2008a) have considered the continuous time threshold-θ contact process on a tree in which each vertex has degree $b + 1$, and they have shown that if b is large enough, then $\lambda_c < \infty$. Our result improves their result by removing the restriction that b is large.

Dying out from small initial density

If we set the death rate = 0 in the threshold-θ contact process, then we can without loss of generality set the birth rate equal to 1 and the process reduces to bootstrap percolation (with asynchronous updating). Balogh and Pittel (2007) have studied bootstrap percolation on random regular graphs. They have identified an interval $[p_-(n), p_+(n)]$ so that the probability that all sites end up active goes sharply from 0 to 1. The limits $p_\pm(n) \to p_*$ and $p_+ - p_-$ is of order $1/\sqrt{n}$. If bootstrap percolation cannot fill up the graph, then it seems that our process with deaths will be doomed to extinction. The next result proves this, and more importantly extends the result to arbitrary initial conditions with a small density of occupied sites.

Here, since processes with larger θ have fewer survivals, it is enough to prove the result when $\theta = 2$.

Theorem 5.9.3 *Suppose $\theta \geq 2$ and $p_2 < 1$. There are constants $0 < \epsilon_2(p_2), C_2(p_2) < \infty$, and a good set of graphs $\mathcal{G}_n$ with $\tilde{\mathbb{P}}(G_n \in \mathcal{G}_n) \to 1$ so that if $G_n \in \mathcal{G}_n$, then for any $p \leq p_2$, and any subset $A \subset V_n$ with $|A| \leq \epsilon_2 n$,*

$$P_{p,\theta}^{G_n}\left(\xi_{\lceil C_2 \log n \rceil}^A \neq \emptyset\right) \leq 2/n^{1/6} \text{ for large enough } n.$$

The density of 1's $\rho(p,\theta) := P_{p,\theta}(\zeta_\infty^1(\mathbf{0}) = 1)$ in the stationary distribution on the homogeneous r-tree is a nondecreasing function of p. The next result shows that the threshold-θ contact process on the r-tree has a discontinuous phase transition.

Corollary 5.9.4 *Suppose $\theta \geq 2$, let p_1 be the constant from Theorem 5.9.1, and let $\epsilon_2(\cdot)$ be as in Theorem 5.9.3. $\rho(p,\theta)$ never takes values in $(0, \epsilon_2(p_1))$.*

This result, like Theorem 5.9.3, does not require the assumption $r \geq \theta + 2$. On the other hand, if $\rho(p,\theta) \equiv 0$ for $r \leq \theta + 1$, the result is not very interesting in that case. Again Fontes and Schonmann (2008a) have proved that the threshold-θ contact process has a discontinuous transition when the degree $b + 1$ is large enough.

Fontes and Schonmann (2008b) have studied θ-bootstrap percolation on trees in which each vertex has degree $b + 1$ and $2 \leq \theta \leq b$. They have shown that there is a critical value p_f so that if $p < p_f$, then for almost every initial configuration of product measure with density p, the final bootstrapped configuration does not have any infinite component. This suggests that we might have $\epsilon_2(p)$ bounded away from 0 as $p \to 1$.

Stability of 0

The previous pair of results are the most difficult in the paper. From their proofs one easily gets results for the process with spontaneous births with probability β, i.e., after the threshold-θ dynamics has been applied to the configuration at time t, we independently make vacant sites occupied with probability β. For this new process, we denote the set of occupied vertices at time t starting with all 0's by $\hat{\xi}_t^0$ and its distribution conditioned on the graph G_n by $P_{p,\theta,\beta}^{G_n}$ to have the following:

Theorem 5.9.5 *Suppose* $\theta \geq 2$. *There is a good set of graphs* $\mathcal{G}_n$ *with* $\tilde{\mathbb{P}}(G_n \in \mathcal{G}_n) \to 1$ *so that if* $G_n \in \mathcal{G}_n$ *and* $p < 1$, *then there are constants* $C_3(p), \beta_3(p), \gamma_3(p, \beta) > 0$ *so that for* $\beta < \beta_3$,

$$P_{p,\theta,\beta}^{G_n}\left(\sup_{t \leq \exp(\gamma_3 n)} \frac{|\hat{\xi}_t^0|}{n} > C_3\beta\right) \leq 2\exp(-\gamma_3 n).$$

Let $\hat{\zeta}_\infty^0$ be the limiting distribution for the process on the homogeneous tree, which exists because of monotonicity.

Corollary 5.9.6 *If* $\theta \geq 2$ *and* $p < 1$, *then* $\lim_{\beta \to 0} P_{p,\theta,\beta}(\hat{\zeta}_\infty^0(\mathbf{0}) = 1) = 0$.

5.9.2 Key Ideas for the Proof

We now describe the "isoperimetric inequalities" that are the keys to the proofs of our results. Let $\partial U := \{y \in U^c : y \sim x \text{ for some } x \in U\}$ be the boundary of U, and given two sets U and W, let $e(U, W)$ be the number of edges having one end in U and the other end in W. Given an $x \in V_n$ let $n_U(x)$ be the number of neighbors of x that are in U, and let

$$U^{*j} = \{x \in V_n : n_U(x) \geq j\}.$$

The estimation of the sizes of $e(U, U^c)$ is key to the study of random walks on graphs, see Chapter 6, and especially Section 6.2. Here we are interested in studying the sizes of U^{*j} and having better constants by restricting to small sets. The last remark will make more sense after reading Section 6.2.

Proposition 5.9.7 *Let* $E^{*1}(m, \leq k)$ *be the event that there is a subset* $U \subset V_n$ *with size* $|U| = m$ *so that* $|U^{*1}| \leq k$. *There are constants* C_0 *and* Δ_0 *so that for any* $\eta > 0$, *there is an* $\epsilon_0(\eta)$ *which also depends on* r *so that for* $m \leq \epsilon_0(\eta)n$,

$$\mathbb{P}\left[E^{*1}(m, \le (r-1-\eta)m)\right] \le C_0 \exp\left(-\frac{\eta^2}{4r} m \log(n/m) + \Delta_0 m\right).$$

This result yields the next proposition, which we need to prove Theorems 5.9.1 and 5.9.3. For Theorem 5.9.1, note that if $W = V_n \setminus \xi_t$ is the set of vacant vertices at time t, then at time $t+1$ the vertices in $W^{*(r-\theta+1)}$ will certainly be vacant and the vertices in its complement will be vacant with probability $1-p$. So having an upper bound for $|W^{*(r-\theta+1)}|$ will be helpful. On the other hand for Theorem 5.9.3, if U is the set of occupied vertices at time t, then at time $t+1$ the vertices in $U^{*\theta}$ will be occupied with probability p and the vertices in its complement will certainly be vacant. So having an upper bound for $|U^{*\theta}|$ will be helpful.

Keeping these in mind, it is easy to see from the definitions that if $j > 1$ and $|Z| = m$, then

$$rm \ge \sum_{y \in Z^{*1}} e(\{y\}, Z) \ge |Z^{*1} \setminus Z^{*j}| + j|Z^{*j}| = |Z^{*1}| + (j-1)|Z^{*j}|.$$

So for any set Z of size m, if $|Z^{*j}| \ge k$, then $|Z^{*1}| \le rm - (j-1)k$. Taking $k = m(1+\eta)/(j-1)$ so that $rm - (j-1)k = (r-1-\eta)m$ and using Proposition 5.9.7 we get

Proposition 5.9.8 *Let $E^{*j}(m, \ge k)$ be the event that there is a subset $Z \subset V_n$ with size $|Z| = m$ so that $|Z^{*j}| \ge k$. For the constants C_0, Δ_0, and $\epsilon_0(\eta)$ given in Proposition 5.9.7, if $j > 1$ and $m \le \epsilon_0(\eta)n$, then*

$$\mathbb{P}\left[E^{*j}\left(m, \ge \left(\frac{1+\eta}{j-1}\right) m\right)\right] \le C_0 \exp\left(-\frac{\eta^2}{4r} m \log(n/m) + \Delta_0 m\right).$$

The reasoning that led to Proposition 5.9.8 depends on the fact that all vertices have the same degree. Danny Nam has developed new results to cover the case of variable degree.

References

Allen LS (2003) An introduction to stochastic processes with applications in biology. Prentice Hall, Pearson

Balogh J, Pittel BG (2007) Bootstrap percolation on the random regular graph. Rand Struct Alg 30:257–276

Berger N, Borgs C, Chayes JT, Saberi A (2005) On the spread of viruses on the internet. In: Proceedings of the 16th symposium on discrete algorithms

Bezuidenhout C, Grimmett G (1990) The critical contact process dies out. Ann Probab 18:1462–1482

Bhamidi S, Nam D, Nguyen O, Sly A (2021) Survival and extinction of epidemics on random graphs with general degrees. Ann Probab 49:244–286

Bollobás B, Chung FRK (1988) The diameter of a cycle plus a random matching. SIAM J Discrete math 1:328–333
Cator E, Don H (2021) Explicit bounds for critical infection rates and expected extinction times of the contact process on finite random graphs. Bernoulli 27:1556–1582
Chatterjee S, Durrett R (2009) Contact processes on random graphs with power law degree distributions have critical value 0. Ann Probab 37:2332–2356
Chatterjee S, Durrett R (2013) A first order phase transition in the threshold $\theta \geq 2$ contact process on random r-regular graphs and r-trees. Stoch Proc Appl 123(2013):561–578
Cranston M, Mountford T, Mourrat JC, Valesin D (2014) The contact process on finite trees revisited. ALEA Lat Am J Probab Math Stat 11:385–408
Durrett R, Jung R (2007) Two phase transitions for contact processes on small worlds. Stoch Proc Appl 117:1910–1927
Durrett R, Liu X (1988) The contact process on a finite set. Ann Probab 16:1158–1173
Durrett R, Schonmann RHS (1988) The contact process on a finite set, II. Ann Probab 16:1570–1583
Fontes LRG, Schonmann RH (2008) Threshold $\theta \geq 2$ contact processes on homogeneous trees. Probab Theory Rel Fields 141:513–541
Fontes LRG, Schonmann RH (2008) Bootstrap percolation on homogeneous trees has two phase transitions. J Stat Phys 132:839–861
Griffeath D (1979) Additive and Cancellative Interacting Particle Systems. Springer
Harris TE (1974) Contact interactions on a lattice. Ann Probab 2:969–988
Harris TE (1978) Additive set-valued Markov processes and percolation methods. Ann Probab 6:355–378
Hoeffding W (1963) Probability inequalities for sums of bounded random variables. J Amer Statist Assoc 58:13–30
Huang X (2020) Exponential growth and continuous phase transition for the contact process on trees. Electron J Probab 25. https://doi.org/10.1214/20-EJP483
Huang X, Durrett R (2020) The contact process on random graphs and Galton-Watson trees. ALEA Lat Am J Probab Math Stat 17(1):159–182
Lalley S, Su W (2017) Contact processes on random regular graphs. Ann Appl Probab 27:2061–2097
Liggett TM (1995) Improved upper bounds for the contact process critical value. Ann Probab 23:697–723
Liggett TM (1996) Multiple transition points for the contact process on the binary tree. Ann Probab 24:1455–1490
Ménard L, Singh A (2016) Percolation by cumulative merging and phase transition for the contact process on some random graphs. Ann Sci École Norm Supér 49:1189–1238
Mountford TS (1993) A metastable result for the finite multidimensional contact process. Canad Math Bull 36:216–226
Mountford TS (1999) Existence of a constant for finite system extinction. J Statist Phys 96:1331–1341
Mountford T, Mourrat J-C, Valesin D, Yao Q (2016) Exponential extinction time of the contact process on finite graphs. Stoch Proc Appl 126:1974–2013
Mountford T, Valesin D, Yao Q (2013) Metastable densities for the contact process on power law random graphs. Electron J Probab 18, paper 103
Mourrat J-C, Valesin D (2016) Phase transitions of the contact process on random regular graphs. Electron J Probab 21, paper 31
Nam D (2019) A discontinuous phase transition in the threshold $\theta \geq 2$ contact process on random graphs. arXiv:1907.05005
Nam D, Nguyen O, Sly A (2022) Critical value asymptotics for the contact process on random graphs. Trans Amer Math Soc 375:3899–3967
Pastor-Satorras R, Castellano C, van Mieghem, Vespignani A (2015) Epidemic processes on complex networks. Rev Mod Phys 87:925–979
Pastor-Satorras R, Vespignani A (2001) Epidemic spreading in scale-free networks. Phys Rev Lett 86:3200–3203

Pastor-Satorras R, Vespignani A (2001) Epidemic dynamics and endemic states in complex networks. Phys Rev E 63:066117
Pastor-Satorras R, Vespignani A (2002) Epidemic dynamics in finite size scale-free networks. Phys Rev E 65:035108(R)
Pemantle R (1992) The contact process on trees. Ann Probab 20:2089–2116
Schapira B, Valesin D (2015) Extinction time for the contact process on general graphs. arXiv:1509.04133
Schapira B, Valesin D (2018) Exponential rate for the contact process extinction time. arXiv:1806.04491
Stacey AM (1996) The existence of an intermediate phase for the contact process on trees. Ann Probab 24:1711–1726
Stacey AM (2001) The contact process on finite homogeneous trees. Probab Theory Rel Fields 121:551–576
Liggett TM (1999) Stochastic interacting systems: contact, voter and exclusion processes. Springer, New York
Watts DJ, Strogatz DH (1998) Collective dynamics of 'small world' networks. Nature 393:440–442

Chapter 6
Random Walks, Mixing Times

In this chapter we will investigate how the time for a random walk on a graph to reach equilibrium depends on the geometry of the graph. In Sect. 6.1 we review basic results about Markov chains and state two general results on the rate of convergence to equilibrium that are based on the size of the spectral gap. In Sect. 6.2 we introduce the notion of **conductance for reversible Markov chains** and prove **Cheeger's inequality** which relates the spectral gap to the "bottleneck ratio."

In Sect. 6.3 we begin to apply these results to bounding convergence times of random walks on random graphs. The first result, which is useful for many examples, is that if the minimum degree is at least three then the convergence time is $O(\log n)$. Section 6.4 shows that when there are also vertices of degree two then convergence is slowed to $O(\log^2 n)$ because there are paths of length $\geq \epsilon \log n$ and if a random walk starts in the middle of one of them then time to escape from the path is $\geq c\epsilon^2 \log^2 n$.

Section 6.5 investigates how the rate of convergence to equilibrium on Erdős-Rényi graphs depends on the mean degree. When the graph is connected, i.e., the mean degree is $c \log n$ with $c > 1$ then convergence occurs in time $O(\log n)$, but as the mean decreases to $\lambda > 1$ the time increases to $O(\log^2 n)$. At the end of that section we will see that having vertices of degree one, which are dead ends, only changes the constant. Of course vertices of degree zero don't do much unless you start on one. It is interesting that $O(\log^2 n)$ is a worst case result. If the walk starts from a vertex chosen at random then the mixing time is $O(\log n)$

Section 6.6 introduces the **cutoff phenomenon**, which occurs when t_n is the time to converge to equilibrium and the distance from the stationary distribution goes from 1 to 0 over a window of size $o(t_n)$. In Sect. 6.7 we describe results of Lubetzky and Sly (2010) on random d-regular graphs. They show that if $d \geq 3$ the mixing time is $\sim c_d \log_{d-1} n$ where $c_d = d/(d-2)$ and there is cutoff with a window of size $\sqrt{\log_{d-1} n}$. It is easy to see this is the right answer. By results in Chap. 1, the largest distance on the graph is $\sim \log_{d-1} n$. The random walk moves away from its starting point with probability $(d-1)/d$ and closer to it with probability $1/d$ so the drift is

R. Durrett, *Dynamics on Graphs*, Probability Theory and Stochastic Modelling 110,
https://doi.org/10.1007/978-3-032-21908-4_6

$(d-2)/d$. However, proving this is not so easy even when you get the hint that "the covering space of a random regular graph is a regular tree."

6.1 General Convergence Results

Given a finite graph (random or not) one can define a random walk by its **transition kernel** or **transition probability**

$$K(x,y) = 1/d(x) \quad \text{if } y \sim x,$$

where $y \sim x$ is short for y is a neighbor of x and $d(x)$ is the degree of x. If we let $D = \sum_x d(x)$ then $\pi(x) = d(x)/D$ is a stationary distribution that satisfies the **detailed balance condition**

$$\pi(x)K(x,y) = \frac{1_{x\sim y}}{D} = \pi(y)K(y,x).$$

Stationary distributions with this property are also called **reversible**, since when the chain is started in state π the process has the same distribution going backwards in time. To explain this note that

$$\begin{aligned} P_\pi(X_0 = y|X_1 = x) &= \frac{P_\pi(X_0 = y, X_1 = x)}{P_\pi(X_1 = x)} \\ &= \frac{\pi(y)K(y,x)}{\pi(x)} \\ &= \frac{\pi(x)K(x,y)}{\pi(x)} = K(x,y). \end{aligned}$$

To avoid the problem of periodicity, we will often make the random walk **lazy**

$$\bar{K}(x,x) = 1/2 \qquad \bar{K}(x,y) = (1/2)K(x,y) \quad \text{when } x \neq y.$$

If the graph is connected then the kernel is irreducible. If the random walk is lazy, it is aperiodic, so it converges to equilibrium, i.e. $K^n(x,y) \to \pi(y)$. Here we will be interested in how fast the convergence occurs. To measure the distance from equilibrium we can use the **total variation distance**. If μ and ν are probability distributions then

$$\|\mu - \nu\|_{TV} = \max_A |\mu(A) - \nu(A)| = \frac{1}{2}\sum_x |\mu(x) - \nu(x)|.$$

We will usually drop the subscript TV. The distance from equilibrium at time t can be defined as

$$d(t) = \sup_x \|K^t(x, \cdot) - \pi\| = \sup_\mu \|\mu K^t - \pi\|,$$

where μ is a probability distribution and $\mu K^t(y) = \sum_x \mu(x) K^t(x, y)$. It is useful to also define

$$\bar{d}(t) = \sup_{x,y} \|K^t(x, \cdot) - K^t(y, \cdot)\| = \sup_{\mu,\nu} \|\mu K^t - \nu K^t\|.$$

$\bar{d}$ has the useful property that it is a **submultiplicative function**

$$\bar{d}(s + t) \leq \bar{d}(s) \cdot \bar{d}(t). \tag{6.1.1}$$

It is easy to see that

$$d(t) \leq \bar{d}(t) \leq 2d(t). \tag{6.1.2}$$

The time to get within ϵ in total variation distance is

$$t_{mix}(\epsilon) = \min\{t : d(t) < \epsilon\}.$$

We define the **mixing time** by

$$t_{mix} = t_{mix}(1/4). \tag{6.1.3}$$

To see why we choose $\epsilon = 1/4$ note that using (6.1.2) and (6.1.1)

$$d(\ell t_{mix}) \leq \bar{d}(\ell t_{mix}) \leq \bar{d}(t_{mix})^\ell \leq (2\epsilon)^\ell = 2^{-\ell}$$

when $\epsilon = 1/4$.

6.1.1 Bounds on Convergence

Consider a Markov chain transition kernel $K(i, j)$ on $\{1, 2, \ldots n\}$ with reversible stationary distribution π_i, i.e., $\pi_i K(i, j) = \pi_j K(j, i)$. For the next result we will measure the distance from equilibrium using the **relative pointwise distance**

$$\Delta(t) = \max_{i,j} \left| \frac{K^t(i, j)}{\pi_j} - 1 \right|,$$

which is larger than two times the total variation distance

$$\Delta(t) \geq \max_i \sum_j \left| \frac{K^t(i, j)}{\pi_j} - 1 \right| \pi_j = \max_i \sum_j |K^t(i, j) - \pi_j| = 2\|K^t(i, \cdot) - \pi(\cdot)\|_{TV}.$$

Let D be a diagonal matrix with entries $\pi_1, \pi_2, \ldots, \pi_n$ and $a = D^{1/2} K D^{-1/2}$. We like a because it is symmetric. Since

$$\begin{aligned} a(i, j) &= \pi_i^{1/2} K(i, j) \pi_j^{-1/2} = \pi_i^{-1/2} \cdot \pi(i) K(i, j) \cdot \pi_j^{-1/2} \\ &= \pi_i^{-1/2} \cdot \pi(j) K(j, i) \cdot \pi_j^{-1/2} = \pi_j^{1/2} K(j, i) \pi_i^{-1/2} = a(j, i), \end{aligned}$$

matrix theory tells us that $a(i, j)$ has real eigenvalues $1 = \lambda_0 \geq \lambda_1 \geq \cdots \geq \lambda_{n-1} \geq -1$. Let $\lambda_{max} = \max\{\lambda_1, |\lambda_{n-1}|\}$ be the eigenvalue with largest magnitude. The lazy chain has transition kernel $\bar{K} = (I + K)/2$ so all $\bar{\lambda}_i = (1 + \lambda_i)/2 \geq 0$ and we do not have to worry about $|\lambda_{n-1}|$.

To explain the interest in eigenvalues near -1, note that if the chain is periodic with period two then we can find a set S such that $p(x, S^c) = 1$ for $x \in S$ and $p(x, S) = 1$ for $x \in S^c$. The function that has $f(x) = 1$ for $x \in S$ and $f(x) = -1$ for $x \in S^c$ has $Pf = -f$, i.e., it is an eigenvector with eigenvalue -1. Thus $|\lambda_{n-1}|$ measures how close the chain is to having period 2. In general, Markov chains can have states with any period $d \geq 3$ but that is impossible in the reversible case.

The next result is from Sinclair and Jerrum (1989), but similar results can be found in many other places.

Theorem 6.1.1 *Let K be the transition matrix of an irreducible reversible Markov chain on $\{1, 2, \ldots, n\}$ with stationary distribution π and let $\pi_{min} = \min_j \pi_j$. Then*

$$\Delta(t) \leq \frac{\lambda_{max}^t}{\pi_{min}} \quad \textit{and} \quad |K^t(i, j) - \pi_j| \leq \lambda_{max}^t \sqrt{\pi_j / \pi_i}.$$

Proof. Since a is symmetric, we can select an orthonormal basis e_m, $0 \leq m < n$, of eigenvectors of a, and a has spectral decomposition:

$$a = \sum_{m=0}^{n-1} \lambda_m e_m e_m^T.$$

The matrix $B_m = e_m e_m^T$ has $B_m^2 = B_m$, and $B_\ell B_m = 0$ if $\ell \neq m$ so

$$a^t(i, j) = \sum_{m=0}^{n-1} \lambda_m^t B_m(i, j) = \sum_{m=0}^{n-1} \lambda_m^t e_m(i) e_m(j).$$

$e_0(i) = \pi_i^{1/2}$ so

$$K^t(i, j) = (D^{-1/2} a^t D^{1/2})_{i,j} = \pi_j + \sqrt{\frac{\pi_j}{\pi_i}} \sum_{m=1}^{n-1} \lambda_m^t e_m(i) e_m(j). \tag{6.1.4}$$

From this it follows that

$$\Delta(t) = \max_{i,j} \frac{\left|\sum_{m=1}^{n-1} \lambda_m^t e_m(i) e_m(j)\right|}{\sqrt{\pi_i \pi_j}} \le \lambda_{max}^t \frac{\max_{i,j} \sum_{m=1}^{n-1} |e_m(i)||e_m(j)|}{\pi_{min}}.$$

The Cauchy-Schwarz inequality implies

$$\sum_{m=1}^{n-1} |e_m(i)||e_m(j)| \le \left(\sum_{m=1}^{n-1} |e_m(i)|^2 \sum_{m=1}^{n-1} |e_m(j)|^2 \right)^{1/2} \le 1. \tag{6.1.5}$$

To see that $\sum_{m=1}^{n-1} |e_m(i)|^2 \le 1$ note that if δ_i is the vector with 1 in the ith place and 0 otherwise then expanding in the orthonormal basis $\delta_i = \sum_{m=0}^{n-1} e_m(i) e_m$, so the desired result follows by taking the L^2 norm of both sides of the equation. To get the second result combine (6.1.4) and (6.1.5). □

6.1.2 Continuous Time Chains

If jumps occur at rate one then there are a Poisson mean t jumps by time t so the transition probability is

$$H_t(x, y) = e^{-t} \sum_{m=0}^{\infty} \frac{t^m}{m!} K^m(x, y). \tag{6.1.6}$$

If λ_i is an eigenvalue of K then $e^{-t(1-\lambda_i)}$ is an eigenvalue of H_t. Thus there are no negative eigenvalues to worry about and we have

Theorem 6.1.2 *If* $\gamma = 1 - \lambda_1$ *is the spectral gap of* K *and* $\Delta(t) = \max_{i,j} |H_t(i, j)/\pi(j) - 1|$ *then*

$$\Delta(t) \le \frac{e^{-\gamma t}}{\pi_{min}} \quad \textit{and} \quad |H_t(i, j) - \pi_j| \le e^{-\gamma t} \sqrt{\pi_j / \pi_i}.$$

Proof. This follows immediately from (6.1.6) and Theorem 6.1.1. □

Given a reversible Markov transition kernel $K(x, y)$ we define an inner product by

$$\mathcal{E}(f, g) = \frac{1}{2} \sum_{x,y} (f(x) - f(y))(g(x) - g(y)) \pi(x) K(x, y),$$

and the **Dirichlet form** by

$$\mathcal{E}(f) = \frac{1}{2} \sum_{x,y} (f(x) - f(y))^2 \pi(x) K(x, y). \tag{6.1.7}$$

Introducing the inner product $\langle f, g\rangle_\pi = \sum_x f(x)g(x)\pi(x)$, a little algebra shows

$$\mathcal{E}(f, f) = \langle f, (I - K)f\rangle_\pi.$$

If we define the variance by $\text{var}_\pi(f) = E_\pi(f - E_\pi f)^2$ then the spectral gap can be computed from the **variational formula**

$$1 - \lambda_1 = \min\{\mathcal{E}(f, f) : \text{var}_\pi(f) = 1\}. \tag{6.1.8}$$

To see this note that $\mathcal{E}(f, f)$ is not affected by subtracting a constant from f so

$$1 - \lambda_1 = \min\{\langle f, f\rangle_\pi - \langle f, Kf\rangle_\pi : E_\pi f = 0, \langle f, f\rangle_\pi = 1\}$$

and the result follows from the usual variational formula for λ_1 for the nonnegative symmetric matrix $a_{i,j} = \pi(i)K(i, j)$, i.e.,

$$\lambda_1 = \max\left\{\sum_{i,j} x_i a_{i,j} x_j : \sum_i x_i^2 = 1\right\}.$$

Up to this point we have obtained the continuous time results from the discrete case. They can also be proved directly without too much effort. The developments here follow Chapter 2 of Saloff-Coste (1996). Let $\pi(f) = \sum_x f(x)\pi(x)$.

Lemma 6.1.3 *Let K be a Markov kernel with spectral gap γ. Then the semigroup $H_t = e^{-t(I-K)}$ satisfies*

$$\|H_t f - \pi(f)\|_2^2 \le e^{-2\gamma t} var_\pi(f) \quad \textit{for all } f \in \ell^2(\pi).$$

Proof. Set $u(t) = \text{var}_\pi(H_t f) = \|H_t(f - \pi(f))\|_2^2 = \|H_t(f) - \pi(f)\|_2^2$. Then

$$\begin{aligned} u'(t) &= -2\mathcal{E}(H_t(f) - \pi(f), H_t(f) - \pi(f)) \\ &\le -2u(t). \end{aligned}$$

Integrating gives $u(t) \le e^{-2\gamma t}u(0)$, which gives the desired result. □

This leads to a slightly weaker version of Theorem 6.1.2

Theorem 6.1.4 *Let $h_t^x = H_x^t(\cdot)/\pi(\cdot)$. If $\gamma = 1 - \lambda_1$ is the spectral gap of K then*

$$\begin{aligned} &\|h_t^x - 1\|_2 \le \sqrt{1/\pi(x)}e^{-\gamma t} \quad \textit{and} \\ &|h_t(i, j) - \pi_j| \le e^{-\gamma t}\sqrt{\pi_j/\pi_i}. \end{aligned}$$

Proof. Let $g_x(y) = 1/\pi(x)$ if $y = x$, 0 otherwise. Lemma 6.1.3 implies

$$\|h_t^x - 1\|_2 \le \sqrt{\frac{1-\pi(x)}{\pi(x)}} e^{-\gamma t}.$$

To prove the second result note that

$$\begin{aligned}|h_t(i,j) - \pi_j| &= \left|\sum_z (h_{t/2}(x,z) - 1)(h_{t/2}(z,y) - 1)\pi(z)\right| \\ &\le \|h_{t/2}^x\|_2 \|h_{t/2}^y\|_2 \le e^{-\gamma t}\sqrt{\pi_j/\pi_i},\end{aligned}$$

where on the second step we have used a Cauchy-Schwarz inequality. □

Relaxation Time

There are many ways to quantify the amount of time needed to reach equilibrium. Earlier we defined the mixing time in (6.1.3) in terms of

$$d(t) = \sup_x \|K^t(x,\cdot) - \pi(\cdot)\|.$$

One can define the **relaxation time** in terms of the absolute spectral gap

$$\gamma_* = \max\{1 - |\lambda_i| : \lambda_i \ne 1\},$$

where the λ_i are the eigenvalues of the transition matrix, by setting

$$t_{rel} = 1/\gamma_*. \tag{6.1.9}$$

Combining Theorems 6.1.1 and 6.1.2 we have that in discrete or continuous time the relative pointwise distance

$$|\Delta(\ell t_{rel})| \le \frac{e^{-\ell}}{\pi_{min}}.$$

6.2 Conductance

Let $Q(x,y) = \pi(x)K(x,y)$ be the flow across the edge $\{x,y\}$ in equilibrium, let $Q(S,S^c) = \sum_{x\in S, y\in S^c} Q(x,y)$ and define

$$\Phi(S) = \frac{Q(S,S^c)}{\pi(S)}, \qquad \Phi_\star = \min_{\pi(S)\le 1/2} \Phi(S).$$

Levin and Peres (2017) call $\Phi_\star$ the **bottleneck ratio**. To see what this has to do with the rate of convergence to equilibrium we will state and prove their Theorem 7.4. The mixing time was defined in (6.1.3).

Theorem 6.2.1 $t_{mix}(1/4) \geq 1/4\Phi_\star$.

Proof. Suppose that X_t is stationary, i.e., the distribution is π for all t.

$$P_\pi(X_0 \in A, X_t \in A^c) \leq \sum_{s=1}^{t} P_\pi(X_{s-1} \in A, X_s \in A^c) = tQ(A, A^c).$$

Dividing by $\pi(A)$, we get $P_\pi(X_t \in A^c | X_0 \in A) \leq t\Phi(A)$ so there is an $x \in A$ with

$$P^t(x, A) \geq 1 - t\Phi(A),$$

and we conclude $d(t) \geq 1 - t\Phi(A) - \pi(A)$. If $\pi(A) \leq 1/2$ and $t < 1/(4\Phi(A))$ then $d(t) > 1/4$. This shows $t_{mix} \geq 1/\Phi(A))$ and maximizing over A gives the desired result. □

Example 6.2.2 Take two complete graphs on n vertices, call them L and R. Pick $\ell \in L$ and $r \in R$ and connect them by an edge. See Fig. 6.1. To simplify the computation we formulate the chain in continuous time with jumps across each edge occur at rate $1/n$. Let P_R denote the law of the Markov chain starting with all states in R having probability $1/n$ and $P_R(X_0 \in L) = 0$. If we let $u(t) = P_R(X_t \in R)$, then

$$\begin{aligned} u'(t) &= \frac{P_R(X_t = r) - P_R(X_t = \ell)}{n} \\ &\approx \frac{P_R(X_t \in R) - P_R(X_t \in L)}{n^2} = \frac{u(t) - (1 - u(t))}{n^2}, \end{aligned}$$

where in the second step we have used the approximation that all sites in R at time t have the same probability. (6.2.1) provides support for this but we leave the tedious details of writing a rigorous proof to the reader. Letting $v(t) = 2u(t) - 1$ we have $v'(t) = 2u'(y) = 2v(t)/n^2$. Since $v(0) = 1$, we have $v(t) = \exp(-2t/n^2)$ and

$$u(t) = \frac{1 + v(t)}{2} = \frac{1 + e^{-2t/n^2}}{2}.$$

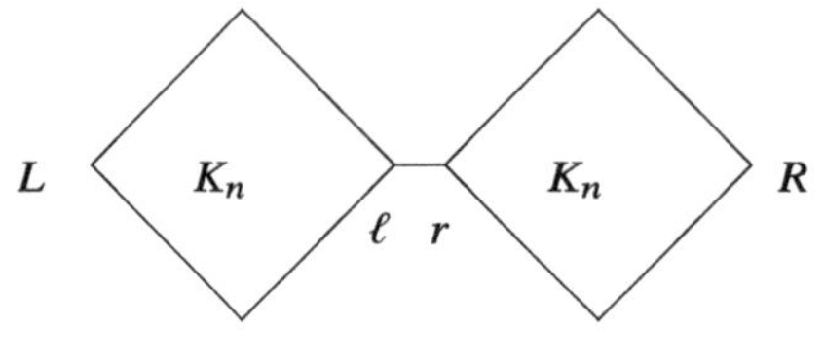

Fig. 6.1 Picture of Example 6.2.2

To compute $\Phi(R, L)$ in this example we must first compute $\pi(x)$. To do this we note that ℓ and r have degree n while all the other vertices that we will call x have degree $n - 1$ so

$$\pi(\ell) = \pi(r) = \frac{n}{2(n + (n-1)^2)}, \qquad \pi(x) = \frac{n-1}{2(n + (n-1)^2)}, \tag{6.2.1}$$

and for the other vertices x in equilibrium we have $\pi(x) = \pi(\ell) \cdot (n-1)/n$. Simplifying the exact formula to $\pi(y) \approx 1/2n$ for all y.

$$\Phi(R, L) = \frac{(1/2n) \cdot (1/n)}{1/2} = n^{-2}.$$

Since we start with all of the mass on R and in equilibrium $\pi(R) = 1/2$ it follows that the time to equilibrium is at least $O(n^2)$.

6.2.1 *Cheeger's Inequality*

Saloff-Coste attributes the result to Diaconis and Stroock (1991), who in turn named the result Cheeger's inequality in honor of the eigenvalue bound in differential geometry. Levin and Peres (2017) cite Sinclair and Jerrum (1989) and Lawler and Sokal (1988). To make things easier to write let $h = \Phi_* = \min_{\pi(S) \le 1/2} Q(S, S^c)/\pi(S)$.

Theorem 6.2.3 *The spectral gap has*

$$\frac{h^2}{2} \le 1 - \lambda_1 \le 2h.$$

Proof. Taking $f = 1_S$ in the variational formula (6.1.8) we have

$$\mathcal{E}(1_S, 1_S) = Q(S, S^c)$$

and $\text{var}_\pi(1_S) = \pi(S)(1 - \pi(S))$, so $1 - \lambda_1 \le Q(S, S^c)/\pi(S)(1 - \pi(S))$. The right-hand side is the same for S and S^c, so we can restrict our attention to $\pi(S) \le 1/2$. Since $1 - \pi(S) \ge 1/2$, we have $1 - \lambda_1 \le 2h$.

For the other direction, fix a function f, let $F_t = \{x : f(x) \ge t\}$ and let f_t be the indicator function of the set F_t. Since only differences $f(x) - f(y)$ appear in $\mathcal{E}(f, f)$, defined in (6.1.7), we can without loss of generality suppose that the median of f is 0, i.e., $\pi(F_t) \le 1/2$ for $t > 0$, and $\pi(F_t^c) \le 1/2$ for $t < 0$. Our next step is to compute something that would be the Dirichlet form if we had squared the increment.

$$\frac{1}{2}\sum_{x,y}|f(x)-f(y)|Q(x,y) = \sum_{f(x)>f(y)}(f(x)-f(y))Q(x,y)$$
$$= \sum_{x,y}\int_{-\infty}^{\infty} 1_{\{f(y)<t<f(x)\}}Q(x,y)\,dt$$
$$= \int_0^{\infty}|\partial F_t|\,dt + \int_{-\infty}^{0}|\partial F_t^c|\,dt$$
$$\geq h\left(\int_0^{\infty}\pi(F_t)\,dt + \int_{-\infty}^{0}\pi(F_t^c)\,dt\right) = h\pi(|f|).$$

Continuing to suppose that the median of f is 0, let $g = f^2\mathrm{sgn}\,(f)$, where $\mathrm{sgn}\,(x) = 1$ if $x > 0$, $\mathrm{sgn}\,(x) = -1$ if $x < 0$, and $\mathrm{sgn}\,(0) = 0$. $|g| = f^2$ so the last inequality implies

$$2h\pi(f^2) \leq \sum_{x,y}|g(x)-g(y)|Q(x,y) \leq \sum_{x,y}|f(x)-f(y)|(|f(x)|+|f(y)|)Q(x,y).$$

To check the last inequality, we can suppose without loss of generality that $f(x) > 0$ and $f(x) > f(y)$. If $f(y) \geq 0$ we have an inequality, while if $f(y) < 0$ we have $f^2(x) + f^2(y) < (|f(x)| + |f(y|)^2$. Using the Cauchy-Schwarz inequality now the above is

$$\leq \left(\sum_{x,y}(f(x)-f(y))^2Q(x,y)\right)^{1/2}\cdot\left(\sum_{x,y}(|f(x)|+|f(y)|)^2Q(x,y)\right)^{1/2}$$
$$\leq (2\mathcal{E}(f,f))^{1/2}(4\pi(f^2))^{1/2}.$$

Rearranging gives $(2\mathcal{E}(f,f))^{1/2} \geq h(\pi(f^2))^{1/2}$. Squaring we have

$$\mathcal{E}(f,f) \geq \frac{h^2}{2}\pi(f^2) \geq \frac{h^2}{2}E_\pi(f - E_\pi f)^2,$$

which proves the desired result. □

Example 6.2.4 (Random walks on graphs) Let G be a finite connected graph, $d(x)$ be the degree of x, and define a transition kernel by $K(x,x) = 1/2$, $K(x,y) = 1/2d(x)$ if $x \sim y$ and $K(x,y) = 0$ otherwise. Our K can be written $(I+p)/2$ where p is another transition probability, so all of the eigenvalues of K are in $[0,1]$, and $\lambda_{max} = \lambda_1$. $\pi(x) = d(x)/D$, where $D = \sum_{y\in G} d(y)$, defines a reversible stationary distribution since $\pi(x)K(x,y) = 1/2D = \pi(y)K(y,x)$. Letting $e(S,S^c)$ is the number of edges between S and S^c, and $\mathrm{vol}(S)$ be the sum of the degrees in S, we have

$$h = \frac{1}{2} \min_{\pi(S) \le 1/2} \frac{e(S, S^c)}{\text{vol}(S)}.$$

When $d(x) \equiv d$, $h = \iota/2d$ where

$$\iota = \min_{|S| \le n/2} \frac{e(S, S^c)}{|S|}$$

is the **edge isoperimetric constant.**

The next two examples show that both inequalities in Theorem 6.2.3 are "sharp."

Example 6.2.5 (Random walk on the circle) To illustrate the use of Theorem 6.2.3 and to show that one cannot get rid of the power 2 from the lower bound, consider the random walk on the circle $\mathbb{Z}$ mod n in which we stay put with probability 1/2 and jump from x to $x \pm 1$ with probability 1/4 each. Taking $S = \{1, 2, \ldots, n/2\}$ we see that

$$\iota = \frac{2}{n/2} = 4/n.$$

To bound the spectral gap, we let $f(x) = \sin(\pi x/n)$. Since $\sin(a + b) = \sin a \cos b + \sin b \cos a$ we have

$$(I - K)f(x) = f(x)(1 - \cos(\pi/n))/2,$$

and $1 - \lambda_1 \le (1 - \cos(\pi/n))/2 \sim \pi^2/4n^2$ as $n \to \infty$. Using Theorem 6.1.1 gives an upper bound on the convergence time of order $O(n^2 \log n)$. However using the local central limit theorem for a random walk on $\mathbb{Z}$ it is easy to see that $\Delta(t) \le \epsilon$ at a time $K_\epsilon n^2$.

Example 6.2.6 (n-dimensional hypercube) Consider the lazy random walk on the hypercube $\{0, 1\}^n$ and let $S = \{x : x_1 = 0\}$. Since $\pi(S) = 1/2$.

$$\Phi(S) = 2 \sum_{x \in S, y \in S^c} 2^{-n} p(x, y) = 2^{-n+1} \cdot 2^{n-1} \cdot \frac{1}{2n} = \frac{1}{2n}.$$

To determine the rate of convergence to equilibrium note that the lazy chain may be formulated as: pick a coordinate at random and then flip a fair coin to determine its state. From this it is clear that when we have touched all the coordinates the system is in equilibrium. To compute the spectral gap note that the coordinates are independent so $\gamma = 1/n$. It is not hard to verify this by hand but the details can be found in Sect. 12.4 of Levin and Peres (2017) which concerns product chains.

6.2.2 Mixing Times and the Conductance Profile

In some examples, given an initial state i, it is possible to define stopping times T (which are called **stationary times**) so that X_T has the stationary distribution. Define $H(i,\pi)$ the minimum value of ET for all such stopping times and let $\mathcal{H} = \max_i H(i,\pi)$. For more on stationary times (and strong stationary times) see Chapter 6 in Levin and Peres (2017).

To be completely accurate we should redefine the mixing time by

$$T_{mix} = \max_i \min\{t : d_{TV}(K^t(i,\cdot),\pi) < 1/e\}.$$

Earlier, see (6.1.3), we used $1/4$ instead of $1/e$. The choice of threshold is not important as long as it is small enough, but it does affect the C_i in the next result. Aldous (1989) has shown, see also Aldous et al. (1997), that

$$C_1\mathcal{H} \le T_{mix} \le C_2\mathcal{H}.$$

Define the **conductance profile** by

$$\Phi(x) = \min_{S:0<\pi(s)\le x} \frac{Q(S,S^c)}{\pi(S)\pi(S^c)}.$$

Lovász and Kannan (1999) have shown that

$$\mathcal{H} \le 32 \int_{\pi_{min}}^{1/2} \frac{dx}{x\Phi(x)^2}.$$

Morris and Peres (2005) used their notion of evolving sets to sharpen this result to

$$\text{If } \; n \ge \int_{\pi(i)\wedge\pi(j)}^{4/\epsilon} \frac{4\,dx}{x\Phi(x)^2} \quad \text{then} \quad \left|\frac{K^n(i,j)}{\pi(i)} - 1\right| \le \epsilon.$$

These results are useful for improving rate of convergence results in some examples. However in some of our favorite examples the worst conductance occurs for small sets. In these cases, we will instead use a recent result of Fountoulakis and Reed (2007).

Theorem 6.2.7 *If $\Phi_c(x)$ is the minimum $Q(S,S^c)/\pi(S)\pi(S^c)$ over all connected sets S with $x/2 \le \pi(S) \le x$ then*

$$T_{mix} \le 32 \int_{\pi_{min}}^{1/2} \frac{dx}{x\Phi_c(x)^2}.$$

6.3 Fixed Degree Distribution, Minimum Degree 3

Gkantsidis et al. (2003) have proved the following:

Theorem 6.3.1 *Consider a random graph with a fixed degree distribution in which the minimum degree is $r \ge 3$. There is a constant $\alpha_0 > 0$ so that with probability tending to 1 as $n \to \infty$*

$$\min_{\pi(S)\le 1/2} \frac{e(S, S^c)}{vol(S)} \ge \alpha_0,$$

and hence by Theorem 6.1.1 it follows that the mixing time is $\le C \log n$.

The diameters of these graphs are $\Theta(\log n)$ so their random walks cannot mix at a faster rate. The condition $r \ge 3$ is necessary since if there is a positive density of vertices of degree 2 then there will be paths of length $O(\log n)$ in which each vertex has degree 2 and if we start in the middle of the path then the mixing time will be $\ge O(\log^2 n)$. We will consider that case in the next section.

Proof. We say that a set of vertices S is **bad** if $e(S, S^c)/\text{vol}(S) \le \alpha$. Our goal is to show that the probability there is a bad set $\to 0$. There are at most $\binom{D/r}{k/r}$ sets that have volume k. Let $f(m)$ be the number of ways of dividing m objects into pairs.

$$f(m) = \frac{m!}{(m/2)!2^{m/2}}. \tag{6.3.1}$$

Let $P(k, \ell)$ be the probability that there is a set S with $\text{vol}(S) = k$ has $e(S, S^c) = \ell$.

$$P(k, \ell) \le \binom{k}{\ell}\binom{D-k}{\ell}\ell! f(k-\ell) f(D-k-\ell)\frac{1}{f(D)}. \tag{6.3.2}$$

To see this, recall that in the configuration model we pair the D half-edges at random, which can be done in $f(D)$ ways. We pick ℓ of k half-edges in S and ℓ of those $D-k$ in S^c. The ℓ which will make up $e(S, S^c)$ can be paired in $\ell!$ ways. Then the remaining $k-\ell$ half-edges in S can be paired in $f(k-\ell)$ ways and the $D-k-\ell$ in S^c in $f(D-k-\ell)$ ways.

To make it easier to compare with the argument in Gkantsidis et al. (2003) we change values $\ell = \alpha k$. Taking into account the number of choices of S, the probability of a bad set with volume k and $e(S, S^c) = \alpha k$ is

$$\binom{D/r}{k/r}\binom{k}{\alpha k}\binom{D-k}{\alpha k}\frac{(\alpha k)! f(k-\alpha k) f(D-k-\alpha k)}{f(D)}. \tag{6.3.3}$$

Their formula (10) is this with $s! = (\alpha k)!$ replaced by the larger $f(2\alpha k)$. They also have a factor αk to account for $1 \le s \le \alpha k$. □

To bound the binomial coefficients, the following lemma is useful.

Lemma 6.3.2

$$\binom{n}{m} \le \frac{n^m}{m!} \le \frac{n^m}{m^m e^{-m}}.$$

Proof. The first inequality follows from $n(n-1)\cdots(n-m+1) \le n^m$. For the second we note that the series expansion of e^m has only positive terms so $e^m > m^m/m!$. □

From Lemma 6.3.2, we see that the three binomial coefficients in (6.3.3) are

$$\le \left(\frac{De}{k}\right)^{k/r} \left(\frac{e}{\alpha}\right)^{2\alpha k} \left(\frac{D-k}{k}\right)^{\alpha k}. \tag{6.3.4}$$

Here, to prepare for a later step, we have transferred part of the bound for the third term into the second.

To bound the f's in (6.3.3) we use Stirling's formula to conclude

$$f(m) = \frac{m!}{(m/2)!2^{m/2}} \sim C \frac{m^{m+1/2}e^{-m}}{(m/2)^{m/2+1/2}e^{-m/2}2^{m/2}} = C(m/e)^{m/2}.$$

From this we see that the fraction in (6.3.3) is

$$\le Ck^{1/2} \frac{(\alpha k/e)^{\alpha k}(k(1-\alpha)/e)^{k(1-\alpha)/2}((D-(1+\alpha)k)/e)^{(D-(1+\alpha)k)/2}}{(D/e)^{D/2}} \tag{6.3.5}$$

$$= Ck^{1/2}(\alpha k)^{\alpha k} D^{-\alpha k} \left(\frac{k(1-\alpha)}{D}\right)^{k(1-\alpha)/2} \left(1 - \frac{(1+\alpha)k}{D}\right)^{(D-(1+\alpha)k)/2}, \tag{6.3.6}$$

since the exponents in the numerator sum to $D/2$.

Combining (6.3.4) and (6.3.6) gives an upper bound

$$\le Ck^{1/2} \left(\frac{De}{k}\right)^{k/r} \left(\frac{e^2}{\alpha}\right)^{\alpha k} \left(\frac{D-k}{D}\right)^{\alpha k}$$
$$\cdot \left(\frac{k(1-\alpha)}{D}\right)^{k(1-\alpha)/2} \left(1 - \frac{(1+\alpha)k}{D}\right)^{(D-(1+\alpha)k)/2}.$$

Ignoring the $Ck^{1/2}$'s, the first term is the first term from (6.3.4), the second and third terms come from combining the second and third terms of (6.3.4) and with the first and second terms of (6.3.6), while the remainder of the formula comes from (6.3.6). Using $\alpha > 0$ and $D - k < D$ and rearranging we have

$$\leq Ck^{1/2}e^{k/r}\left(\frac{e^2}{\alpha}\right)^{\alpha k}\cdot\left(\frac{k}{D}\right)^{k(1-\alpha)/2-k/r}\left(1-\frac{(1+\alpha)k}{D}\right)^{(D-(1+\alpha)k)/2}.$$

Setting $\beta = e^2/\alpha$ and $\gamma = (1-\alpha)/2 - 1/r$ we have

$$\leq Ck^{1/2}e^{k/r}\beta^{\alpha k}\cdot\left(\frac{k}{D}\right)^{\gamma k}\left(1-\frac{(1+\alpha)k}{D}\right)^{(D-(1+\alpha)k)/2}. \tag{6.3.7}$$

Comparing with formula (17) in Gkantsidis et al. (2003), we see that apart from the differences that result from our use of $(\alpha k)!$ instead of $f(2\alpha k)$, they are missing the $e^{k/r}$ and we have retained an extra term to compensate for the error. Let

$$G(k) = e^{k/r}\beta^{\alpha k}\cdot\left(\frac{k}{D}\right)^{\gamma k}\left(1-\frac{(1+\alpha)k}{D}\right)^{(D-(1+\alpha)k)/2}.$$

Since $Ck^{1/2} \leq Cn^{1/2}$, we can show $h \geq \alpha_0$ by showing that for $0 \leq \alpha \leq \alpha_0$

$$\sup_{1\leq k\leq D/2} G(k) = o(n^{-5/2}),$$

because then we can sum our estimate over $k \leq D/2$ and $s = \alpha k$ with $\alpha \leq \alpha_0$ and end up with a result that is $o(1)$.

$\beta^{\alpha k} = \exp(\eta k)$ where $\eta = \alpha\log(e^2/\alpha) \to 0$ as $\alpha \to 0$. Ignoring this term, and setting $k = D/2$, $\alpha = 0$

$$G(D/2) = e^{D/2r}(1/2)^{[1/2-1/r]D/2+D/4} = \left(e^{1/3}(1/2)^{2/3}\right)^{D/2},$$

when $r = 3$, the worst case. Since $4 > e$, the quantity in parentheses is < 1 when $\alpha = 0$ and hence also when $0 \leq \alpha \leq \alpha_0$, if α_0 is small.

To extend this result to other values of k, let

$$H(k) = \log G(k) = \frac{k}{r} + k\alpha\log\beta + k\gamma\log(k/D) + \frac{D-(1+\alpha)k}{2}\log\left(1-\frac{(1+\alpha)k}{D}\right).$$

Since $G(k) = \exp(H(k))$, differentiating gives $G'(k) = G(k)H'(k)$ where

$$H'(k) = \frac{1}{r} + \alpha\log\beta + \gamma\log(k/D) + \gamma - \frac{(1+\alpha)}{2}\log\left(\frac{D-(1+\alpha)k}{D}\right)$$
$$+\frac{D-(1+\alpha)k}{2}\cdot\frac{D}{D-(1+\alpha)k}\cdot\left(\frac{-(1+\alpha)}{D}\right).$$

Differentiating again $G''(k) = G(k)(H'(k)^2 + H''(k))$ where

$$H''(k) = \frac{\gamma}{k} - \frac{(1+\alpha)}{2}\cdot\frac{D}{D-(1+\alpha)k}\cdot\left(\frac{-(1+\alpha)}{D}\right) > 0.$$

From the last calculation we see that $G(k)$ is convex. We have control of the value for $k = D/2$. It remains then to inspect the values for small k. Dropping the last factor which is < 1

$$G(k) \leq e^{k/r} \beta^{k\alpha} \left(\frac{1}{\alpha D} \right)^{\gamma k}.$$

When $0 \leq \alpha \leq \alpha_0 \leq 1/24$, $\gamma \geq 7/48$ and hence $G(24) \leq Cn^{-7/2}$. Since $e(S, S^c) \geq 1$ there is nothing to prove for $k \leq 1/\alpha_0 = 24$ and the proof is complete. □

6.4 Effect of Degree 2 Vertices

In the previous section we saw that if the minimum degree on a connected graph with n vertices is 3 then the random walk mixes in time $O(\log n)$. This conclusion is false for configuration model graphs if vertices of degree 2 have positive probability. It is easy to see that such graphs will have paths of length $\geq \delta \log n$ in which all the vertices have degree 2. If we start in the middle of such a path, it will with positive probability take time $\geq \delta^2 \log^2 n$ to escape from the path.

I knew this when I was writing the first edition of this book, which was published in 2007. I wanted to prove an upper bound of the same order, but I could not cope with the complexities of a general graph, so I created a simple example. Start with a random 3-regular graph H with $(1 - a)n$ vertices, and hence $3(1 - a)n/2$ edges. We produce a new graph G_{23} by replacing each edge by a path with a geometric number of edges with success probability r, i.e., with probability $(1 - r)^{j-1}r$ we have j edges. The number of vertices of degree 2 in one of these paths has mean $(1/r) - 1$ so if we pick r so that

$$\frac{3(1 - a)}{2} \cdot ((1/r) - 1) = a,$$

then we asymptotically have n vertices and $p_2 = a$.

Our main result is that

Theorem 6.4.1 *The mixing time of the lazy random walk on G_{23} is $\Theta(\log^2 n)$.*

To prove an upper bound, we will use mixing time result in Theorem 6.2.7. I appreciate the fact that on a trip to Montreal (circa 2006) Bruce Reed (a) spent a long time explaining the proof to me, and (b) gave me a tutorial on the wonders of Belgian beer.

Proof To begin we need a simple combinatorial result.

Lemma 6.4.2 *The number of connected subsets of H of size k containing a fixed vertex v_0 is $\leq 3^{3k}$.*

Proof. Given a connected set V of vertices of H, define the set $W = \{(x, y) : x, y \in V, x \neq y\}$. Note that if $(x, y) \in W$ then $(y, x) \in W$ and think of these as two oriented edges between x and y. We will show that there is a Hamiltonian path starting from v_0 that traverses each oriented edge at most once. The number of edges in W is at most $3k$. At each stage we have at most 3 choices so the number of such paths is $\leq 3^{3k}$ and this proves the desired result.

To construct the path start at v_0 and pick an outgoing edge. When we are at a vertex $v \neq v_0$ we have used one more incoming edge than outgoing edge so we have at least one way out. This procedure may terminate by coming back to v_0 at a time when there are no more outgoing edges. If so, and we have not exhausted the graph, then there is some vertex v_1 on the current path with an outgoing edge. Repeat the construction starting from v_1 using edges not in the current path. We will eventually come back to v_1. We can combine the two paths by using the old path from v_0 to the first visit to v_1, using the new path to go from v_1 to v_1, and then the old path to return from v_1 to v_0. Repeating this construction we will eventually exhaust all of the edges. □

Let B be a connected subset of G, and let $A = B \cap H$. It is easy to see that A is a connected subset of H. By the isoperimetric inequality for random regular graphs, there is an $\alpha > 0$ so that $|\partial A| \geq \alpha |A|$, where ∂A is the set of edges (x, y) with $x \in A$ and $y \notin A$. From the construction of the graph it is easy to see that $|\partial B| = |\partial A|$.

It remains to see how big $|B|/|A|$ can be. When $|A| = 1$ we can have $|B| = O(\log n)$. The key to the proof is to show that the ratio cannot be big when $|A|$ is. Let X_i be i.i.d. with $P(X_i = j) = (1-r)^j r$ and let $S_m = X_1 + \cdots + X_m$.

Lemma 6.4.3 *There are constants β and γ so that*

$$P(S_m \geq \beta \log n + \gamma m) \leq n^{-2}(2/81)^m.$$

Proof. The moment generating function

$$\psi(\theta) = E e^{\theta X_i} = \sum_{j=0}^{\infty} (e^{\theta}(1-r))^j r = \frac{r}{1 - e^{\theta}(1-r)},$$

when $e^{\theta}(1-r) < 1$. If we pick $\theta > 0$ so that $e^{\theta}(1-r) = 1 - r/2$ then $\psi(\theta) = 2$. Markov's inequality implies

$$P(S_m \geq \beta \log n + \gamma m) \leq \psi(\theta)^m \exp(-\theta[\beta \log n + \gamma m]).$$

Letting $\beta = 2/\theta$ and $\gamma = 81/\theta$ the desired result follows. □

If $|A| = k$ then the number of edges adjacent to some point in A is $\geq k + 2$, the value for a tree, and $\leq 3k$. Since the number of connected sets of size k is $\leq n 27^k$ it follows that with probability $1 - O(n^{-1})$ we have $|B| \leq \beta \log n + 3\gamma |A|$ for connected sets B. From this it follows that

$$|\partial B| = |\partial A| \geq \alpha |A| \geq \frac{\alpha}{\gamma}(|B| - \beta \log n),$$

if $|B| \geq 2\beta \log n$ then $|\partial B|/|B| \geq c$, while for $|B| \leq 2\beta \log n$, $|\partial B|/|B| \geq 2/|B|$.

To evaluate $\int_{1/3n}^{1} dx/(x\Phi(x)^2)$ up to a constant factor we note that

$$\int_{2\beta \log n/n}^{1/2} \frac{dx}{x} = O(\log n),$$

while changing variables $y = nx$, $dy = n\,dx$ shows

$$\int_{1/3n}^{2\beta \log n/n} \frac{dx}{x(2/xn)^2} = \int_{1/3}^{2\beta \log n} (y/2)\,dy = O(\log^2 n),$$

and completes the proof. □

6.5 Erdős-Rényi Graphs

In this section we will explore how the time for a random walk to converge to equilibrium depends on the mean degree of the Erdős-Rényi graph.

6.5.1 Connected Graphs

We begin with random walk on Erdős-Rényi$(n, (c \log n)/n)$ with $c > 1$, which Theorem 1.4.1 has shown is connected with high probability for large n. The word "connected" in the title of this subsection is a bit of a red herring because we make this assumption on the value of λ not to avoid the trivial obstruction to convergence that comes from lack of irreducibility, but to have the reduction in randomness that occurs when all vertices have large degree.

Let $d(x)$ be the degree of x, write $x \sim y$ if x and y are neighbors. To avoid problems associated with periodicity, we run the lazy random walk with

$$K(x,x) = 1/2, \qquad K(x,y) = 1/2d(x) \quad \text{if } x \sim y,$$

and $K(x, y) = 0$ otherwise. Our analysis of the mixing time follows Cooper and Frieze (2003), who were primarily interested in the cover time C_G, i.e., the time to visit all the vertices.

Theorem 6.5.1 *Suppose that* $np = c_n \log n$ *where* $c_n = O(1)$ *and* $(c_n - 1)\log n \to \infty$. *Then whp*

$$\mathcal{C}_G \sim c \log\left(\frac{c}{c-1}\right) n \log n.$$

To prove this result they needed to establish a number of properties of the Erdős-Rényi graph $G_{n,p}$ including

Theorem 6.5.2 *Consider $ER(n, (c \log n)/n)$ with $c > 1$. The lazy random walk mixes in time $O(\log n)$.*

Proof. We begin by estimating the maximum and minimum degrees of vertices. □

Lemma 6.5.3 *There is a constant $\delta > 0$ so that if n is large then*

$$\delta c \log n \le d(x) \le 4c \log n \qquad \textit{for all } x.$$

Proof. By the large deviations result in Lemma A.1.8, if $X = \text{Binomial}(n, p)$ then

$$P(X \ge np(1+y)) \le \exp(-npy^2/2(1+y)).$$

Taking $p = (c \log n)/n$, and $y = 3$

$$P(X \ge 4c \log n) \le \exp(-9(c \log n)/8) = n^{-9c/8}.$$

Since we have assumed $c > 1$ then with high probability, the maximum degree in the graph is $\le 4c \log n$.

To get a lower bound, we need the more precise result in Lemma A.1.6. The function

$$H(a) = a \log(a/p) + (1-a) \log((1-a)/(1-p))$$

defined there has $H(0) = -\log(1-p)$, which is sensible since $P(X = 0) = (1-p)^n$. When $p = c \log n$

$$(1 - (c \log n)/n)^n \le n^{-c}.$$

Taking $a = (\delta c \log n)/n$, we have

$$H(a) = \frac{\delta c \log n}{n} \log \delta + \left(1 - \frac{\delta c \log n}{n}\right) \log\left(\frac{1 - \delta c \log n/n}{1 - c \log n/n}\right).$$

The logarithm in the second term is

$$\log\left(1 + \frac{(1-\delta) c \log n/n}{1 - c \log n/n}\right) \sim (1-\delta) c \log n/n,$$

as $n \to \infty$. As $\delta \to 0$, $\delta \log \delta \to 0$, so if δ is small enough then $H(\delta) \sim (b \log n)/n$ with $b > 1$ as $n \to \infty$ and we conclude that with probability that tends to one, the minimum degree in the graph is $\ge \delta c \log n$. □

To prove the theorem we will estimate the conductance h introduced in Sect. 6.2. By considering the number of edges we see that

$$\text{vol}(G) = 2 \cdot \text{binomial}\left(\binom{n}{2}, \frac{c \log n}{n}\right),$$

which has mean $\sim cn \log n$ and variance $\sim cn \log n$, so $\text{vol}(G) \sim cn \log n$. The maximum degree $\leq 4c \log n$ with high probability for large n, so if $|S^c| \leq n/10$ then

$$\text{vol}(S^c) \leq 4c \log n \cdot n/10,$$

and hence for large n no set with $|S| \geq 9n/10$ will have $\pi(S) \leq 1/2$.

Case 1. Consider $B = \{S : n/(c \log n) \leq |S| \leq 9n/10\}$ and let $s = |S|$. Using Lemma 6.3.2 we have

$$\binom{n}{s} \leq (n/s)^s e^s = \exp(s[\log(n/s) + 1]). \tag{6.5.1}$$

Lemma 1.9.5 says that if $X = \text{binomial}(n, p)$ then

$$P(X \leq np(1-y)) \leq \exp(-npy^2/2).$$

The number of edges from S to S^c is binomial$(s(n-s), c \log n)/n)$. Using the large deviations result with $n = s(n-s)$, $p = p$, and $y = 1/2$ gives that the binomial probability

$$\leq \exp(-s(n-s)(c \log n)/8n). \tag{6.5.2}$$

Combining the last two estimates and using $n - s \geq n/10$, $s \geq n/(c \log n)$ we see that there is an $S \in B$ with $e(S, S^c) \leq s(n-s)p/2$ so that the binomial probability

$$\leq n \cdot \exp\left(-\frac{n}{c \log n}\left[\frac{c \log n}{80} - \log(c \log n) - 1\right]\right),$$

which goes to 0 exponentially fast as $n \to \infty$. To finish up now we note that

$$s(n-s)p/2 = s(n-s)(c \log n)/2n \geq sc(\log n)/20,$$

while $\text{vol}(S) \leq 4sc \log n$, so for sets in B we have $e(S, S^c)/\text{vol}(S) \geq 1/80$.

Case 2. Finally we have to deal with the small sets $A = \{S : 1 \leq |S| \leq n/(c \log n)\}$. In this case we upper bound $e(S, S)$ in order to conclude $e(S, S^c)$ is large. $E|e(S, S)| \leq (s^2/2)p \leq s/2$ so

$$P(\exists S \in A : e(S, S) \geq s \log \log n) \leq C\binom{n}{s}\binom{s^2/2}{s \log \log n} p^{s \log \log n}.$$

The right-hand side is the probability $e(S, S) = s \log\log n$, ignoring the fact that this may not be an integer. However in this part of the tail, the probabilities decay exponentially fast. Bounding the binomial coefficients using Lemma 6.3.2, and filling in the value of p

$$\leq C \left(\frac{ne}{s}\right)^s \frac{(s^2/2)^{s \log\log n} p^{s \log\log n}}{(s \log\log n)^{s \log\log n} e^{-s \log\log n}}$$
$$= C \left(\frac{ne}{s}\right)^s \left(\frac{s}{2 \log\log n} \cdot \frac{ec \log n}{n}\right)^{s \log\log n}.$$

Reorganizing we have

$$= C \exp\left(s[\log(ne) - \log s] + s \log\log n \left[\log s + \log\left(\frac{ec \log n}{2n \log\log n}\right)\right]\right).$$

Differentiating the exponent with respect to s we have

$$\log(ne) - \log s - 1 + \log\log n[\log s + \log(ec \log n) - \log(2n \log\log n)] + \log\log n.$$

When $1 \leq s \leq n/(c \log n)$ this is negative, so the worst case is $s = 1$. In this case the quantity of interest is

$$= \exp\left(\log(ne) + \log\log n[\log(ec \log n) - \log(2n \log\log n)]\right),$$

which tends to 0 as $n \to \infty$.

To bound $e(S, S^c)$ we note that $e(S, S^c) = d(S) - e(S, S) \geq s\delta \log n - s \log\log n \geq s(\delta/2) \log n$ when n is large. $\text{vol}(S) \leq 4s \log n$, so for sets in A we have $e(S, S^c)/\text{vol}(S) \geq \delta/8$. □

6.5.2 *Mixing Time Changes from Connected to Sparse*

Fountoulakis and Reed (2008) proved results about the mixing times of random walks on Erdős-Rényi graphs that cover the entire range from $(1 + \epsilon)/n$ to $(2 \log n)/n$. In the first result $p(n)$ is the probability an edge is present and $d(n) = np(n)$ is the average degree.

Theorem 6.5.4 *For every $p = p(n)$ with $d(n) - \log n \to \infty$ we have*

$$\left| T_{mix}(G_{n,p}) - \frac{\log n}{\log d} \right| \leq 3.$$

Intuitively the mixing time is the diameter of the graph. On a high degree Erdős-Rényi graph almost all the steps take us away from the starting point.

As the degree decreases the mixing time is a competition between the diameter and the amount of time to get stuck in long paths. Letting $H_{n,p}$ be the giant component of $G_{n,p}$ and writing a.a.s. (asymptotically almost surely) instead of w.h.p. we have the following result.

Lemma 6.5.5 *For $p < (\log n)/5n$, $H_{n,p}$ a.a.s. contains paths of length more than $(\log n)/d$ all of whose interior vertices have degree two.*

To state the next two results we need some notation. Let

$$\bar{P}^t_{x_0} = (1/t)\sum_{m=1}^{t} P^m(x_0,\cdot) \quad \text{and} \quad T'_{mix}(G) = \sup_{x_0} \min\{t : d_{TV}(\bar{P}^t_{x_0},\pi) < 1/e\}.$$

Theorem 6.5.6 *For every $p(n)$ with $\sqrt{(\log n)\log\log n} \le np(n) \le 2\log n$ we have*

$$\left| T'_{mix}(G) - \frac{\log n}{\log d} \right| = O((\log n/d)^2).$$

Theorem 6.5.7 *For every $p(n)$ with $1+\epsilon \le np(n) \le \sqrt{(\log n)\log\log n}$ we have*

$$\left| T'_{mix}(G) \right| = O((\log n/d)^2).$$

To explain why the transition between the last two results happens at $p = \sqrt{(\log n)\log\log n}$ we note that for this value of p

$$(\log n/d)^2 = \frac{\log n}{\log\log n} \sim \frac{\log n}{\log d}.$$

To really understand what is going on here you should go to the paper, which has many more details.

6.5.3 *Sparse Erdős-Rényi Graphs*

In Sect. 6.4 we saw that if we have a random graph generated by the configuration model in which vertices of degree 2 have positive probability then the mixing time of the random walk on the giant component is at least $C\log^2 n$ since there are paths of length $\log n$ in which all vertices have degree 2. If the walk starts in the middle of such a path then it takes time $O(\log^2 n)$ to escape from the set.

Benjamini, Kozma, and Wormald (2014) studied the Erdős-Rényi graph $G(n,m)$ with a fixed number of edges m and $m/n \to c > 1$ and showed that the mixing time was $\Theta(\log^2 n)$. They did this by showing that (i) these graphs are *decorated expanders*: there is an expander subgraph B which they call the *strong core*, which leaves only small components when it is deleted, and (ii) proving a mixing time result for decorated expanders

Berestycki, Lubetzky, Peres, and Sly (2018) have shown that if we start at a randomly chosen vertex in the giant component then the mixing time becomes $O(\log n)$ and there is cutoff with a window of $(\log n)^{1/2+o(1)}$. (The notion of cutoff will be explained in the next section.)

Theorem 6.5.8 *Let $\mathcal{C}_1$ be the giant component of an Erdős-Rényi$(n, \lambda/n)$ random graph with $\lambda > 1$ fixed. For any $0 < \epsilon < 1$ whp the mixing time of the random walk starting from a randomly chosen vertex $x \in \mathcal{C}_1$ satisfies*

$$|t^{v}_{mix}(\epsilon) - (\nu\mathbf{d})^{-1} \log n| \leq (\log n)^{1/2+o(1)}.$$

Here ν is the speed of a random walk on a Poisson(λ) Galton-Watson tree and $\mathbf{d}$ is the dimension of harmonic measure on the ends of the tree. For an explanation of this you will need to consult Lyons, Pemantle, and Peres (1995). In addition to the result for Erdős-Rényi graphs they have a result for a graph G with general degree distribution p_k. Condition (6.5.3) is strong but weaker than having finite exponential moments.

Theorem 6.5.9 *Suppose that for fixed $\delta > 0$ the random variable Z given by $P(Z = k - 1) \propto kp_k$ satisfies*

$$P(Z > \Delta_n) = o(1/n), \qquad \Delta_n \equiv \exp((\log n)^{1/2-\delta}), \tag{6.5.3}$$

and let $t_\star = (\nu\mathbf{d})^{-1} \log n$ where ν and $\mathbf{d}$ are the speed of the random walk and the dimension drop for the Galton-Watson tree with offspring distribution Z.

(i) Set $w_n = (\log n)^{1/2}(\log\log n)^3$. If $p_2 < 1 - \delta$ and $1 + \delta < EZ < K$ for all n then whp on the event that z is in the largest component of G

$$d^z_{TV}(t_\star - w_n) > 1 - \epsilon) \quad and \quad d^z_{TV}(t_\star + w_n) < \epsilon.$$

(ii) Set $w_n = (\log n)^{1/2}$. If $Z > 2$ and $EZ < K$ for all n then for any $\epsilon > 0$ there is a $\gamma < \infty$ so that with probability at least $1 - \epsilon - o(1)$

$$d^z_{TV}(t_\star - w_n) > 1 - \epsilon) \quad and \quad d^z_{TV}(t_\star + w_n) < \epsilon.$$

The structure theorem of Ding et al. (2014) gives an intuitive explanation. From page 3 of BLPS: "A contiguous model for the giant component is given by (i) choosing a kernel uniformly over graphs on i.i.d. Poisson truncated to be at least 3, (ii) subdividing every edge into i.i.d. geometric variables, and (iii) hanging i.i.d. Poisson Galton-Watson trees on every vertex. Observe that steps (ii) and (iii) introduce i.i.d. delays with an exponential tail for the walk; thus starting from a uniform random vertex would essentially eliminate all but the typical $O(1)$ delays and it should mix on the kernel (whp an expander) in time $O(\log n)$"

6.6 Cutoff Phenomenon

A sequence of finite state Markov chains is said to exhibit **cutoff** if its distance from the stationary distribution π drops from 1 to 0 over a window that is $o(t_n)$ where t_n is the time to converge to equilibrium. To make this notion precise we need some more definitions. Let

$$d_n(t) = \max_{x \in \mathcal{S}} \| P_x(X_t \in \cdot) - \pi \|_{TV}$$

be the distance from equilibrium and let

$$t^n_{mix}(\epsilon) = \min\{t : d_n(t) < \epsilon\}.$$

We say there is **cutoff** if for any $\epsilon > 0$

$$\lim_{n \to \infty} t^n_{mix}(\epsilon)/t^n_{mix}(1 - \epsilon) \to 1.$$

We say there is **cutoff at** t_n **with window size** w_n if

$$\lim_{\lambda \to \infty} \liminf_{n \to \infty} d_n(t_n - \lambda w_n) = 1,$$

$$\lim_{\lambda \to \infty} \limsup_{n \to \infty} d_n(t_n + \lambda w_n) = 0.$$

A nice introduction to cutoff is given in Diaconis (1996), an "Inaugural Article" written on the occasion of his election to the National Academy of Science in April 1995. What follows is our rendition of his work, which lacks much of the information presented there and some of the style of the original. We begin with an example of a Markov chain that does not have cutoff.

Example 6.6.1 (Simple random walk on the integers modulo n) To avoid problems with periodicity the transition probability has

$$p(i, i + 1) = p(i, i) = p(i, i - 1) = 1/3,$$

where the arithmetic is done modulo n so that $p(0, n - 1) = p(n - 1, 0) = 1/3$. The stationary distribution in uniform $\pi(i) = 1/n$. It is easy to show, see e.g., Diaconis (1986).

Theorem 6.6.2 $\|P^k_0 - \pi\| \sim Ck/n^2$.

Here and in the next three examples we have changed notation so that $P^k_x = P_x(X_k \in \cdot)$. It is clear from the central limit theorem that the total variation distance is ≈ 1 when $k = o(n^2)$, and that we need $k \gg n^2$ for the normal distribution of the position of the walk on $\mathbb{Z}$ to flatten out enough so that when reduced modulo n it is almost uniform.

It is surprising (to me at least) that cutoff occurs in

Example 6.6.3 (Ehrenfest chain) There are two urns and d balls. In the simplest version on each step one ball is chosen at random and transferred to the other urn. However to avoid periodicity, we do nothing with probability $1/(d+1)$.

$$p(i, i-1) = \frac{i}{d+1}, \qquad p(i, i) = \frac{1}{d+1}, \qquad p(i, i+1) = \frac{d-i}{d+1}.$$

Theorem 6.6.4 *For the Ehrenfest chain started at 0, if $k = \frac{1}{4}(d+1)(\log d + \theta)$ then*

$$\|P_0^k - \pi\| \le \frac{1}{\sqrt{2}}(\exp(e^{-\theta}) - 1)^{1/2},$$

while if $k = (1/4)d(\log d - \theta)$ then the total variation distance tends to 1 as $d, \theta \to \infty$.

Example 6.6.5 (Random transpositions) This example, considered by Diaconis and Shahshahani (1981), was perhaps the first example where a sharp cutoff was demonstrated. Picture n cards at random on a table. At each time your left and right hands choose cards independently (so left=right with probability $1/n$) and then the cards' positions are switched. The stationary distribution is uniform over all the $n!$ arrangements

Theorem 6.6.6 *Let $k = (1/2)n(\log n + \theta)$ with $\theta > 0$, For any starting position x,*

$$\|P_x^k - \pi\| \le Ae^{-\theta/2}.$$

To explain the answer, suppose we start with the identity permutation. The expected number of fixed points of a randomly chosen permutation is 1, and the distribution is roughly Poisson. By results for the coupon collectors problem, see Example 6.6.10 for more detail, it takes time $\sim (n/2)\log(n)$ until we have moved every card at least once. If there are a lot of cards that have not moved the chain is not in equilibrium because there are too many fixed points.

Example 6.6.7 (Riffle shuffles) are an approximation of how humans, that are not magicians by training, shuffle cards. The deck is cut into two pieces according to a symmetric binomial distribution. Then the two halves are riffled together according to the following rule. If the pile in the left hand has A cards and the one in the right hand has B cards then the next card drops from the left pile with probability $A/(A+B)$ and from the right pile with probability $B/(A+B)$. Let $p(x, y)$ be the transition probability of this shuffling method, which is called the **Gilbert-Shannon-Reeds riffle shuffle**. Bayer and Diaconis (1992) proved

Theorem 6.6.8 *Let $k = (3/2)\log_2 n + \theta$. Then*

$$\|P_x^k - \pi\| = 1 - 2\Phi(-2^{-\theta}/4\sqrt{3}) + O(1/\sqrt{n}),$$

where Φ is the standard normal distribution function.

There are by now many examples. We will add two more from Levin and Peres (2017) that are closely related to the first four and then leave it to the reader to explore the subject further. For the next two example references such as the one to Proposition 7.14 are to their book.

Example 6.6.9 (Random walk on the hypercube) The state space is $\{0, 1\}^n$. The dynamics are simple. We pick a coordinate at random and flip it. To avoid the problems of periodicity we consider the lazy walk that does nothing with probability 1/2.

This example is used throughout the book by Levin and Peres (2017). Their Proposition 7.14 shows that

$$t_{mix}(1-\epsilon) \geq \frac{1}{2}n\log n - c_\ell(\epsilon)n.$$

In their Example 12.19 using the known eigenvalues and eigenfunctions for the chain, it was proved that

$$t_{mix}(\epsilon) \leq \frac{1}{2}n\log n - c_u(\epsilon)n.$$

Thus the lazy random walk on the hypercube has cut off at $(1/2)n\log n$ with a window of size $O(n)$.

Example 6.6.10 (Top to random shuffle) Suppose to simplify the discussion that initially the Ace of spades (A♠) is at the bottom of the deck. It will gradually rise to the top as cards are inserted below it. It is easy to show by induction that when there are k cards below the A♠ they are in random order. Thus if we let τ_n be the time at which the A♠ is inserted into a random position in the deck, the system is in equilibrium.

In the terminology of Chapter 6 of Levin and Peres (2017) τ_n is a **stationary time**. It has the additional property that

$$P_x(\tau_n = t, X_\tau = y) = P_x(\tau_n = t)\pi(y),$$

so it is a **strong stationary time**, i.e., the chain is stationary even if we condition on when τ occurs. Proposition 6.11 says that if τ is a strong stationary time then

$$d_n(t) \equiv \|P^t(x,\cdot) - \pi\|_{TV} \leq P(\tau_n > t).$$

In the case under consideration τ_n is a coupon collectors random variable. It is a sum of independent geometric(k/n) random variables with $k = 1, \ldots, n$. It is well known (see e.g., Example 2.2.7 in PTE5) that

$$E\tau_n \sim n\log n \quad \text{and} \quad \text{var}(\tau_n) \sim N^2 \sum_{m=1}^{\infty} m^{-2}.$$

Proposition 2.4 in Levin and Peres (2017) says that

$$P(\tau_n > n \log n + cn) \le e^{-c}. \tag{6.6.1}$$

The proof is so easy that we go ahead and give it.

Proof. Let A_i be the probability that card i has not been drawn by time $n \log n + cn$.

$$\begin{aligned} P(\tau > n \log n + cn) &= P\left(\cup_{i=i}^{n} A_i\right) \le \sum_{i=1}^{n} P(A_i) \\ &= n \left(1 - \frac{1}{n}\right)^{n \log n + cn} \le n \exp(-\log n - c) = e^{-c}, \end{aligned}$$

which is the desired result. □

In Sect. 7.4.1 they look at the time it takes for the jth card from the bottom to reach the top in order to prove (see their Proposition 7.15)

Theorem 6.6.11 *Given $\epsilon > 0$ there is an $\alpha(\epsilon)$ so that for large n*

$$d_n(n \log n - \alpha n) \ge 1 - \epsilon.$$

This shows that there is cut off at $n \log n$ with a window of size $O(n)$.

6.7 Random Regular Graphs

Let $\mathcal{G}(n, d)$ be the collection of random regular graphs in which all vertices have degree d. The case $d = 1$ is boring: each vertex has degree 1, so the graph is a collection of isolated edges. The case $d = 2$ is not much better: the graph is a union of circles of random sizes. In what follows, we restrict our attention to $d \ge 3$ even if we do not state that explicitly. Theorem 6.3.1 implies that when $d \ge 3$ the mixing time on a random d-regular graph is $\le C \log n$. Figure 6.2 compares the analytic result in (6.7.1) with simulation.

Lubetzky and Sly (2010) have identified the constant C in the mixing time and shown that there is cutoff with a window of size $O(\sqrt{\log_{d-1} n})$.

Theorem 6.7.1 *(Theorem 1.) Let $G \sim \mathcal{G}(n, d)$ be a random regular graph in which all vertices have degree $d \ge 3$. Then the simple random walk exhibits cutoff at $t_n = (d/(d-2)) \log_{d-1} n$ with a window of size $w_n = O(\sqrt{\log n})$. Furthermore*

$$t_{mix}(s) = \frac{d}{d-2} \log_{d-1} n - \Phi^{-1}(s)(\Lambda + o(1))\sqrt{\log_{d-1} n}, \tag{6.7.1}$$

where $\Lambda = 2\sqrt{d(d-1)}/(d-2)^{3/2}$ and Φ is the cumulative distribution function of the normal.

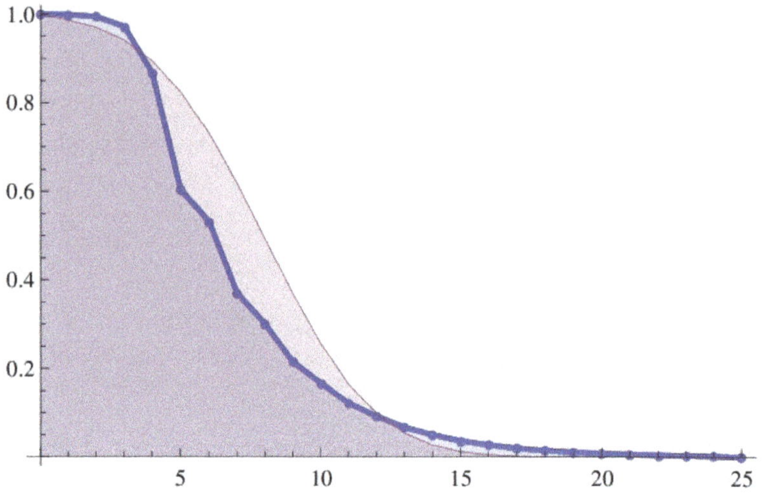

Fig. 6.2 Distance from stationarity for a random 6-regular graph with 5000 vertices compared with theoretical result. Picture from Lubetzky and Sly (2010)

It is easy to guess that there is cutoff at time t_n with a window of width w_n. Seen from a fixed vertex that we will call 0, the graph G looks locally like a tree in which the root 0 has degree d while the other vertices have one edge pointing back toward 0 and $d-1$ pointing away. If we assume the approximate picture is exactly correct, then the diameter of the graph is $\log_{d-1} n$ and the distance of the random walk from the root, X_m, when it is positive, changes by $+1$ with probability $(d-1)/d$ and by -1 with probability $1/d$.

The mean of one step of the random walk (ignoring the times it is at 0) is $(d-2)/d$ so by the law of large numbers the time to distance $\log_{d-1} n$ is asymptotically

$$T_n \equiv \frac{d}{d-2} \log_{d-1} n.$$

Clearly we cannot be in equilibrium at time $(1-\epsilon)T_n$ since there are only $O(n^{1-\epsilon})$ vertices reachable in that time and the stationary distribution is uniform.

The more precise result given in (6.7.1) comes by reasoning that we cannot be in equilibrium until the random walk has moved by more that $\log_{d-1} n$ and using the central limit theorem to evaluate that probability. The funky formula for Λ should indicate that there is more to the sharp result in the displayed formula than just computing the variance of the steps of the random walk, so my advice to the reader is to wait until the formula is done for you in (6.7.7).

Main Ideas of the Proof

In what follows the Lemmas are numbered as in Lubetzky and Sly (2010). The hard part of the proof of Theorem 1 is to show that once the walker has moved a distance larger than $\log_{d-1} n$ it is in equilibrium. The proof is based on two ideas.
I. The first important idea, which is widely known, is that the graph is **locally tree-like**. To prove a result to quantify this, let

$$B_t(u) = \{v \in V : \text{dist}(u, v) \leq t\} \quad \text{and} \quad \partial B_t(u) = B_t(u) - B_{t-1}(u).$$

Let $\text{tx}(B_t)$ be the maximum number of edges that can be deleted from B_t and still keep it connected. Here tx is short for tree excess. Sharpening the conclusion in our Theorem 1.2.3 they show the following.

Lemma 2.1. *Let* $G \sim \mathcal{G}(n, d)$*for a fixed* $d \geq 3$*and let* $s = \lfloor (1/5) \log_{d-1} n \rfloor$*then with high probability* $\text{tx}(B_s(u)) \leq 1$*for all* $u \in V(G)$.

"Proof." When we grow the clusters out to distance s there are (ignoring constants) $n^{1/5}$ vertices. In running the exploration process for time $m = n^{1/5}$ the number of collisions will be smaller than binomial$(m, m/n)$. The probability of more than 2 collisions is

$$\leq \binom{m}{2} \cdot (m/n)^2 \leq m^4/n^2 = O(n^{-6/5}),$$

so the expected number of vertices for which this occurs is $n^{-1/5}$ and the probability of the existence of such a vertex $\to 0$.

Theorem 1.2.3 shows that for a given vertex the number of collisions before the cluster grows to size n^α is $n^{2\alpha-1}$. If $\alpha = 2/3$ then this is $O(n^{1/3})$. While there are a large number of these, vertices with this problem, which we will call bad vertices, are rare. Let A_k be the probability that the random walk has stepped on a bad vertex by time k. Berestycki and Durrett (2008), who studied random walks on a three-regular graph, proved that $P(A_{3(1-\epsilon)\log n}) \to 0)$, so with high probability the walk will get to time $3(1-\epsilon)\log n$ without noticing that it is not on a 3-regular tree. While the last picture is useful it breaks down badly when the walk reaches a distance equal to the diameter.

II. To analyze what happens beyond distance $\log_{d-1} n$ becomes much easier one uses the notion of the **cover tree of a graph**. Let $G = (V, E)$ be a d-regular graph and let $a \in V$. The cover tree is a mapping $\phi : \mathcal{T} \to V$ where $\mathcal{T}$ is a d-regular tree with root ρ so that $\phi(a) = \rho$, which preserves the neighborhood relationship:

$$\mathcal{N}_G(\phi(x)) = \{\phi(y) : y \in \mathcal{N}_\mathcal{T}(x)\},$$

where for $H = G, \mathcal{T}, \mathcal{N}_H(y)$ are the neighbors of y in H.

To prove Theorem 1 we need several results about the structure of the cover tree. In order to conserve the reader's energy and to make it easier to see the "big picture" we state these without proof, and refer the reader to the paper for details. To state these lemmas we need four definitions

$$K = \lfloor \log_{d-1} \log n \rfloor, \qquad R = \lfloor (4/7) \log_{d-1} n \rfloor, \qquad T = \lfloor (1/2) \log_{d-1} n \rfloor.$$

A vertex $a \in V$ is a **K-root** if and only if the induced subgraph on $B_K(a)$ is a tree.

Lemma 3.2. *Suppose that every* $u \in V$ *has* $\text{tx}(B_{5K}(u)) \leq 1$.*Then for any* $v \in V$ *the simple random walk of length 4K ends in a* K*-root with probability* $1 - o(1)$.*In particular, the number of* K*-roots is* $n - o(n)$.

Lemma 3.3. *With high probability every* K*-root* u*satisfies*

$$|\partial B_t(u)| \geq (1 - o(1))d(d-1)^{t-1} \quad \text{for all } t < R.$$

Let $\partial B_t^*(u)$ denote the set of vertices in $\partial B_t(u)$ with a single (simple) path of length t to u.

Lemma 3.4. *With high probability any two* K*-roots with* $\text{dist}(u, v) > 2K$*satisfy*

$$|\partial B_t^*(u) - B_{t+1}(v)| = (1 - o(1))d(d-1)^{t-1} \quad \text{for all } t < R - 1.$$

Let $\mathcal{S}_k$ be the number of simple paths of length k between u and v.

Lemma 3.5. *With high probability any two* K*-roots with* $\text{dist}(u, v) > 2K$*satisfy*

$$\mathcal{S}_{2T+t}(u, v) \geq (1 - o(1))\frac{1}{n}d(d-1)^{2T+t-1},$$

for all $2K \leq \ell \leq (1/20) \log_{d-1} n$.

Proof of Theorem 6.7.1. By Lemma 3.2 after K steps whp the random walk is at a K-root. Since we are only seeking to establish t_{mix} to an accuracy of $o(\sqrt{\log_{d-1} n})$ and since $K = o(\sqrt{\log_{d-1} n})$ it is enough to suppose we are starting at a K-root.

By Lemma 3.5

$$\mathcal{S}_{2T+t}(u, v) \geq (1 - o(1))\frac{1}{n}d(d-1)^{2T+t-1} \qquad \text{for all } 2K \leq \ell \leq (1/20) \log_{d-1} n.$$

Since each simple path in G corresponds to a simple path in the cover tree $\mathcal{T}$

$$\begin{aligned} &|\{w \in \mathcal{T} : \phi(w) = v, \text{dist}(\rho, w) = 2T + \ell\}| \\ &\geq \mathcal{S}_{2T+\ell}(u, v) \geq \frac{1 - o(1)}{n} d(d-1)^{2T+t-1}, \end{aligned} \tag{6.7.2}$$

when $2K \leq \ell \leq (1/20) \log_{d-1} n$.

Let X_t be a simple random walk on $\mathcal{T}$ started from ρ and let $W_t = \phi(X_t)$ be the corresponding SRW on G started from u. Note that by symmetry, the random walk conditioned on $\text{dist}(\rho, X_t) = k$ is uniform on the $d(d-1)^{k-1}$ points at distance k from ρ in $\mathcal{T}$

A random walk on $\mathcal{T}$ with $d \geq 3$ is transient so it returns to ρ only finitely many times. If $X_t \neq \rho$ then on the next step the change the distance from ρ increases by 1 with probability $(d-1)/d$ and decreases by 1 with probability $1/d$. Let ξ be a random variable with this distribution. $E\xi = (d-2)/d$ and

$$\text{var}\,(\xi) = 1 - (E\xi)^2 = 1 - \frac{d^2 - 4d + 4}{d^2} = \frac{4(d-1)}{d^2}.$$

Therefore the central limit theorem gives

$$\text{dist}(X_t, \rho) = \frac{t(d-2)/d}{\sqrt{t} \cdot 2\sqrt{d-1}/d} \Rightarrow \text{normal}(0, 1). \tag{6.7.3}$$

Let A be the set of vertices that are K roots and whose distance from u is $\geq 2K$. Since the number of vertices with distance $2K$ is $o(n)$ it follows from Lemma 3.2 that $|A| \geq n - o(n)$. If $v \in A$ and

$$t = \left\lfloor \frac{d}{d-2} \log_{d-1} n + k\sqrt{\log_{d-1} n} \right\rfloor, \tag{6.7.4}$$

there are many points $w \in \mathcal{T}$ with $\phi(w) = v$ so

$$P(W_t = v) = \sum_{j=0}^{t} P(\text{dist}(\rho, X_t) = j) \frac{|\{w \in \mathcal{T} : \phi(w) = v, \text{dist}(\rho, w) = j\}|}{d(d-1)^{j-1}}.$$

Recalling $2T = 2\lfloor (1/2) \log_{d-1} n \rfloor \approx \log_{d-1} n$ and using (6.7.2) the above

$$\begin{aligned}
&= \sum_{\ell=2K}^{(1/20)\log_{d-1} n} P(\text{dist}(\rho, X_t) = 2T + \ell) \cdot \frac{1 + o(1)}{n} \\
&= \frac{1 + o(1)}{n} \cdot P(\log_{d-1} n + 2K \leq \text{dist}(\rho, X_t) \leq 1.05 \log_{d-1} n) \\
&= \frac{1 + o(1)}{n} \cdot P(\text{dist}(\rho, X_t) - \log_{d-1} n \geq 0),
\end{aligned}$$

since in the second line $K = o(\sqrt{\log_{d-1} n})$ and the other inequality holds with probability $\to 1$. Substituting in the value of t

$$\text{dist}(\rho, X_t) - t(d-2)/d = \log_{d-1} n + k\sqrt{\log_{d-1} n} \cdot (d-2)/d$$

which implies $P(\text{dist}(\rho, X_t) - \log_{d-1} n \geq 0)$ is

$$= P\left(\frac{\text{dist}(\rho, X_t) - t(d-2)/d}{\sqrt{t} \cdot 2\sqrt{d-1}/d} \geq \frac{-k\sqrt{\log_{d-1} n} \cdot (d-2)/d}{\sqrt{t} \cdot 2\sqrt{d-1}/d}\right). \tag{6.7.5}$$

Using the central limit theorem in (6.7.3) we have

$$P(\text{dist}(\rho, X_t) - \log_{d-1} n \geq 0) \approx 1 - \Phi(-k/\Lambda), \tag{6.7.6}$$

where Φ is the distribution function of the standard normal and

$$\Lambda = \sqrt{\frac{d}{d-2}} \cdot \frac{2\sqrt{d-1}/d}{(d-2)/d} = \frac{2\sqrt{d(d-1)}}{(d-2)^{3/2}}. \tag{6.7.7}$$

To get to this easily start by flipping the right-hand side of (6.7.5) and then use the definition of t in (6.7.4).

To complete the proof now recall that A is the set of vertices that are K-roots and use (6.7.6)

$$\begin{aligned}
\|P(W_t \in \cdot) - \pi\| &= \sum_{v \in V} \max\left(\frac{1}{n} - P(W_t = v), 0\right) \\
&\leq \frac{n - |A|}{n} + \sum_{v \in A} \max\left(\frac{1}{n} - P(W_t = v), 0\right) \\
&= o(1) + (1 + o(1)) \cdot \frac{1}{n}\Phi(-k/\Lambda).
\end{aligned}$$

At this point the proof of the matching lower bound is not very hard, but again we refer the reader to the paper for details.

References

Aldous D (1989) The Poisson clumping heuristic. Springer, New York

Aldous D, Lovász L, Winkler P (1997) Mixing times for uniformly ergodic Markov chains. Stoch Proc Appl 125:408–420

Bayer D, Diaconis P (1992) Trailing the dovetail shuffle to its lair. Ann Appl Probab 2:37–50

Benjamini I, Kozma G, Wormald N (2014) The mixing time of the giant component of a random graph. Rand Struct Alg 45(3):383–407

Berestycki N, Durrett R (2008) Limit behavior for the distance of a random walk. Electron J Probab 13:44

Berestycki N, Lubetzky E, Peres Y, Sly A (2018) Random walks on the random graph. Ann Probab 46:456–490

Cooper C, Frieze A (2003) The cover time of sparse random graphs. In: Proceedings of 14th annual ACM-SIAM symposium on discrete algorithms, pp 140–147

Diaconis P (1986) Group representations in probability and statistics. In: IMS lecture notes monographs series
Diaconis P (1996) The cut-off phenomenon in finite Markov chains. Proc Natl Acad Sci USA 93:1659–1664
Diaconis P, Shahshahani M (1981) Generating a random permutation with random transpositions. Probab Theory Rel Fields 57(1981):159–179
Diaconis P, Stroock D (1991) Geometric bounds for the eigenvalues of Markov chains. Ann Appl Probab 1:36–61
Ding J, Lubetzky E, Peres Y (2014) Anatomy of a giant component: the strictly supercritical regime. Eur J Combin 35:155–168
Fountoulakis N, Reed B (2007) Faster mixing and small bottlenecks. Probab Theory Rel Fields 137:475–486
Fountoulakis N, Reed B (2008) The evolution of the mixing rate of a simple random walk on the giant component of a random graph. Rand Struct Alg 33:68–86
Gkantsidis C, Mihail M, Saberi A (2003) Conductance and congestion in power law graphs. In: Proceedings of the 2003 ACM SIGMETRICS international conference on measurement and modeling of computer systems, pp 148–159
Lawler G, Sokal A (1988) Bounds on the L^2 spectrum for Markov chains and Markov processes: a generalization of Cheeger's inequality. Trans Am Math Soc 309:557–580
Levin DA, Peres Y (2017) Markov chains and mixing times, 2nd edn. American Mathematical Society, Providence, RI
Lovász L, Kannan R (1999) Faster mixing via average conductance. Symp Theory Comput 1999:282–287
Lubetzky E, Sly A (2010) Cutoff phenomena for random walks on random regular graphs. Duke Math J 153:475–510
Lyons R, Pemantle R, Peres Y (1995) Ergodic theory on Galton-Watson trees: speed of random walk and dimensions of harmonic measure. Ergodic Theory Dyn Syst 15:593–619
Morris B, Peres Y (2005) Evolving sets, mixing and heat kernel bounds. Probab Theory Relat Fields 133:245–266
Saloff-Coste L (1996) Lectures on finite Markov chains. Ecole d'eté de probabilités de Saint-Flour XXVI, vol 1665. Springer
Sinclair A, Jerrum M (1989) Approximate counting, uniform generation and rapidly mixing Markov chains. Inf Comput 82:93–133

Chapter 7
Voter Models, Coalescing RWs

The voter model was introduced independently by Clifford and Sudbury (1973) and Holley and Liggett (1975) on the d-dimensional integer lattice. It is a very simple model for the competition of two opinions. Each voter at the times of a rate one Poisson process wakes up and imitates the opinion of one of its neighbors chosen at random. While no one could think that this is an accurate model of how people form opinions, as a mathematical model, it has the desirable properties that it is simple and has interesting behavior.

The key to the analysis of the voter model is a duality similar to the one used to study the contact process, which is introduced in Sect. 7.1. Here the dual, which traces the origin of the opinion at x at time t, is a **coalescing random walk**. Particles in the dual move like independent random walks until they hit, at which point they coalesce to one. In the case of a homogeneously mixing population, each dual particle gives the genealogy an individual. The collection of dual particles is an instance of **Kingman's coalescent**.

The chapter could have been named coalescing random walks (CRW) because they are the main focus of our investigation. In Sect. 7.2 we consider the behavior of CRW on the torus in dimensions $d \geq 3$. In Sect. 7.3 we begin the study of CRW on graphs by examining the hitting times for two random walks. **Aldous's Poisson Clumping Heuristic** allows us to compute the asymptotic behavior of the hitting time, including the constant.

Another result of Aldous and Fill shows that given a sequence of graphs G_n, the hitting time T_n for walks starting from two randomly chosen starting points divided by its mean converges to an exponential distribution if we have

(H) the mixing time for the random walk on the graph is of smaller order of magnitude than the time for the two walks to hit.

A major theme in the chapter is that if (H) holds then a coalescing random walk when properly rescaled converges to Kingman's coalescent. In Sect. 7.4, we describe the result of Cooper et al. (2012) for random regular graphs. Section 7.5 is devoted to

R. Durrett, *Dynamics on Graphs*, Probability Theory and Stochastic Modelling 110,
https://doi.org/10.1007/978-3-032-21908-4_7

work of Oliveira on CRW on general graphs that confirms a conjecture of Aldous and Fill.

In Sect. 7.6 we turn our attention to the short time behavior of CRW starting from one particle at each site. The combined work of a half dozen people from 1977–1980 showed that on $\mathbb{Z}^d$ the density of occupied sites p_t obeyed

$$p_t \sim \begin{cases} c_1/t^{1/2} & d = 1 \\ c_2(\log t)/t & d = 2 \\ c_d/t & d \geq 3. \end{cases}$$

In the second half of the section we describe work of Hermon et al. (2022) who show that for random graphs generated by the configuration model the decay of the density of CRW is like that on the torus in $d \geq 3$.

In the last section of this chapter the focus returns to the voter model. We consider three classes of finite graphs: the complete graph, the d-dimensional torus in $d \geq 3$, and graphs generated by the configuration model. On a finite graph the voter model will eventually enter an absorbing state that represents consensus. Results for CRW tell us how long it takes for this to occur: it is the same as time s_N for the coalescing random walk starting from one particle at each site to reduce to one particle. Results of Cox with Greven (1990) and with Chen and Choi (2016) describe how the behavior of the voter model starting from product measure with density p evolves. At times that are large but $o(s_N)$ the system looks like the voter model equilibrium ν_p. At times $O(s_N)$ it is a mixture of these equilibrium states where the mixture is given by the distribution of a Wright-Fisher diffusion. As Fermat once said, "there is a wonderful proof, but it won't fit in the margin."

7.1 On $\mathbb{Z}^d$ and on Graphs

As was the case for the contact process in Chap. 5, we construct the voter model using a **graphical representation**.

> For each site x there is an independent rate 1 Poisson process, T_n^x, $n \geq 1$. At the times T_n^x, $n \geq 1$ the individual at x picks a neighbor y_n^x at random, and at time $t = T_n^x$ we set $\xi_t(x) = \xi_t(y_n^x)$.

To construct the process, we will think about 1's as occupied and 0's as vacant rather than two opinions. We put a • at x at times T_n^x (which causes a death as in the contact process) and draw an arrow from $(y_n^x, T_n^x) \to (x, T_n^x)$ (which causes a birth as in the contact process). The arrow arrives just after the dot (or else it would do nothing).

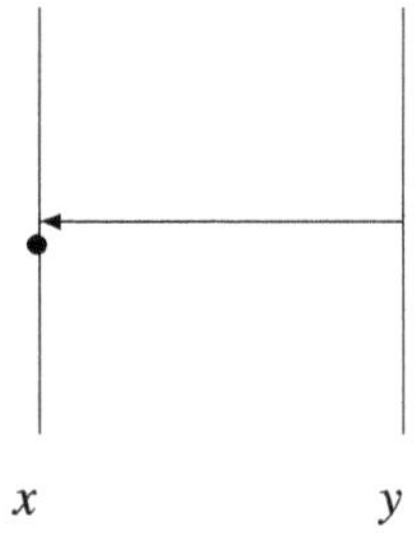

A little thought reveals that the combination of these two events causes the following changes to occur. Here y is short for y_n^x.

Before x	y	After x	y
0	0	0	0
0	1	1	1
1	0	0	0
1	1	1	1

To check the table, note that the • kills any particle at x while the arrow gives birth from y to x if there is a particle at y. On the last line the • kills the particle at x, but then the arrow restores it. No change occurs at y. The result of these two mechanisms is a voter event: x imitates the value at y.

This construction allows us to define a **dual process** (see Fig. 7.1) by working down the graphical representation and crossing arrows in the direction opposite their

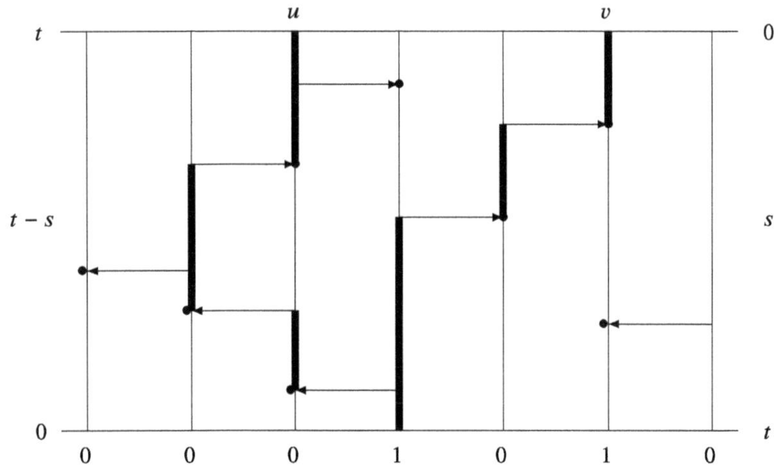

Fig. 7.1 A realization of the dual process for the voter model. Particles start at u and v performing random walks that jump at rate 1 and to a neighbor chosen at random. After they hit the walks coalesce and we know that the two individuals have the same opinion

orientation. The combination of the two events causes the following changes in the dual process.

Before x	y	After x	y
0	0	0	0
0	1	0	1
1	0	0	1
1	1	0	1

To check the table, note that if x is in state 0 nothing happens, while if x is in state 1 then the arrow from y to x, which is crossed in the direction opposite its orientation, gives birth to a particle at y and then the • kills the 1 at x. The result is a **coalescing random walk** step: the particle at x jumps to y, and if y is occupied the two particles coalesce.

As in the contact process, the duality generates an almost sure equality

$$\xi_t(x) = \xi_{t-s}(\zeta_s^x) \quad \text{for} \quad 0 \le s \le t. \tag{7.1.1}$$

In words, the opinion at x at time t is the same as the opinion of ζ_s^x at time $t-s$. If $A = \{x : \xi_0(x) = 1\}$ then we let ξ_t^A be the set of sites in state 1 at time t. If we start with coalescing random walk particles on sites B in the dual then we let $\zeta_t^B = \cup_{x \in B} \zeta_t^x$. In this notation the duality becomes

$$P(\xi_t^A \cap B \neq \emptyset) = P(A \cap \zeta_t^B \neq \emptyset). \tag{7.1.2}$$

Remark. The graphical representation construction also works for the voter model on graphs.

If we consider the voter model on $\mathbb{Z}^d$ with the usual nearest neighbors, then as Holley and Liggett (1975) have shown.

Theorem 7.1.1 *In $d \le 2$ the voter model approaches complete consensus, i.e., $P(\xi_t(x) = \xi_t(y)) \to 1$. In $d \ge 3$ if we start from product measure with density p (i.e., we assign opinions 1 and 0 independently to sites with probabilities p and $1-p$) then as $t \to \infty$, ξ_t^p converges in distribution to ν_p, a stationary distribution in which a fraction p of the sites have opinion 1.*

Proof. In $d \le 2$ recurrence of random walk implies that $P(\zeta_t^x \neq \zeta_t^y) \to 0$ so $P(\xi_t(x) \neq \xi_t(y)) \to 0$. Figure 7.2 gives simulations of the voter model in dimensions 1 and 2 showing the clustering. If we let ξ_t^p be the set of sites occupied by 1's at time t then the duality Eq. (7.1.2) and the fact that sites are independently occupied with probability p implies that

$$P(\xi_t^p \cap B = \emptyset) = P(\xi_0^p \cap \zeta_t^B = \emptyset) = E(1-p)^{|\zeta_t^B|}.$$

Since the size of the dual never increases, $|\zeta_t^B| \downarrow$ a limit.

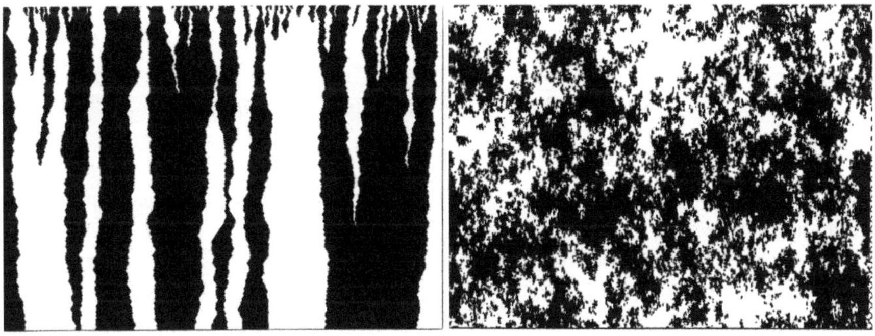

Fig. 7.2 Simulations of the one and two-dimensional voter models starting from a random mix of 0's and 1's showing clustering. That is, the regions of voters with the same opinion grow as $t \to \infty$

$$\lim_{t\to\infty} P(\xi_t^p \cap B = \emptyset) = \lim_{t\to\infty} E(1-p)^{|\zeta_t^B|} \quad \text{exists.}$$

Knowing the probabilities $P(\xi_\infty^p \cap B \neq \emptyset)$ determines the distribution of ξ_∞^p. Since ξ_∞^p is the limit in distribution of ξ_t^p as $t \to \infty$ it must be a stationary distribution for the voter model.

7.1.1 *On the Complete Graph*

We consider the complete graph with vertex set $\{1, 2, \dots, n\}$ and edges connecting each pair of vertices. For computations it is convenient to have particles jump with equal probability to all vertices, including the one they are on. If we start the coalescing random walk with k walkers, each jumping at rate one, then the time until the first coalescence is approximately exponential with rate $k(k-1)/n$. To see this note that our k particles jump at rate k and when one does it hits another particle with probability $(k-1)/n$. The resulting Markov chain on the positive integers is called **Kingman's coalescent**.

Let T_m be the first time the coalescing random walk has only m particles. By the calculation in the previous paragraph

$$E(T_{m-1} - T_m) = n/(m(m-1)).$$

We will follow a common practice in population genetics and "let τ_m be the time T_m measured time in units of n generations," i.e., $\tau_m = T_m/n$, so we have

$$E(\tau_{m-1} - \tau_m) = 1/(m(m-1)) = \frac{1}{m-1} - \frac{1}{m}.$$

Telescoping the series

$$E_m\tau_1 = \sum_{k=2}^{m}\left(\frac{1}{k-1} - \frac{1}{k}\right) = 1 - 1/m,$$

so (after scaling by the population size) the time to go from m particles to 1 stays bounded as $m \to \infty$.

The interest in this process in population genetics is that it gives the genealogy of the sample of m individuals in the Wright-Fisher model. As we work backwards in discrete time each individual chooses their parent at random, so two lineages will coalesce with probability $1/n$. If k stays constant then as $n \to \infty$ the genealogy of the sample converges to the continuous time coalescent. For more on this see the original paper of Kingman (1982) or the nice book by Wakeley (2009).

When a system is considered on the complete graph the result is often called the **mean-field version** of the system because each site interacts equally with all the others. In this chapter, we will prove results which show that on many large random graphs coalescing random walks exhibit mean-field behavior, i.e., all pairs of particles have an equal chance of being the next to coalesce and the elapsed time between coalescing events converges to

$$E(\tau_{m-1} - \tau_m) = c/(m(m-1)) \quad \text{as} \quad n \to \infty,$$

where the constant depends on the graph under consideration.

7.1.2 On Non-regular Graphs

There are two versions of the voter model, which coincide on a regular graph, but can have much different behavior in general. In either version each site x has an opinion $\xi_t(x)$ at time t. The first formulation matches the one we have given above.

Vertex Voter Model. For each vertex x, at the times T_n^x, $n \geq 1$ of a rate 1 Poisson process x decides to change its opinion. To do this it picks a neighbor y_n^x at random, and at time $t = T_n^x$ we set $\xi_t(x) = \xi_t(y_n^x)$.

Edge Voter Model. For each oriented edge (x, y) at the times of a rate 1 Poisson process $T_n^{(x,y)}$ the voter at x imitates the voter at y. This can also be formulated as; if x has degree $d(x)$ then at the times $\bar{T}_n^x$, $n \geq 1$ of a Poisson process rate $d(x)$, x picks a neighbor y_n^x at random, and at time $t = T_n^x$ we set $\xi_t(x) = \xi_t(y_n^x)$.

In the vertex case the dual random walk jumps from x to a neighbor $y \sim x$ at rate $q(x, y) = 1/d(x)$ so if we let $a(x, y)$ be the adjacency matrix of the graph then

$$d(x)q(x, y) = a(x, y) = a(y, x) = d(y)q(y, x),$$

and the measure that assigns mass $d(x)$ to vertex x is a reversible stationary measure. If the graph is finite we can convert the measure to a stationary distribution by dividing by the sum of the degrees: $\pi(x) = d(x)/D$ where $D = \sum_x d(x)$. If $\pi(x)$ is not uniform then the density defined as the fraction of sites in state 1 is not preserved in time but in Sect. 7.7.3 we will see that on finite graphs the weighted sum $p_1(\xi) = \sum_x \pi(x)\xi(x)$ is a martingale.

In the edge case the rate of jumps from x to y and from y to x are equal so the dual random walk is reversible with respect to the uniform stationary distribution. In the words of Suchecki, Eguíluz and Miguel (2005): the system displays "conservation of the global magnetization." In terms more familiar to probabilists, the number of voters with a given opinion is a time change of simple random walk and hence is a martingale. To see this note that at each edge between a 0 and 1 the 0 changes to 1 at the same rate that the 1 changes to 0.

7.2 Coalescing Random Walk on the Torus in $d \geq 3$

Cox (1989) studied the nearest neighbor voter model on a finite torus $(\mathbb{Z} \bmod N)^d$. Since the state space is finite the chain will eventually be absorbed in a state in which all individuals have the same opinion, an event we call **consensus**. Let τ_N be the time to reach consensus. We are interested in the asymptotic behavior of τ_N as $N \to \infty$

This question is easiest to study if we imagine each site starts in a different state, so we let $\bar{\xi}_0(x) = x$. Using duality we see that the time to reach consensus, is the same as the time it takes a coalescing random walk to be reduced to one particle if we start with one particle at each site. Cox (1989) has shown, see his Theorem 1.

Theorem 7.2.1 *As $N \to 0$, τ_N/s_N converges in distribution to a limit we call τ where*

$$s_N = \begin{cases} N^2 & d = 1, \\ N^2 \log N & d = 2, \\ N^d & d \geq 3. \end{cases}$$

The case $d = 1$ of Theorem 7.2.1 is easy. Boundaries between types move like random walks and annihilate when they hit, so invoking the central limit theorem it takes time of order N^2 for all the boundaries on $Z \bmod N$ to disappear. One of the main results of this chapter is:

Metatheorem. *If the time for a random walk on a graph to converge to equilibrium is much smaller than the time for random walks from two randomly chosen sites to hit, then the time for these two walks to hit has an exponential distribution and the coalescing random walk behaves like Kingman's coalescent.*

Since the time to converge to equilibrium on the torus is, by the central limit theorem, of order N^2, the metatheorem suggests the convergence to Kingman's coalescent holds in $d = 2$. The $d = 2$ case is delicate (and interesting), since the recurrence of two-dimensional random walk produces a log N factor in the asymptotics. However this phenomenon is very different from the behavior on random graphs, so we will restrict our attention to the case $d \geq 3$, where there are strong similarities.

Coalescence Time of Two Particles

Let S_n be the discrete time random walk on the d-dimensional lattice that jumps to the $2d$ nearest neighbors with equal probability, let p_n be its n-step transition probability, and let

$$G = \sum_{n=0}^{\infty} p_n(0, 0) = 1/\beta_d,$$

where $\beta_d = P(S_n \neq 0 \text{for all} n \geq 1)$. To check the last equality note that the number of visits to 0 (including the one at time 0) has a geometric distribution with "success probability" β_d. We have put success probability in quotes since success is failing to return to 0.

Theorem 4 in Cox (1989) gives the asymptotic behavior of the hitting time of the origin $T^0 = \inf\{t \geq 0 : X_t = 0\}$. His theorem contains a result for $d = 2$ but we will ignore that result in order concentrate on $d \geq 3$, which is more closely analogous to the behavior on random graphs.

Lemma 7.2.2 *Assume $d \geq 3$, and let $a_N \to \infty$. Then uniformly in all starting points x with distance $d(x, 0) \geq a_N$*

$$P(T^0 > tN^d) \to \exp(-t/G) \quad as \quad N \to \infty.$$

It follows immediately from this result that if τ_2 is the first hitting time of two independent random walks starting from x and y with $d(x, y) \geq a_N$ on the torus with side N then

$$P(\tau_2 > tN^d) \to \exp(-2t/G), \tag{7.2.1}$$

since the difference in the two locations is a random walk that jumps at rate 2.

Before turning to the coalescence times of $k \geq 3$ particles we will prove an estimate which implies that the hypotheses of the metatheorem hold for random walks on the torus in $d \geq 3$. The mixing time of random walk on the torus (in any dimension) is $O(N^2)$. With the metatheorem in mind the first thing to do is to show that the hitting time of 0, T_0, is larger than the mixing time if the particle is not too close to 0. The result here is more crude than the previous result but the proof is simple and we will need some of estimates later.

Lemma 7.2.3 *Suppose $a_N \to \infty$ and let $b_N = B \log N$ then $P_x(T_0 \leq b_N N^2) \to 0$ uniformly for x with $d(0, x) \geq a_N$.*

Proof. Let $\epsilon \leq 0.1$. The proof is accomplished in three steps. We first list the steps and then prove them.

(i) If the initial point x has $|x| \geq a_N$ then with high probability the particle will reach a distance $N^{1-\epsilon}$ from 0 before hitting 0.
(ii) A random walk on the torus requires time $\geq r(N) = N^{2(1-2\epsilon)}$ to go a distance $N^{1-\epsilon}$. This lower bounds the additional amount of time to hit 0 if goal (i) is achieved.
(iii) The local central limit theorem implies that for $t = r(N)$

$$\max_x P(S_t = x) \leq CN^{-d(1-2\epsilon)}. \tag{7.2.2}$$

The Markov property extends this to $t \geq r(N)$ so the time to hit 0 after goal (ii) is achieved is $\geq N^{d(1-3\epsilon)}$ with high probability.

It suffices to prove (i) and (ii).

Proof of (i). If $0 < \alpha < d - 2$ then there is a K_α so that the function $1/|x|^\alpha$ is superharmonic for the random walk when $|S_t| \geq K_\alpha$, see e.g., Example 5.3 in Varadhan's (2001) book. The proof is not hard. Look at $|x + y|^{-\alpha}/|x|^{-\alpha}$ for $|y| = 1$, expand in power series assuming $|x|$ is large, and then compute the expected value of the result of one jump.

Using this with the optional stopping theorem for nonnegative supermartingales (Theorem 4.8.4 in PTE5) we see that if $|S_0| = a_N$, $T = \inf\{n : S_n \notin (K_\alpha, N^{1-\epsilon})\}$, and $q = P(S_T = K_\alpha\}$ then ignoring the fact that $N^{1-\epsilon}$ is not an integer

$$a_N^{-\alpha} \geq qK_\alpha^{-\alpha} + (1-q)N^{-(1-\epsilon)\alpha} \quad \text{so} \quad q \leq Ca_N^{-\alpha}.$$

Proof of (ii). If $S_0 = 0$ then $E|S_t|^2 = t$, so if $r(N) = N^{2(1-2\epsilon)}$ then the L^2 maximal inequality for submartingales (Theorem 4.4.4 in PTE5) implies that

$$P\left(\max_{0 \leq t \leq r(N)} |S_t|^2 \geq N^{2(1-\epsilon)}/2\right) \leq 4 \cdot \frac{N^{2(1-2\epsilon)}/2}{N^{2(1-\epsilon)}} \leq 2N^{-2\epsilon}.$$

This completes the proof of (ii) and hence of Lemma 7.2.3.

Coalescence time for k particles

Lemma 7.2.4 *Run $k \geq 3$ independent random walks starting from points separated by distances at least a_N. (a) If we let $\tau_{i,j}$ be the first time i and j hit and let $\tau^k = \min_{1 \leq i < j \leq k} \tau_{i,j}$ then*

$$P(\tau^k > tN^d) \to \exp(-tk(k-1)/G). \tag{7.2.3}$$

(b) The probability that at time τ_{ij} there are random walks (other than ith and jth) separated by distance $\leq a_N$ tends to 0.

Proof. To prove (b), we note that by the proof of Lemma 7.2.3 the probability of a pair hitting by time $\leq b_N N^2$ tends to 0. Using (7.2.2) we conclude that the probability that two pairs of particles $\{i_1, j_1\}$ and $\{i_2, j_2\} \neq \{i_1, j_1\}$ are both within distance a_N at some time $t \in [b_N N^d, N^d \log N]$ is

$$\leq k^4 \cdot N^d \log N \left(\frac{2a_N}{N^{d(1-2\epsilon)}} \right)^2 \to 0,$$

if $\epsilon < 1/4$.

To prove (a) now, we need to argue that the $\tau_{i,j}$ are approximately independent. This ingenious calculation is taken from Cox and Griffeath (1986). Let $c_N = GN^d/2$. Let

$$H_t(i,j) = \{\tau_{ij} \leq tc_N\}, \quad F_t(i,j) = \{\tau = \tau_{ij} \leq tc_N\}, \quad \text{and} \quad q(t) = P(\tau \leq tc_N).$$

Then

$$P(H_t(i,j)) = P(F_t(i,j)) + \sum_{\{\ell,m\}\neq\{i,j\}} \int_0^{tc_N} P(\tau = \tau_{\ell.m} = s, \tau_{ij} \leq tc_N)\, ds, \quad (7.2.4)$$

where the quantity being integrated on the right is the density of the hitting time $\tau_{\ell.m}$. To evaluate the $\ell.m$ term in the sum we break things down according to the locations X_s^i and X_s^j. By (b) we can ignore the possibility that $|X_s^i - X_s^j| < a_N$. When the distance between the particles is $\geq a_N$ the argument for Lemma 7.2.2 shows that the positions will become randomized before they hit (i) so each hitting time $\tau_{i,j}$ will be exponentially distributed in the limit as $N \to \infty$, and each pair of particles has an equal probability to be the first to collide.

Writing ϵ_N for a quantity that goes to 0 as $N \to \infty$

$$\int_0^{tc_N} P(\tau = \tau_{\ell,m} = s, \tau_{ij} \leq tc_N)\, ds$$

$$= \int_0^{tc_N} P(\tau = \tau_{k\ell} = s) \left[1 - \exp\left(-\frac{tc_N - s}{c_N} \right) \right] ds + \epsilon_N.$$

Integrating by parts and then changing variables $s = u/c_N$, the above is

$$= \int_0^{tc_N} \frac{1}{c_N} \exp(-(tc_N - s)/c_N)) P(\tau = \tau_{k\ell} \leq s)\, ds + \epsilon_N$$

$$= \int_0^{t} \exp(-(t-u)) P(\tau = \tau_{k\ell} \leq uc_N)\, du + \epsilon_N.$$

Using this in our initial decomposition (7.2.4), with the convergence of the hitting time $\tau_{i,j}/c_N$ to the exponential(1) distribution, we get

$$1 - e^{-t} = P(F_t(i, j)) + \sum_{\{\ell,m\} \neq \{i,j\}} e^{-t} \int_0^t e^u P(F_u(\ell, m))\, du + \epsilon_N. \qquad (7.2.5)$$

Summing over all $\binom{k}{2}$ pairs $\{i, j\}$, and letting $q(t) = P(F_t(i, j))$ we have

$$\binom{k}{2}(1 - e^{-t}) = q(t) + \left[\binom{k}{2} - 1\right] e^{-t} \int_0^t e^u q(u)\, du + \epsilon_N.$$

It follows that as $N \to \infty$, $q(t)$ converges to $r(t)$, the solution of

$$\binom{k}{2}(1 - e^{-t}) = r(t) + \left[\binom{k}{2} - 1\right] e^{-t} \int_0^t e^u r(u)\, du.$$

Multiplying both sides by e^t and rearranging we have

$$e^t r(t) - \binom{k}{2}(e^t - 1) = -\left[\binom{k}{2} - 1\right] \int_0^t e^u r(u)\, du.$$

Differentiating we have

$$e^t r(t) + e^t r'(t) - \binom{k}{2} e^t = -\left[\binom{k}{2} - 1\right] e^t r(t).$$

The first term on the left cancels the last term on the right. Dividing by e^t and rearranging gives

$$\binom{k}{2}(1 - r(t)) = r'(t) = -\frac{d}{dt}(1 - r(t)),$$

which has solution $1 - r(t) = \exp(-k(k-1)t/2)$. Since $q(t) \to r(t)$ it follows that

$$P(\tau = \tau_{ij} \leq t/c_N) \to 1 - \exp(-k(k-1)t/2).$$

Recalling $c_N = GN^2/2$ the desired result follows.

Remark 7.2.5 To prepare for later arguments, note that the proof of (a) only requires that $\tau_{i,j}/c_N$ converges to a mean one exponential, and that the time to reach equilibrium is $o(c_N)$.

Using Lemma 7.2.4 and induction gives,

Theorem 7.2.6 *Suppose that we start a coalescing random walk on* $(\mathbb{Z} \bmod N)^d$ *in* $d \geq 3$ *with* m *particles that are separated by* $a_N \to \infty$ *then the number of particles at time* tGN^d *converges to Kingman's coalescent in which jumps from* k *to* $k-1$ *occur at rate* $k(k-1)$.

Since the expected time to jump from k to $k-1$ is $1/k(k-1)$, it is easy to see that we can start Kingman's coalescent from ∞ and the number of particles will immediately become finite. In our study of coalescing random walk, we will want to start with all sites on the torus occupied, and show that when time is run at rate GN^d then as $N \to \infty$ the coalescing random walk converges to Kingman's coalescent starting from infinitely many particles. To do this we will need to show that at time tGN^d the number of particles on the torus is finite. This is an event that many years ago I named the Big Bang. We will delay addressing this point until we can use a nice argument of Cooper et al. (2012) in Sect. 7.4.3 to deal with it.

7.3 Hitting Times for Two Random Walks

When the degrees of vertices in the graph are not constant the random walk does not spend an equal amount of time in all parts of the space. If $a(i,j)$ is the adjacency matrix of the graph, then the transition probability

$$p(i,j) = a(i,j)/d(i)$$

where $d(i)$ is the degree of i, so if we let $\pi(i) = d(i)/D$ where $D = \sum_i d(i)$ then the walk is reversible with respect to π

$$\pi(i)p(i,j) = a(i,j)/D = a(j,i)/D = \pi(j)p(j,i).$$

The first step in understanding the time to reach consensus in the voter model is to pick two starting points x_1 and x_2 at random according to the stationary distribution π for the random walk and investigate the time it takes for independent random walks starting from these sites to hit. It is natural to consider coalescing random walks in continuous time, but, for simplicity, we consider the discrete time version of the two particle chain $(\bar{X}_n^1, \bar{X}_n^2)$ in which $\bar{X}_0^i = x_i$ and at each step we pick a particle at random and let it jump. Let $A = \{(x,x) : x \in G\}$ be the "diagonal" and let

$$T_A = \inf\{n \geq 1 : X_n^1 = X_n^2\}$$

be the first hitting time of A by (X_n^1, X_n^2), and write P_A for $P_\pi(\cdot|(X_0^1, X_0^2) \in A)$.

7.3.1 Asymptotics for $E_\pi T_A$

Theorem 7.3.1 (Aldous' (1989a) Poisson Clumping Heuristic) *Let t_{mix} be the mixing time for the Markov chain defined in (6.1.3) and suppose that $t_{mix} \ll E_\pi T_A$, then*

$$E_\pi(T_A) = \frac{1}{\pi(A)} \cdot \frac{1}{P_A(T_A \gg t_{mix})}. \tag{7.3.1}$$

Proof. Kac's recurrence theorem (Theorem 6.3.3 in PTE5) implies that

$$E_A(T_A) = 1/\pi(A).$$

The expected value on $T_A \le t_{mix}$ makes a negligible contribution to $E_A T_A$ so

$$1/\pi(A) \approx P_A(T_A \gg t_{mix}) E_\pi(T_A).$$

Rearranging gives the desired formula.

To connect with Aldous' (1989a) clumping heuristic, we note that the naive guess for the waiting time is $1/\pi(A)$ but this must be corrected for by multiplying by the clump size, i.e., the expected number of hits that occur soon after the first one. In nice cases, e.g. the torus or random regular graphs, the number of hits in a clump is geometric and hence has mean $1/P_A(T_A \gg t_{mix})$. However, a geometric distribution of return times is not necessary for (7.3.1) to hold.

Example 7.3.2 (d-dimensional torus, $d \ge 3$) $(\mathbb{Z} \bmod N)^d$ has $n = N^d$ sites. $\pi(x, x) = N^{-2d}$ so $\pi(A) = N^{-d}$. By the central limit theorem the mixing time $t_{mix} = O(N^2)$ in any dimension. In $d \ge 3$, $P_A(T_A \gg t_{mix}) \to \beta_d = P_0(S_n \ne 0$ for all $n \ge 1\}$, so

$$E_\pi T_A \sim N^d/\beta_d.$$

Example 7.3.3 (Random regular graphs) Locally these graphs look like a tree in which each vertex has degree r. The probability that two random walkers that start from the origin on an r-regular tree will hit after they separate is the same as the probability that a single random walk will return to the origin, which is $1/(r-1)$. To check this note that $\phi(x) = 1/(r-1)^x$ is a harmonic function for the distance S_n from the root of the tree, i.e., $\phi(S_n)$ is a martingale when $S_n > 0$. Since $P_A(T_A \gg t_{mix}) \approx (r-2)/(r-1)$, we should have

$$E_\pi T_A \sim \frac{r-1}{r-2} n. \tag{7.3.2}$$

Note that the right-hand side is ∞ when $r = 2$, since a random 2-regular graph is a union of circles.

Example 7.3.4 (Power law random graphs) If the graph is generated by the configuration model the diagonal has probability

$$\pi(A) = \sum_{i=1}^{n} d_i^2/D^2 \quad \text{where} \quad D = \sum_{j=1}^{n} d_j.$$

If the degree distribution has finite second moment then $\pi(A) = O(1/n)$. When the degree distribution has infinite variance $\pi(A)$ will go to 0 more slowly than $1/n$. Consider random graphs with a power law degree distribution $p_k \sim Ck^{-\gamma}$. When $\gamma > 3$ the distribution has finite variance. For $2 \le \gamma \le 3$ results of Sood and Redner (2005) for the consensus time of the voter model suggest that

$$E_\pi T_A \approx \begin{cases} n/\log n & \gamma = 3, \\ n^{(2\gamma-4)/(\gamma-1)} & 2 < \gamma < 3, \\ (\log n)^2 & \gamma = 2, \end{cases} \tag{7.3.3}$$

where $\approx$ should be read "is of order." For simplicity we will only consider the case $2 < \gamma < 3$. The first step is to show.

Lemma 7.3.5 *When $2 < \gamma < 3$ the asymptotic behavior of $1/\pi(A)$ is given by the formulas above.*

Proof. When $p_k \sim Ck^{-\gamma}$ we have

$$P(d_i > k) \sim C_1 k^{-(\gamma-1)} \quad \text{and} \quad P(d_i^2 > k) \sim C_2 k^{-(\gamma-1)/2}.$$

To prepare for later, note that the maximum degree is $O(n^{1/(\gamma-1)})$ when $\gamma > 2$.

When $2 < \gamma < 3$, d_i^2 is in the domain of attraction of a stable law index $\beta = (\gamma-1)/2$ and asymmetry $\kappa = 1$. See page 165 in PTE5. It follows that

$$\frac{1}{n^2} \sum_{i=1}^{n} d_i^2 \sim Y_{\beta,1} n^{2/(\gamma-1)}/n^2 = Y_{\beta,1} n^{(2\gamma-4)/(\gamma-1)},$$

where $Y_{\beta,1}$ has the indicated stable law.

If we assume that the minimum degree is 3 then the mixing time $t_{mix} = O(\log n)$, and in the example $T_A/(\log n) \to \infty$ in probability.

7.3.2 Asymptotic Exponential Distribution

Proposition 3.23 in Aldous and Fill (2002) implies

$$\sup_t |P_\pi(T_A > t) - \exp(-t/E_\pi T_A)| \le t_{rel}/E_\pi T_A, \tag{7.3.4}$$

where t_{rel} is the **relaxation time**, which is defined to be 1 over the spectral gap, see 6.1.9. In the random regular case if $r \geq 3$ then $t_{rel} \leq C$ and, as we have seen,

$$E_\pi T_A \sim \frac{r-1}{r-2} n, \tag{7.3.5}$$

so the hitting time is approximately exponential.

The proof of (7.3.4) is based on a result of Mark Brown (1983) for IMRL (increasing mean residual life) distributions. If one is willing to give up on the explicit error bound, it is fairly easy to give a proof based on the idea that since convergence to equilibrium occurs much faster than the two particles hitting, then subsequential limits of $T_A/E_\pi T_A$ must have the **lack of memory property**, and hence the sequence converges to a mean 1 exponential.

Theorem 7.3.6 *If the mixing time $t_{mix} = o(1/\pi(A))$, and $E_\pi T_A \leq C/\pi(A)$, then under P_π, $T_A/E_\pi T_A$ converges weakly to an exponential with mean 1.*

Proof. Since $t_{mix} \ll 1/\pi(A) \leq E_\pi T_A$, we can find $\epsilon_n \to 0$ with $(\epsilon_n E_\pi T_A)/t_{mix} \to \infty$. The expected time that $X_t^1 = X_t^2$ in $[rE_\pi T_A, (r+\epsilon_n)E_\pi T_A + 1]$ is equal to $\pi(A)(\epsilon_n E_\pi T_A + 1)$. If $X_s^1 = X_s^2$ then with probability $\geq e^{-2}$ neither jumps before time $s+1$ and hence they agree on $[s, s+1]$. From this we conclude that

$$P_\pi(T_A/E_\pi T_A \in [r, (r+\epsilon_n)]) \leq \delta_n = \pi(A)(\epsilon_n E_\pi T_A + 1) \to 0. \tag{7.3.6}$$

Notice that δ_n does not depend on r. Using this result with $r = s+t$ and writing $X_t = (X_t^1, X_t^2)$, $x = (x_1, x_2)$

$$\begin{aligned} P_\pi(T_A/E_\pi T_A > (s+t)) &= P_\pi(T_A/E_\pi T_A > (s+t+\epsilon_n)) + \delta_n \\ &= \sum_{x,y} P_\pi(T_A/E_\pi T_A > s, X(sE_\pi T_A) = x) \\ &\qquad P_x(X(\epsilon_n E_\pi T_A) = y) P_y(T_A/E_\pi T_A > t). \end{aligned}$$

Subtracting $\pi(y)$ from $P_x(X(\epsilon_n E_\pi T_A) = y)$ and adding $\pi(y)$ gives two terms. The second is

$$\begin{aligned} &\sum_{x,y} P_\pi(T_A/E_\pi T_A > s, X(sE_\pi T_A) = x)\pi(y) P_y(T_A/E_\pi T_A > t) \\ &\quad = P_\pi(T_A/E_\pi T_A > s) P_\pi(T_A/E_\pi T_A > t). \end{aligned}$$

The absolute value of the first term is bounded by

$$P_\pi(T_A/E_\pi T_A > s) \sup_x \sum_y |P_x(X(\epsilon_n E_\pi T_A) = y) - \pi(y)| \to 0, \tag{7.3.7}$$

due to the definition of the mixing time.

The sequence of random variables $T_A/E_\pi T_A$ has mean one, so it is tight. Let F denote a subsequential limit. From the calculation above we see that if s, t, and $s+t$ are continuity points of F then

$$1 - F(s+t) = (1 - F(s))(1 - F(t)).$$

F can have at most countably many discontinuity points, so there is a $\theta > 0$ so that F is continuous at all points $m/(\theta 2^n)$ where m and n are positive integers. Define λ by $e^{-\lambda} = 1 - F(1/\theta)$. It follows from the equation that if $t = m/(\theta 2^n)$ then $1 - F(t) = e^{-\lambda t}$. To conclude that λ is independent of the subsequential limit, note that for all n, $P_\pi(T_A/E_\pi T_A > 2) \le 1/2$, so using the remark about δ_n and (7.3.7), we see that if $\gamma > 0$ and n is large

$$P(T_A/E_\pi T_A > 2(k+1)) \le P(T_A > 2k)(\gamma + P(T_A > 2)).$$

This gives an exponential bound on the tail of the distribution, which enables us to conclude that every subsequential limit has mean 1.

7.4 A Bound on the Limiting Coalescence Time

Suppose we run a coalescing random walk starting with one particle at each site of a graph. Let **C** be the time until only one random walk remains. In this section we describe a result of Cooper et al. (2012) on the expected coalescence time. These authors use lazy discrete time simple random walks, but the result is easier to prove and more closely related to the voter model if the walks have continuous time, so we study that case here.

Theorem 7.4.1 *Let G be a connected graph with n vertices, average vertex degree $\bar{d}$ and maximum degree $\Delta = O(n^{1-\epsilon})$. Let $\nu = (\sum_{v \in V} d^2(v))/(\bar{d}^2 n)$. Let $\mathbf{C}(n)$ be the expected coalescence time for a system of particles making a continuous time random walk starting from each vertex that jumps at rate 1. Then*

$$\mathbf{C}(n) = O\left(\frac{n}{\nu(1-\lambda_1)}\right),$$

where λ_1 is the second largest eigenvalue.

Remark. Noting that $\nu/n = \pi(A)$, we see that if the minimum degree is 3 and hence by Theorem 6.4.1 the spectral gap $1 - \lambda_1$ is bounded away from 0, this result gives an upper bound for the coalescence time which is accurate in the case of power law degree distributions $p_k \sim Ck^{-\gamma}$ with $2 < \gamma < 3$. See Lemma 7.3.5.

7.4.1 Random Walk Properties

Let $G = (V, E)$ be a connected graph with $|V| = n$ and let $d(v)$ be the degree of the vertex v. Let $W_u(t)$ denote a continuous time random walk that starts from u, jumps at rate 1, and jumps from v to each of its neighbors with equal probability. Let $P_u^t(x) = P(W_u(t) = x)$. It follows from Theorem 6.1.1 that

$$|P_u^t(x) - \pi_x| \le (\pi_x/\pi_u)^{1/2} e^{-(1-\lambda_1)t}. \tag{7.4.1}$$

Define the time to reach equilibrium, T_G, so that for $t \ge T_G$

$$\max_{u,x} |P_u^t(x) - \pi_x| = o(1/n^2). \tag{7.4.2}$$

Let $E_\pi(H_w)$ denote the expected hitting time of w starting from the stationary distribution.

Lemma 7.4.2 *Let F be a graph with eigenvalue gap $1 - \lambda_1$. Then*

$$E_\pi H_v \le \frac{1}{1-\lambda_1} \cdot \frac{1}{\pi_v}.$$

Proof. According to Chap. 2 of Aldous and Fill (2002)

$$E_\pi(H_v) = Z_{v,v}/\pi_v, \tag{7.4.3}$$

where Z is the recurrent potential kernel

$$Z_{v,v} = \int_{t=0}^{\infty} P_v^t(v) - \pi_v \, dt.$$

They prove this result in discrete time in Lemma 11, but as they observe in Sect. 2.3 the result is the same in continuous time. Using (7.4.1) with $x = u = v$

$$|P_v^t(v) - \pi_v| \le e^{-(1-\lambda_1)t}.$$

Integrating gives the desired result.

Let $A_v(t; u)$ be the event that W_u does not visit vertex v in $[0, t]$.

Lemma 7.4.3 $P(A_v(t; u)) \le \exp(-\lfloor t/(T_G + 3E_\pi H_v)\rfloor)$.

Proof. To simplify formulas write T_G as T. Let $\rho = P_u^T$ be the distribution of W_u after T steps. Then (7.4.2) and the fact that $\pi_x \ge 1/n^2$ for any connected graph (since $1 \le d_y \le n-1$ for all $y \in V$) imply

$$E_\rho(H_v) = (1 + o(1))E_\pi(H_v),$$

and it follows that

$$E_u(H_v) \le T_G + (1 + o(1))E_\pi(H_v). \tag{7.4.4}$$

If we let $H_v(\rho)$ be the hitting time of v starting from ρ and $\tau = T + 3E_\pi(H_v)$ then

$$\begin{aligned} P(A_v(\tau; u)) &= P(A_v(T; u), H_v(\rho) \ge 3E_\pi H_v) \\ &\le P(H_v(\rho) \ge eE_\pi H_v) \le e^{-1}. \end{aligned}$$

Iterating this bound $\lfloor t/(T_G + 3E_\pi H_v)\rfloor$ times gives the desired result.

7.4.2 Multiple Random Walks

To study the coalescence of $k \ge 2$ walks on a graph $G = (V_G, E_G)$, we replace the k walks by one walk on a new graph Q_k with vertex set V^k. Since we work in continuous time where only one particle jumps at a time, our edge set is much simpler than the one in Cooper et al: vertices $v, w \in Q_k$ are adjacent if there is a j so that $v_i = w_i$ for $i \ne j$ and $\{v_j, w_j\} \in E_G$.

Although we are interested in coalescence, our k random walks will be independent. For any starting positions $u = (u_1, \dots, u_k)$ for the walks, let M_k be the time of the first meeting in G. Let

$$S_k = \{(v_1, v_2, \dots, v_k) : v_i = v_j \text{ for some} 1 \le i < j \le n\}.$$

To use results from the previous section it is convenient to contract the set S_k to be a single vertex γ making a new graph Γ_k. On contraction all edges including those that have become loops or parallel edges are retained.

Let π_k and $\hat\pi_k$ be the stationary distributions on Q_k and Γ_k respectively. For any vertex $v = (v_1, v_2, \dots, v_k) \in Q_k$, $\pi_k(v) = \pi(v_1)\cdots\pi(v_k)$. If $v \notin S_k$, $\hat\pi_v = \pi_v$, while $\hat\pi_\gamma = \sum_{x\in S}\pi_x$. The next order of business is to find a lower bound for $\hat\pi_\gamma$.

Lemma 7.4.4 *There is a constant $c_k > 0$ so that if $k \le \log^4 n$*

$$\hat\pi_\gamma \ge \frac{c_k k^2 \nu}{2n}.$$

Proof. For $1 \le x < y \le k$ let $S_{x,y} = \{(v_1, \dots, v_k) : v_x = v_y\}$ and note $S = \cup_{1\le x<y\le k} S_{x,y}$.

$$\pi(S_{k-1,k}) = \sum_{v_1,\dots v_{k-1}} \pi(v_1)\cdots\pi(v_{k-2})\pi^2(v_{k-1}) = \sum_v \left(\frac{d(v)}{n\bar d}\right)^2 = \frac{\nu}{n}.$$

The inclusion exclusion formula implies

$$\pi(S_k) \geq \sum_{x,y} \pi(S_{x,y}) - sum_{\{x,y\}\neq\{p,q\}} \pi(S_{x,y} \cap S_{p,q}).$$

Taking the cases $\{x, y, p, q\} = 3, 4$ separately (the second one does not exist if $k = 3$)

$$\pi(S_{k-2,k-1} \cap S_{k-1,k}) = \sum_v \pi(v)^3 \leq \Delta \sum_v \frac{d(v)^2}{n^3 \bar{d}^3} \leq \frac{\Delta \nu}{n^2 \bar{d}},$$

$$\pi(S_{1,2} \cap S_{k-1,k}) = \sum_{u,v} \pi(u)^2 \pi(v)^2 \leq \frac{\nu^2}{n^2}.$$

Combining the last two results

$$\pi(S) \geq \binom{k}{2}\frac{\nu}{n} - 3\binom{k}{3}\frac{\Delta\nu}{n^2\bar{d}} - 3\binom{k}{4}\frac{\nu^2}{n^2}$$
$$\geq \binom{k}{2}\frac{\nu}{n}\left[1 - \frac{k\Delta\nu}{n\bar{d}} - \frac{k^2\nu}{n}\right].$$

We have assumed $\Delta = O(n^{1-\epsilon})$ and $k \leq \log^4 n$, so the middle term inside the square brackets is $o(1)$. Noting that

$$\nu \leq \Delta \sum_v \frac{d(v)}{\bar{d}^2 n} = \frac{\Delta}{\bar{d}}, \tag{7.4.5}$$

the last term is also $o(1)$ and the proof is complete.

It follows from (7.4.4) that if T_Γ is the time to reach equilibrium satisfying (7.4.2) then

$$E(M_k) \leq T_\Gamma + (1 + o(1))E_{\hat{\pi}}(H_\gamma).$$

To bound the first quantity we use

Lemma 7.4.5 *For the random walks on G, Q, and Γ there are mixing times*

$$T_G \leq \frac{C \log n}{1 - \lambda_1(G)}, \quad T_Q = O(kT_G), \quad \text{and} \quad T_\Gamma = O(kT_G).$$

Proof. The bound on T_G follows from (7.4.1). In the jargon of Markov processes the random walk on Q_k is the tensor product chain. The eigenvectors are the k-wise products of those of G, so $\lambda_1(Q_k) = \lambda_1(G)$. See page 168 of Levin, Peres, and Wilmer's book for more details.

In the notation of Aldous and Fill (2002), Chap. 3, the random walk on Γ_k is the random walk on Q_k with S_k collapsed to γ_k. Their Corollary 27 implies that if a subset A of vertices is collapsed to one then the second eigenvalue is not increased.

We get the factor of k in the bounds on the mixing times because we need to have $|P_u^T(x) - \pi_x| = o(1/n^{2k})$ and $\pi_x/\pi_u \le n^{2k}$ by the trivial bound we used in the proof of Lemma 7.4.3.

Lemma 7.4.6 *If* $k \le \log^4 n$ *then*

$$E_u(M_k) = O\left(\frac{1}{1-\lambda_1(G)}\left(k\log n + \frac{n}{\nu k^2}\right)\right).$$

Proof. Using Lemmas 7.4.2 and 7.4.4

$$E_{\hat{\pi}}(H_\gamma) \le \frac{1}{\hat{\pi}(\gamma)} \cdot \frac{1}{1-\lambda_1(\Gamma)} \le \frac{1}{c_k k^2} \cdot \frac{n}{\nu(1-\lambda_1(G))}. \tag{7.4.6}$$

Using (7.4.4) we have $E_u(M_k) \le O(kT_G) + (1+o(1))E_{\hat{\pi}} H_\gamma$, so the desired result follows from Lemma 7.4.5.

Summing over $k \le \log^4 n$

$$\sum_{k=2}^{\log^4 n} E(M_k) = O\left(\frac{n}{\nu(1-\lambda_1(G))}\right) + O\left(\frac{\log^7 n}{1-\lambda_1(G)}\right).$$

From (7.4.5) we have $\nu \le \Delta/\bar{d}$, and the first term is dominant as long as $\Delta = O(n^{1-\epsilon})$.

7.4.3 *The Big Bang*

The last thing to do is to estimate the time for the coalescing random walk to be reduced from n particles to $\log^3 n$. We prove that

Lemma 7.4.7 *With high probability there cannot be a set of* $k_* = \log^4 n$ *particles that has not had a meeting by time* $t^* = k_*^{3/2}(T_\Gamma + 3E_\pi(H_\gamma))$.

Proof. Suppose that the starting points are $v = (v_1, \ldots, v_k)$. The probability that the particles do not meet by time t is the same as the probability that the random walk in Γ_{k_*} starting from v does not visit γ by time t^*. By Lemma 7.4.3 this probability is

$$\le \exp(-\lfloor t^*/(T_\Gamma + 3E_\pi H_\gamma)\rfloor) \le \exp(-k^{3/2}) = e^{-\log^6 n}.$$

The probability no set of size k does this is

$$\leq \binom{n}{k} e^{-\log^6 n} \leq n^k e^{-\log^6 n} \leq \exp(-\log^6 n + \log^5 n).$$

To complete the result of Theorem 7.4.1 now, we note that Lemma 7.4.5 and (7.4.6) imply

$$t^* = O\left(\frac{1}{1-\lambda_1(G)}\left(k^{5/2}\log n + \frac{n}{\nu}\right)\right)$$

and $k^{5/2}\log n = o(n/\nu)$ as long as $\Delta = O(n^{1-\epsilon})$.

7.5 Mean-Field Behavior of CRW on General Graphs

On the torus and random regular graphs we proved convergence of the rescaled coalescence time to Kingman's coalescent using detailed calculation. In this section we give a general result due to Oliveira (2012, 2013). To set the stage we go back to Chapter 14 of Aldous and Fill's (2002) book.

Open Problem 14.13. Let $\mathbf{C}$ be the coalescent time for a coalescing random walk starting with one particle at each site. Let $T^G_{hit} = \max_{u,v} E_u T_v$. Prove there is a constant K so that

$$E\mathbf{C} \leq K T^G_{hit}.$$

It might seem more natural to phrase the conjecture in terms of $T^G_{meet} = max_{i,j} E M_{i,j}$ where $M_{i,j}$ is the time for two independent random walks starting at i and j to hit. The next two results show that in general this is not possible. Proposition 14.5 in Aldous and Fill shows that $T^G_{meet} \leq T^G_{hit}$. They also show that on a "symmetric" (i.e., vertex transitive) graph $T^G_{meet} = (1/2)T^G_{hit}$.

Example 7.5.1 **The star graph** has center 0 and leaves $1, \ldots, n$.

- $T^G_{meet} = \Theta(1)$. Letting o denote outside and c denote center we can formulate a Markov chain with three states (o, o), (c, o), and (c, c) and conclude that $E_{(o,o)} T_{(c,c)} \to 2$ as $n \to \infty$.
- $T^G_{hit} = E_0 T_1 \geq n$, since particle has to return to the center an average of n times before it will jump to 1. A little more thought gives $E_0 T_1 \sim 2n$.
- The coalescence time $\mathbf{C}$ is $\Omega(\log n)$. To prove the lower bound note that when there are $k+1$ particles coalescence occurs at rate $\leq (k+1)$ since the rate is 0 when there is no particle in the center, and bounded by this when there is. This shows that unlike Kingman's coalescent on the complete graph or on the d-dimensional torus, the coalescent on the star graph does not come down from infinity in finite time.

Oliveira's (2012) Proof of the Conjecture of Aldous and Fill

Let Q be the generator of a reversible, irreducible, continuous time Markov chain on a finite set V. Let $\mathbf{C}_k$ be the time until only k particles remain.

Theorem 7.5.2 *There is a constant $K > 0$ so that for any n and $\mathbf{x} = (x_1, \ldots, x_n) \in V^n$*

$$E_{\mathbf{x}}\mathbf{C}_k \le K \left(T^Q_{hit}/k + T^Q_{mix} \right).$$

This is his Theorem 1.5, which shows that the dependence on k is the same as in Kingman's coalescent. Theorem 1.3 covers the case $k = 1$, where the second term is not needed.

Sketch of Proof of Theorem 1.3. Label the particles $X_t(a)$, $a = 1, \ldots, n$. Instead of a coalescence a particle dies when it collides with a lower numbered particle. As in the TV show the dead continue walking. If we let A_t be the particles alive at time t then in this system

$$P(\mathbf{C} > t) \le \sum_{a=2}^{n} P(a \in A_t).$$

Oversimplification #1. We say that particle a dies if it hits a particle $b < a$ even if that particle is dead. This leads to the formula

$$P(a \in A_t | X_s(a) = h_s, 1 \le s \le t) = \prod_{b=1}^{a-1} P(X_s(b) \neq h_s \text{for} 0 \le s \le t).$$

Oversimplification #2. Start the particles independent and distributed according to the stationary distribution. This allows the proof of an intriguing"meeting time lemma" which is similar to the previous displayed equation. A few lines leads to the simple bound

$$E\mathbf{C} \le (\ln 2 + 1) T^Q_{hit}.$$

This is not a proof because of the oversimplifications, but this is the beginning of his explanation of the proof. The reader should consult the paper for more details.

Oliveira's (2013) Proof of Convergence to Kingman's Coalescent

His first result is for the **transitive case**: i.e., given x and y one can find a permutation of the state space that preserves adjacency and maps x to y. In addition he assumes **reversibility** i.e., the stationary distribution satisfies detailed balance $\pi(x)Q(x, y) = \pi(y)Q(y, x)$. There is a second similar result for a general chain, which assumes that the mixing time is small enough relative to the other parameters of the chain. We refer the reader to the paper for details of the proof of that result.

To state his result for transitive reversible chains we need some notation.

Let X_t be a continuous time Markov chain with generator Q. Let π be the stationary distribution of Q. Let $m(Q)$ be the expected meeting time of two independent copies of the chain Q, each starting from the stationary distribution.

Let $t^Q_{mix}(\alpha)$ be the α-mixing time, i.e., the smallest value of t so that for all initial states x

$$\|P_x(X_t \in \cdot) - \pi(\cdot)\|_{TV} \le \alpha,$$

where TV indicates we are considering the total variation distance. The specific value $t^Q_{mix}(1/4)$ is called the mixing time and denoted by t^Q_{mix}. The constant 1/4 is chosen, see (6.1.3), so that for all $\epsilon < 1/4$,

$$t^Q_{mix}(\epsilon) \le C \ln(1/\epsilon) t^Q_{mix}.$$

To define the sense in which convergence occurs define the **Wasserstein distance** between the distributions of two random variables X and Y with finite means by

$$d_W(X, Y) = \int |P(X > x) - P(Y > x)|\, dx.$$

As is well known $d_W(X, Y) = \sup\{|Ef(X) - Ef(Y)| : f \text{is 1-Lipshitz}\}$.

Finally to define the limiting coalescence time of Kingman's coalescent we let Z_i, $i \ge 2$, be independent exponentials with

$$P(Z_i > t) = \exp\left(-t\binom{i}{2}\right).$$

Theorem 7.5.3 *Start with one copy of the Markov chain at each site and let* $\mathbf{C}$ *be the time needed for all the particles to coalesce to 1. If* $\rho(Q) = t^Q_{mix}/m(Q)$ *then*

$$d_W\left(\frac{\mathbf{C}}{m(Q)}, \sum_{i\ge 2} Z_i\right) = O\left([\rho(Q)\ln(1/\rho(Q))]^{1/6}\right),$$

and hence $\mathbf{C}/m(Q) \Rightarrow \sum_{i\ge 2} Z_i$ *if* $\rho(Q) \to 0$.

The strength of this result is that it provides quantitative bounds. Since he used the Wasserstein distance, he could conclude

$$E\mathbf{C}/m(Q) - 2 = O\left([\rho(Q)\ln(1/\rho(Q))]^{1/6}\right).$$

Ideas that Go Into the Proof. Let $\mathbf{C}_k$ be the time until only k particles remain. $M = \mathbf{C}_1 - \mathbf{C}_2$ is the hitting time of

$$\Delta = \{(x, x) : x \in V\}.$$

$M^{(i+1)} = \mathbf{C}_i - \mathbf{C}_{i+1}$ is the hitting time of

$$\Delta^{(i+1)} = \{(x_1, \dots x_{i+1}) : x(i_1) = x(i_2) \text{ for some } 1 \le i_1 < i_2 \le i+1\}.$$

As Oliveira described in Sect. 1.4, to prove the result we have to face several difficulties:

- At time $\mathbf{C}_{i+1}$ the random walkers are not in equilibrium. This is where the assumption that mixing is faster than hitting comes in.
- The mean-field picture suggests that $M^{i+1}/m(Q)$ should be close in distribution to Z_i. Results of Aldous and Fill suggest that hitting times are exponentially distributed but to control the Wasserstein distance we need better control of the tail of $M^{(i+1)}$.
- We do not expect the exponential approximation to hold when the set $\Delta^{(i+1)}$ is too large, so we need different methods to control the Big Bang phase.

7.6 Asymptotics for CRW Densities

7.6.1 On the Torus in d ≥ 2

By duality between the voter model ξ_t^0 and coalescing random walk ζ_t^1

$$p_t = P(\xi_t^0 \neq \emptyset) = P(0 \in \zeta_t^1).$$

In a sequence of papers Sudbury (1976), Kelly (1977), Sawyer (1979), and Bramson and Griffeath (1980) combined their talents to prove

$$p_t \sim \begin{cases} (\log t)/\pi t & d = 2, \\ 1/\beta_d t & d \ge 3, \end{cases} \tag{7.6.1}$$

where β_d is the probability a d-dimensional simple random walk starting at 0 never returns there. The result in one dimension is easy and is omitted.

There is a simple heuristic that gives the answer in $d \ge 3$. If lattice sites on $\mathbb{Z}^d$ are independent and occupied with probability u, then the instantaneous rate at which collisions occur is u^2. If we assume that the particles are always randomly scattered on the lattice then their density would satisfy the differential equation:

$$\frac{du}{dt} = -u^2 \quad \text{and} \quad u(0) = 1.$$

The solution is $u(t) = 1/(1+t)$. If our random walks do not coalesce when they meet then in $d \geq 3$, two that hit once would hit a geometric number of times with mean $1/\beta_d$. Removing the over counting from the previous equation

$$\frac{du}{dt} = -\beta_d u^2,$$

so $u(t) \sim 1/(\beta_d t)$. To prepare for the next calculation, note that we have multiplied the naive asymptotics, $1/u$, by the expected number of collisions given that one occurred.

In $d = 2$, this reasoning needs to be modified since random walks are recurrent and hence they hit infinitely many times. In $d = 2$ the random walk S_t that jumps to a randomly chosen nearest neighbor at rate 2 has mean 0 and covariance matrix I, where I is the identity matrix, so using the (local) central limit theorem

$$P(S_t = 0) \sim \frac{1}{2\pi t}.$$

When S_t hits 0, it stays there for an amount of time that is exponential with mean 1/2. The expected number of times two random walks hit in $[0, t]$ is thus

$$\sim 2 \int_1^t \frac{1}{2\pi s}\, ds \sim \frac{\log t}{\pi}.$$

Multiplying the naive asymptotic $\sim 1/t$ by this gives the result in $d = 2$.

The rigorous derivation of (7.6.1) involves a number of clever observations. We follow the summary in Bramson and Griffeath (1980). Unfortunately in that paper the voter model is ζ_t and the coalescing random walk is ξ_t. We will write things in our notation. Consider the voter model starting from all sites different and let

$$N_t = |\{x : \xi_t(x) = \xi_t(0)\}|$$

be the size of the patch of particles in the voter model at time t that is the same type as the particle at the origin. Sudbury (1976) observed that

Lemma 7.6.1 *If R_t is the number of points visited by our rate 2 random walk by time t, i.e., $R_t = |\{x : x = S_s \text{ for some } s \leq t\}|$ then*

$$EN_t = ER_t. \tag{7.6.2}$$

Proof. In order for $\xi_t(0) = \xi_t(x)$, the dual processes must have $\zeta_s^{0,t} = \zeta_s^{x,t}$ for some $s \leq t$., which has the same probability as $\zeta_r^{0,t} - \zeta_r^{x,t}$ hitting 0 by time t, which is the same as the probability S_r hits x by time t when it starts from 0.

Combining (7.6.2) with results of Dvoretsky and Erdős (1951) for the range of random walk we have

$$EN_t \sim \begin{cases} \frac{2\pi t}{\log t} & d = 2, \\ 2\beta_d t & d \geq 3. \end{cases} \tag{7.6.3}$$

The reasoning is the same as in our heuristic proof. The rate 2 random walk makes $\sim 2t$ jumps by time t. To find the range we divide this by the average number of times a site is visited given that it is visited at least once.

Kelly (1977) noticed

Lemma 7.6.2 *Let* $n_t = |\xi_t^0|$.

$$P(N_t = k) = kP(n_t = k). \tag{7.6.4}$$

Proof. Suppose someone shows us a large d-dimensional cube of side L from the configuration from time t, not telling us where the origin is located, and then we pick a site at random to be the origin. An opinion with k representatives will be chosen with probability k/L^d. Thus when we pick a cluster containing a given site, we get a size-biased pick from the original distribution. Note that since $En_t = 1$, the right-hand side is a probability distribution.

Combining (7.6.4) with Jensen's inequality and (7.6.3) we have

$$p_t = E(N_t^{-1}) \geq (EN_t)^{-1},$$

which gives an asymptotic lower bound of 1/2 the correct answer in (7.6.1). To close the gap we use the following result of Sawyer (1979).

Theorem 7.6.3 *For* $d \geq 2$*, and* $k = 1, 2, \ldots$

$$\lim_{t\to\infty} E(N_t/EN_t)^k = \frac{(k+1)!}{2^k},$$

and hence

$$\lim_{t\to\infty} P(N_t/EN_t \leq x) = \int_0^x 4ye^{-2y}\, dy.$$

In words N_t/EN_t converges to a limit Y with gamma(2, 2) distribution, the sum of 2 independent exponentials with rate 2. Less formally

$$N_t \sim Y \cdot EN_t.$$

Ignoring the fact that x^{-1} is unbounded near 0, and hence small probabilities can have a large impact on the expected value, we have

$$p_t = EN_t^{-1} \sim \frac{EY^{-1}}{EN_t} \sim \frac{2}{EN_t},$$

and we have the quoted result.

Bramson and Griffeath (1980) make this rigorous by showing $p_t = \Theta((\log t)/t)$ in $d = 2$ and $p_t = \Theta(1/t)$ in $d \geq 3$. Our description makes it seem that all they did was to cross the t's and dot the i's in asymptotics. To complete the proof they had to use an ingenious argument to show that the coalescing random walk had a self-correcting mechanism, i.e., if p_t is larger than it should be then coalescence will be more rapid. A major technical difficulty is that knowing the density does not give one much of an idea about how they are distributed in space.

Open Problem 7.6.4 *It would be nice to have a less devious proof for the asymptotics in* (7.6.1). *Kesten and van den Berg (2000) have given a much more direct proof but unfortunately it only works in dimensions* $d \geq 6$.

We will discuss their result in the next section.

7.6.2 On Graphs

Hermon et al. (2022), or HLYZ for short, study CRW on a finite graph $\mathcal{G} = (V, \mathcal{E})$ starting from one particle at each vertex and in which particles perform independent **edge simple random walks**, i.e., for each edge incident to the current location jumps across that edge occur at rate 1. They also consider the more general situation in which the jump rate is symmetric $r_{x,y} = r_{y,x}$ but depends on the edge. In this case the Markov chain is reversible with respect to counting measure.

They denote by ξ_t the set of elements of V that are occupied by a particle at time t and define the coalescence time

$$\tau_{coal} = \inf\{t : |\xi_t| = 1\}.$$

Interest focuses on the rate of decay of $\bar{u}_t = \sum_x u_t(x)/|V|$ where

$$u_t(x) = P(x \in \xi_t).$$

HLYZ call this $P_t(x)$ but there are already way too many p's in the paper. Being easily confused, we will modify the notation to distinguish the average $\bar{u}_t$, which is a number, from the function $u_t(x)$.

HLYZ are interested in the rate at which $\bar{u}_t \to 0$ for two collections of graphs

(i) Graphs generated by the configuration model with minimum degree 3. In this case it is natural to guess that $\bar{u}_t = \Theta(1/t)$ or even $\lim_{t\to\infty} t\bar{u}_t = c$.
(ii) A family of vertex transitive graphs $\mathcal{G} = (V, \mathcal{E})$ satisfying some general 'transience-like' and 'spectral' conditions.

To formally state their results we need some definitions. Given two independent walks X_t and Y_t they define the **meeting time**, which is a random variable, by

$$\tau_{meet}(X_0, Y_0) = \inf\{t : X_t = Y_t\}.$$

For the graph they define a constant by

$$t_{meet}(\mathcal{G}_n) = E_{\pi,\pi}\tau_{meet}.$$

For each $x \in V$ they define the **neighborhood distribution** ν_x, which is the distribution of the particle after it has jumped from x.

In case (i) it is the uniform distribution on the neighbors of x.

In case (ii) it is $\nu_x(y) = r_{x,y}/r(x)$ where $r(x) = \sum_y r_{x,y}$.

The definition in case (i) is a special case of that in case (ii), so in general we can define

$$\alpha_t(x) = r(x)P_{x,\nu_x}(t_{meet} > t), \tag{7.6.5}$$

where the subscript gives the distributions of X_0 and Y_0 in the definition of the meeting time. Intuitively $\alpha_t(x)$ gives the distribution of the meeting time for two particles that start at x. To have a quantity associated with the graph as a whole, we define

$$\bar{\alpha}_t(\mathcal{G}_n) = \frac{1}{|V|}\sum_{x \in V}\alpha_t(x).$$

t_{meet} and $\bar{\alpha}_t$ depend on the graph $\mathcal{G}_n$, but we will drop the dependence from the notation

Using the newly introduced notation the two main conclusions can be stated succinctly as

(T1) $\bar{u}_t \approx 1/t\alpha_t$.
(T2) $\bar{u}_t \approx 2t_{meet}/nt$.

Their paper is rather lengthy so we will devote most of our effort here to explaining why these results are true. Our next step is to state the results precisely

Configuration Model. Assume that the degree distribution D has $P(3 \le D \le M) = 1$ for some $M < \infty$. Let $\mathcal{G}_n$ be the graph on n vertices generated by the configuration model $\mathbb{CM}_n(D)$. The lower bound $D \ge 3$ implies that the probability G is connected tends to 1. They also consider the local weak limit of $\mathbb{CM}_n(D)$, which is a **unimodular Galton-Watson tree** $\mathcal{G} \sim \mathbb{UGT}(D)$. In this structure the root has offspring distribution D and later generations have offspring the size-biased distribution D^*

$$P(D^* = k) = \frac{(k+1)P(D = k+1)}{ED}.$$

Here, we are thinking of D as a random variable, while HLYZ think of it as a density function, i.e. our ED is their $\sum_i iD(i)$.

Let o be the root of $\mathcal{G}$, let $\bar{\alpha}_\infty(D) = E(\alpha_\infty(o))$ where $\alpha_\infty(x) = \lim_{t\to\infty} \alpha_t(x)$. Theorem 1 in HLYZ is

Theorem 7.6.5 *Let $\mathcal{G}_n$ be sampled from $\mathbb{CM}_n(D)$. For any sequence of times t_n so that $1 \ll t_n \ll n$ we have the following sequences converging in probability.*

$$\lim_{n\to\infty} t_n u(t_n) = \frac{1}{\bar{\alpha}_\infty(D)}, \tag{7.6.6}$$

$$\lim_{n\to\infty} t_n u(t_n) \cdot \frac{n}{2t_{meet}(\mathcal{G}_n)} = 1. \tag{7.6.7}$$

HLYZ prove in their Sect. 6.4 that

$$t_{meet}(\mathcal{G}_n)/n \to 1/(2\bar{\alpha}_\infty(D)).$$

This was proved earlier by Chen (2017) for random regular graphs. Of course, if (7.6.6) and (7.6.7) both hold then the two limits must be equal.

Transitive Markov Chains. Theorem 1.9 in HLYZ is

Theorem 7.6.6 *Suppose $r(x) = \sum_y r_{x,y} = 1$, $t_{meet}(r_n) \le |V_n|$ and $t_{rel}(r_n) \ll t_{meet}(r_n)$. For any sequence of times satisfying $t_{rel}(r_n) \ll t_n \ll t_{meet}(r_n)$ we have*

$$\lim_{n\to\infty} \frac{|V_n|t_n}{2t_{meet}(r_n)} P_{t_n} = 1, \tag{7.6.8}$$

$$\lim_{n\to\infty} t_n \alpha_{t_n} P_{t_n} \to 1. \tag{7.6.9}$$

Their Theorem 4 is a stronger result under a stronger transience condition. We refer the reader to their paper for details.

A Simple Explanation for (T1). Kesten and van den Berg (2000) have given a simple argument for the asymptotics for a coalescing random walk on $\mathbb{Z}^d$, $d \ge 3$. Unfortunately their proof only works in $d \ge 6$ and the proof is 49 pages because they consider very general coalescence rules. Here, inspired by the discussion in Sect. 1.2.2 of HLYZ we generalize the argument to graphs. The starting point is the equality

$$\frac{d}{dt} u_t(x) = -2 \sum_{y:y\sim x} r_{y,x} P(x, y \in \xi_t).$$

If we reverse the direction of time then the jump from y to x that caused coalescence becomes a jump from x to a neighbor y according to the neighborhood distribution ν_x. In order for the jump to be a coalescence in the forward process then the reverse trajectories cannot hit as we work backwards in time, which has probability

$$\alpha_t(x) = r(x)P_{x,\nu_x}(t_{meet} > t).$$

Suppose we follow the trajectories back for a time s such that $t_{mix} \ll s \ll n$. Then the locations of endpoints will be independent and (almost) uniformly distributed, and the density of coalescing random walks will not have changed. Thus using duality, we have

$$P(x, y \in \xi_t) \approx (\bar{u}_t)^2,$$

and it follows that

$$\frac{d}{dt}u_t(x) = -\alpha_\infty(x)(\bar{u}_t)^2.$$

Averaging over x we have

$$\frac{d}{dt}\bar{u}_t = -\bar{\alpha}_\infty(\bar{u}_t)^2.$$

Solving the differential equation we have $\bar{u}_t \sim 1/(t\bar{\alpha}_\infty)$, which is the desired result.

Outline of the Real Proof. It is somewhat surprising (to me at least) that they follow the approach described in the previous section that led to the results for the asymptotics on $\mathbb{Z}^d$. Pick a site at the beginning, call it 0, and let N_t be the number of particles that coalesce with it by time t including itself. If we consider the voter model ζ_t starting from all sites different then

$$N_t = \{x : \zeta_t(x) = \zeta_t(0)\}.$$

Let $\hat{N}_t = \{x : \zeta_t(x) = \zeta_t(\mathcal{U})\}$ where $\mathcal{U}$ is a vertex chosen at random. HLYZ1 show (see Lemma 2.1) that in the finite set-up we also have $\bar{u}_t = E(\hat{N}_t^{-1})$, so it suffices to study the moments of N_t and study the behavior near 0.

As the authors say, Sawyer's result, Theorem 7.6.3, involves a "magical" combinatorial calculation that relies heavily on the structure of $\mathbb{Z}^d$, so they must use a different approach. The starting point is a result that provides rough upper and lower bounds on the decay of $u_t(x)$ which holds for general (V, r). We will state the result given in their Theorem 8 in a moment. They then bootstrap the general bounds into bounds sharp to smaller order terms. To do this they derive the asymptotics of moments of $\hat{N}_t$ by studying the CRW directly. They say this is "the most difficult part of our argument." As a corollary they generalize Sawyer's result for $\hat{N}_t/E(\hat{N}_t)$ to other graphs, see their Sect. 1.1.2.

A Chinese proverb says "A journey of a thousand miles begins with a single step." For coalescing random walk densities on graphs that step is:

Theorem 3.1. If $M_t = \sup_{x\in V} \int_0^t p_s(x,x)\,ds$ and $m_t = \inf_{x\in V} \int_0^t p_s(x,x)\,ds$ then there is a constant C_0 so that for every Markov chain (V, r) and all $y \in V$

$$\frac{m_t}{C_0 t} \le u_t(y) \le \frac{C_0 M_t}{t}.$$

For a transitive chain $M_t = m_t = \int_0^t p_s(0,0)\,ds \equiv I_t^0$. A little thought (see Remark 3.3 in HLYZ) shows that up to a universal constant, t/I_t is the expected number of vertices visited by the chain up to time t.

A simple and elegant bound with an interesting interpretation should have an informative proof but we don't want to spoil the fun by revealing the details.

7.7 Voter Model at Intermediate Times

Our results for coalescing random walk concern the long time behavior of the voter model, i.e., its convergence to consensus. Consider for concreteness the voter model on a d-dimensional torus with n sites and follow the behavior of finitely many particles in a coalescing random walk with the aim of computing the joint distribution of these sites at time t. If $t = o(n^{1/d})$ it is impossible to have coalescences that would not have occurred on $\mathbb{Z}^d$ so the behavior of the model on the torus is the same as on $\mathbb{Z}^d$. If the initial configuration is product measure with density p and t is large (and $o(n^{1/d})$) then the distribution is close to the stationary distribution ν_p. Results for the hitting time of random walks on the torus derived earlier in this chapter, see Sect. 7.3, show that at times that are $o(n)$ there are no coalescence that would not have occurred on $\mathbb{Z}^d$ so the distribution remains close to ν_p.

When the time is $\geq \epsilon n$ then the behavior on the torus begins to deviate from that on $\mathbb{Z}^d$. Cox and Greven (1990) have proved a very nice result that gives a precise description of the behavior of the voter model on the torus.

Theorem 7.7.1 *If the voter model on the torus in $d \geq 3$ starts from product measure with density p then at time Nt it looks locally like a mixture of stationary distributions $\nu_{\theta(t)}$ where the distribution of θ_t changes according to the Wright-Fisher diffusion process*

$$d\theta_t = \sqrt{\beta_d \cdot 2\theta_t(1-\theta_t)}dB_t \tag{7.7.1}$$

and β_d is the probability that two random walks starting from neighboring sites never hit.

Intuitively the distribution is a mixture of equilibria since the density, which is conserved by the voter model dynamics changes on a much slower time scale than the voter model takes to converge to equilibrium.

7.7.1 On the Complete Graph

Here we follow the approach in Lecture 3 of Capannoli and den Hollander (2024). These are notes from the Brazilian School of Probability held August 5–9, 2024. Let the number of 1 opinions at time t be

$$O_t^n = \sum_{i=1}^{n} \xi_t(i).$$

This is not a 0, which would be a crazy notation for the number of 1's, but is a capital letter O, which you should think of as short for occupied. This quantity is a Markov chain on $\{0, \ldots, n\}$ with transition probability

$$\begin{aligned} k &\to k+1 \quad &&\text{at rate} k(n-k)/n, \\ k &\to k-1 \quad &&\text{at rate} k(n-k)/n. \end{aligned}$$

The formulas hold for $0 \le k \le n$ but the rates are 0 when $k = 0$ or n. We have n instead of $n - 1$ in the denominator since we allow the voter to imitate himself, which does nothing but simplifies the formulas.

Theorem 7.7.2 *As $n \to \infty$, the sequence of process $n^{-1} O_{nt}^n$ converges to the Wright-Fisher(2) diffusion χ_s on $[0, 1]$ given by*

$$d\chi_s = \sqrt{2\chi_s(1-\chi_s)}. \tag{7.7.2}$$

The Wright-Fisher diffusions in (7.7.1) and (7.7.2) are two of many that we will see in this section. They are special cases of $d\chi_s = \sqrt{\gamma\chi_s(1-\chi_s)}$.

Before we can prove Theorem 7.7.2 we need to give a brief explanation of what it means. If $x \in (0, 1)$ and f defined on $[0, 1]$ is smooth enough then in the general Wright-Fisher diffusion

$$\frac{d}{dt} E_x f(\chi_t)\bigg|_{t=0} = \frac{1}{2} \cdot \gamma x(1-x) f''(x) = \mathcal{L}_\gamma f.$$

This describes the motion of χ_t while it is in $(0, 1)$. The description is completed by saying that 0 and 1 are absorbing states. By Dynkin's formula

$$f(\chi_t) - \int_0^t \mathcal{L} f(\chi_s)\, ds \quad \text{is a martingale.}$$

A process X_t is a **solution of the martingale problem for** $\mathcal{L}$ if for enough functions f

$$f(X_t) - \int_0^t \mathcal{L} f(X_s)\, ds \quad \text{is a martingale.}$$

In the case of the Wright-Fisher equation we say that the martingale problem is **well-posed** since its distribution is the only solution of the martingale problem. For a comprehensive introduction to generators and martingale problems that will fill in the details that we have been vague about, see Chap. 4 of Ethier and Kurtz (1986).

Proof. Since it is known that the martingale problem for the Wright-Fisher model is well posed, Theorem 4.1 in Chap. 7 of Ethier and Kurtz (1986) implies that is enough to show that the infinitesimal mean and variance converge to those of the limiting diffusion. The infinitesimal mean is 0 since the rate of jumps of O^n_{nt} from $k \to k+1$ is the same as the rate of jumps from $k \to k-1$. Taking into account that time is run at rate n the infinitesimal variance is

$$\frac{1}{n^2} \cdot n \cdot 2k(n-k)/n = (k/n)(1-k/n).$$

The last two calculations show that conditions (4.1) and (4.2) in Theorem 4.1 hold. One must also show that in the limit there are no jumps $> \epsilon$, conditions (4.3)–(4.5), but that is easy since all of the jumps have size $\le 1/n$.

7.7.2 On the d-Dimensional Torus, $d \ge 3$

We have already stated the main result in Theorem 7.7.1. We content ourselves to prove an earlier version of the result from Cox (1989). Using his notation the torus

$$\Lambda(N) = \mathbb{Z}^d \cap [-N/2, N/2)^d,$$

with "periodic boundary conditions." Let $\eta_t^{(N)}$ denote the voter model on $\Lambda(N)$ in which the neighbors of x are its nearest neighbors. Let

$$\Delta_t^{(N)} = N^{-d} \sum_{x \in \Lambda(N)} \eta^{(N)}(x)$$

be the fraction of sites in state 1, and let G be the expected number of returns to 0 for a nearest neighbor simple random walk starting at the origin, 0, on $\mathbb{Z}^d$, including the visit that occurs at time 0. Let Y_t be the Wright-Fisher diffusion with infinitesimal generator $(1/2)x(1-x)d^2/dx^2$ where $\gamma = 1$.

Theorem 7.7.3 *If $d \ge 3$, then time is scaled by $s_N = N^d$ and as $N \to \infty$, $\Delta_{ts_n}^{(N)} \Rightarrow Y_{2t/G}$ where $\Rightarrow$ denotes converge in distribution of the stochastic processes.*

Proof. **Step 1** is to prove convergence of the one-dimensional distributions by using the method of moments

$$E(\Delta_t^{(N)})^m = N^{-md} \sum_{x_1,\ldots x_m}^{all} P(\eta_{ts_N}^{(N)} = 1, \ 1 \le i \le m). \tag{7.7.3}$$

Suppose that the initial state of the voter model is product measure with density θ and let $\xi_{ts_N}^{x_1,\ldots,x_m}$ be the dual process working backwards from the indicated sites at time ts_N.

$$\sum_{x_1,\dots,x_m}^{all} P(\eta_{tsN}^{(N)}(x_i) = 1,\ 1 \le i \le m) = \sum_{x_1,\dots,x_m}^{all} \sum_k P(|\xi_{tsN}^{x_1,\dots,x_m}| = k)\theta^k.$$

We want to use the convergence of the coalescing random walk to Kingman's coalescent proved in Sect. 7.2. Let $a_N = \log N$ and let $\sum^{far}$ be $\sum^{all}$ restricted to $(x_1,\dots,x_m)$ with $dist(x_i,x_j) \ge a_N$ for all $i \ne j$. Since m points chosen from the torus satisfy this constraint with probability tending to 1 as $N \to \infty$

$$N^{-md}\left(\sum_{x_1,\dots x_m}^{all} \sum_k - \sum_{x_1,\dots x_m}^{far} \sum_k \right) \to 0.$$

Using the convergence of coalescing random walks starting from points separated by a_N to Kingman's coalescent, Theorem 7.2.6, we have

$$\sum_{x_1,\dots x_m}^{far} \sum_k P(|\xi_{tsN}^{x_1,\dots x_m}| = k)\theta^k \to \sum_k \theta^k q_{m,k}(2t/G).$$

Step 1 will be complete when we show, in the discussion of Function Based Duality below, that

$$\sum_k \theta^k q_{m,k}(2t/G) = E_\theta(Y_{2t/G}^m). \tag{7.7.4}$$

Since that requires a somewhat lengthy discussion, we will first complete the proof of the theorem

Step 2 is to prove convergence of the finite-dimensional distributions by establishing the convergence of the moments

$$E\left[\prod_{i=1}^m (\Delta_{t_i sN}^{(N)})^{m_i}\right] \to E_\theta\left[\prod_{i=1}^m Y_{2t_i/G}^{m_i}\right].$$

The first step is to convert the moments into sums of the values of the voter model at a collection of space time points. Using duality we can convert this into a statement about the behavior coalescing random walks starting at a number of different points in space-times. To use the convergence of the dual processes to Kingman's coalescent we need to assume that the particles that start at the same time are separated by a_N. As before, this does not significantly change the value of the sum. Finally we need to use the coalescent probabilities to Wright-Fisher moments using the as yet unproved identity.

Step 3 is to prove tightness. Cox (1989) checks this using Corollary 8.7 in Chap. 4 of Ethier and Kurtz (1986). However since $M^N(t) = \Delta_{tsN}^{(N)}$ are martingales this is easier to do using Proposition 1.2 of Aldous (1989b). These arguments are well-known to those skilled in the craft of stochastic analysis, but this topic is far away from the central focus of the book, so we refer the readers to the references cited for details. If you haven't seen Aldous' paper, it is well worth reading.

Function-based duality

Markov processes X_t and Y_t are said to be **in duality with respect to** H if

$$EH(X_t, Y_0) = EH(X_0, Y_t). \tag{7.7.5}$$

If the two processes are set-valued and $H_a(\xi, \zeta) = 1$ when $\xi \cap \zeta \neq \emptyset$ (and 0 otherwise) this gives **additive duality**, which holds for the voter model ξ and coalescing random walk ζ, or for the contact process with the dual $\zeta = \xi$

$$P(\xi_t^A \cap B \neq \emptyset) = P(A \cap \zeta_t^B \neq \emptyset).$$

If the two processes are set-valued and $H_c(\xi, \zeta) = 1$ when $|\xi \cap \zeta|$ is odd (and 0 otherwise) this gives **cancellative duality**

$$P(|\xi_t^A \cap B| \text{ is odd}) = P(|A \cap \zeta_t^B| \text{ is odd}).$$

This holds for a number of models, see Griffeath's (1979) book: annihilating random walks, some nonlinear voter model, and some instances of the Ising model. It is less useful than the additive case since it is harder to check $|\xi_t^A \cap B|$ is odd than it is to check $|\xi_t^A \cap B| > 0$, since in the latter case a few extra particles doesn't hurt.

Moment duality refers to the case: $H_m(y, n) = y^n$ where $y \in [0, 1]$ and n is a positive integer. The most important case for us occurs when Y_t is a Wright-Fisher diffusion and N_t is Kingman's coalescent, provided the constants are chosen correctly.

The Wright-Fisher diffusion generator is $\mathcal{L}^Y f(y) = (1/2)\gamma y(1-y) f''(y)$.

Kingman's coalescent generator is $\mathcal{L}^N f(n) = \gamma \binom{n}{2}(f(n-1) - f(n))$.

Proposition 1.2 in Jansen and Kurt (2014) implies that (7.7.5) holds if

$$\mathcal{L}_y^Y H(y, n) = \mathcal{L}_n^N H(y, n), \tag{7.7.6}$$

where the generator subscripts indicate the variables they act on (with the other variable held constant). To check the Wright-Fisher diffusion-Kingman coalescent duality now we note that if $n \geq 2$

$$\begin{aligned} \mathcal{L}_y^Y H(y, n) &= \frac{1}{2} \gamma n(n-1) y^{n-2} y(1-y) \\ &= \gamma \frac{n(n-1)}{2} (y^{n-1} - y^n) = \mathcal{L}_n^N H(y, n). \end{aligned}$$

If $n = 1$ then all of these quantities are 0.

By our convention $q_{m,k}(t)$ is the transition probability of Kingmans' coalescent when $\gamma = 1$. Theorem 7.2.6 shows that when time is run at rate N^d in $d \geq 3$ coalescing random walk on the torus converges to Kingman's coalescent in which jumps k to $k-1$ occur at rate $k(k-1)/G$, which corresponds to $\gamma = 2/G$, so the transition probability for the limiting coalescent is $q_{m,k}(2t/G)$. Combining their observation with the bounded convergence theorem proves (7.7.4) and completes the proof of Theorem 7.7.3.

7.7.3 Chen et al. (2016)

Chen, Choi, and Cox have developed martingale of this section have developed martingale-based methods to prove convergence of voter density processes to Wright-Fisher diffusions. We will refer to their paper as CCC. The first step is to introduce the voter models they consider. Given a finite set E the dynamics are defined by a q-matrix which has $q(x, y) \geq 0$ when $y \neq x$ and $q(x) = -q(x, x) = \sum_{y \neq x} q(x, y)$. Site x changes its opinion at rate $q(x)$ and when it does it imitates y with probability $q(x, y)/q(x)$.

We will work with a sequence of q-matrices $q^{(n)}$ defined on finite sets E_n with $|E_n| \to \infty$. We will usually take $E_n = \{1, 2, \ldots, n\}$. Letting $\pi^{(n)}$ be the stationary distribution for $q^{(n)}$, and dropping the superscript n's, the density of 1's is defined to be

$$p_1(\xi_t) = \sum_x \pi(x)\xi_t(x).$$

Lemma 7.7.4 *$p_1(\xi_{\gamma t})$ is a martingale with jumps bounded by* $\max_x \pi(x)$.

Proof. Let ξ^x be the configuration with the opinion flipped at x, i.e., $\xi^x(x) = \widehat{\xi}(x) = 1 - \xi(x)$ and $\xi^x(y) = \xi(y)$ when $y \neq x$. The change in density due to a flip at x is

$$p_1(\xi^x) - p_1(\xi) = \pi(x)(\widehat{\xi}(x) - \xi(x)). \tag{7.7.7}$$

Let $\mathcal{L}$ is the generator of the voter model and recall that

$$\mathcal{L}f(\xi) = \sum_x c(\xi, \xi^x)(f(\xi^x) - f(\xi)),$$

$$\text{where} \quad c(\xi, \xi^x) = \sum_y (\xi(x)\widehat{\xi}(y) + \widehat{\xi}(x)\xi(y))\, q(x, y). \tag{7.7.8}$$

Applying $\mathcal{L}$ to $f = p_1$

$$\mathcal{L}p_1(\xi) = \sum_x \pi(x)\widehat{\xi}(x) \sum_y q(x, y)\xi(y) - \sum_x \pi(x)\xi(x) \sum_y q(x, y)\widehat{\xi}(y).$$

Note that the second sum in each term includes $y = x$ but that is OK because $q(x, x) = -q(x)$ so the two terms cancel. Using $\widehat{\xi}(z) = 1 - \xi(z)$ we have

$$= \sum_{x,y} \pi(x)q(x, y)\xi(y) - \sum_{x,y} \pi(x)\xi(x)q(x, y)\xi(y)$$
$$- \sum_{x,y} \pi(x)\xi(x)q(x, y) + \sum_{x,y} \pi(x)\xi(x)q(x, y)\xi(y).$$

The two double sums cancel. For fixed y, $\sum_x \pi(x)q(x, y) = 0$ and for fixed x, $\sum_y q(x, y) = 0$. In both cases it is important $y = x$ is included.

Our next goal is to compute the predictable quadratic variation $\langle p_1(\xi)\rangle_t$. To be able to do this we need to introduce some notation. Let

$$\nu(x, y) = \pi(x)^2 q(x, y)\mathbf{1}_{x\neq y}$$
$$\bar{\nu}(x, y) = \nu(x, y)/\nu(\mathbf{1}) \text{where} \nu(\mathbf{1}) = \sum_{x,y} \nu(x, y)$$
$$p_{10}(\xi) = \sum_{x,y\in E} \bar{\nu}(x, y)\xi(x)\widehat{\xi}(y)$$
$$p_{01}(\xi) = \sum_{x,y\in E} \bar{\nu}(x, y)\widehat{\xi}(x)\xi(y). \tag{7.7.9}$$

Note that p_{10} is the sum of the weights on the oriented edges (x, y) in the graph with state 10.

Example 7.7.5 (Graphs with constant degree r) In these examples, we have a transition probability $p(x, y) = 1/r$ if y is one of the r neighbors of x, and 0 otherwise. Since the underlying structure is a graph $p(x, y) = p(y, x)$ and $\pi(x) = 1/n$ is a reversible stationary distribution. The jump rates of the voter model have $q(x, y) = p(x, y) - I(x, y)$ where I is the identity matrix.

$$\nu(\mathbf{1}) = \sum_x \pi(x)^2 \sum_y q(x, y)\mathbf{1}_{x\neq y} = n \cdot \frac{1}{n^2} \cdot 1 = \frac{1}{n},$$

so we have $\bar{\nu}(x, y) = p(x, y)/n$ when $y \neq x$. The edge weights $\bar{\nu}(x, y) = 1/nr$ or 0.

Example 7.7.6 (Graphs generated by the configuration model) Let $d(x)$ be the degree of x and $p(x, y) = 1/d(x)$ if y is a neighbor of x. If we let D_n be the sum of the $d(x)$ then $\pi(x) = d(x)/D_n$ is a reversible stationary distribution. If $\mu = Ed(x) < \infty$ then $D_n \sim \mu n$ as $n \to \infty$. The jump rates of the voter model have $q(x, y) = p(x, y) - I(x, y)$ where I is the identity matrix. In Sect. 7.3 we introduced $A = \{(x, x) \in V \times V\}$ and wrote $\pi(A)$ for the measure of A under $\pi \times \pi$

$$\nu(\mathbf{1}) = \sum_x \pi(x)^2 \sum_y q(x, y)\mathbf{1}_{x \neq y} = \sum_x \left(\frac{d(x)}{D_n}\right)^2 \cdot 1 = \pi(A),$$

$$\bar{\nu}(x, y) = \frac{\pi(x)^2/d(x) \cdot \mathbf{1}_{y \sim x}}{\nu(\mathbf{1})} = \frac{d(x) \cdot \mathbf{1}_{y \sim x}}{\sum_z d(z)^2}.$$

Lemma 7.7.7 *The predictable quadratic variation of the density process*

$$\langle p_1(\xi) \rangle_t = \nu(\mathbf{1}) \int_0^t [p_{10}(\xi_s) + p_{01}(\xi_s)]\, ds. \tag{7.7.10}$$

Proof. Since $p_1(\xi_t)$ is a martingale it is enough to show

$$p_1(\xi_t) p_0(\xi_t) + \nu(\mathbf{1}) \int_0^t [p_{10}(\xi_s) + p_{01}(\xi_s)]\, ds$$

is a martingale and then subtract the last expression from formula from $p_1(\xi_t)$. A little arithmetic shows

$$\begin{aligned} p_1(\xi^x) p_0(\xi^x) - p_1(\xi) p_0(\xi) &= (p_1(\xi^x) - p_1(\xi)) \cdot (p_0(\xi^x) - p_0(\xi)) \\ &\quad + p_0(\xi)(p_1(\xi_x) - p_1(\xi)) + p_1 \xi)(p_0(\xi_x) - p_0(\xi)). \end{aligned}$$

The first term on the right is $-\pi(x)^2$. To see this consider two cases $\xi(x) = 1$ and $\xi(x) = 0$. Since $\mathcal{L} p_1 = \mathcal{L} p_0 = 0$, using the formula for $c(x, \xi)$ in (7.7.8), we have

$$\begin{aligned} &\sum_x c(x, \xi)(p_1(\xi^x) p_0(\xi^x) - p_1(\xi) p_0(\xi)) \\ &\qquad = -\sum_x \pi(x)^2 (\xi(x)\widehat{\xi}(y) + \widehat{\xi}(x)\xi(y))\, q(x, y) \\ &\qquad = -\nu(\mathbf{1})(p_{10}(\xi) + p_{01}(\xi)), \end{aligned}$$

which completes the proof.

Let $\pi^{(n)}_{diag} = \sum_x \pi^{(n)}(x)^2$. This is called $\pi^{(n)}_{diag}$ because it is the mass on the diagonal in $\pi^{(n)} \times \pi^{(n)}$. Let Y_t be the Wright-Fisher diffusion with generator $\mathcal{L} f(x) = (1/2) x(1 - x) f''(x)$. The next result is Theorem 2.1 in CCC.

Theorem 7.7.8 *Let $u \in (0, 1)$ and let γ_n be a sequence of positive constants. Assume that*

$$\lim_{n \to \infty} \pi^{(n)}_{diag} = 0.$$

Then the convergence of density processes starting from μ_u product measure with density u

$$(p_1(\xi_{\gamma_n \cdot}), P^{(n)}_{\mu_u}) \Rightarrow (Y, P_u) \tag{7.7.11}$$

holds if and only if the following **mean-field condition (MFC)** *holds*

$$\gamma_n \nu_n(\mathbf{1}) \int_0^T [p_{10}(\xi_{\gamma_n s}) + p_{01}(\xi_{\gamma_n s})]\, ds - \int_0^T p_1(\xi_{\gamma_n s})[1 - p_1(\xi_{\gamma_n s})]\, ds \Rightarrow 0. \tag{7.7.12}$$

The reasoning here is similar to the proof on the complete graph. The density process is a martingale, so we will have weak convergence to the Wright-Fisher diffusion if (and only if) the quadratic variations converge.

To develop sufficient conditions for the (7.7.12), let $q_t = \exp(tq)$ be the continuous time transition probability and define the total variation distance from equilibrium by

$$d_E(t) = \max_{x \in E} \|q_t(x, \cdot) - \pi(\cdot)\|_{TV}.$$

Define the mixing time $t^{(n)}_{mix} = \inf\{t \geq 0 : d_E(t) \leq 1/2e\}$ and the relaxation time $t_{rel} = \mathbf{g}^{-1}$ where $\mathbf{g}$ is the spectral gap, see 6.1.9. Let

$$\pi^{(n)}_{max} = \max\{\pi^{(n)}(x) : x \in E\} \quad \text{and} \quad qi^{(n)}_{max} = \max\{q^{(n)}(x) : x \in E\}.$$

Let $M_{U,U'}$ be the meeting time of two independent copies of q_t starting from U and U' chosen independently according to π, and let $t_{meet} = E(M_{U,U'})$. Theorem 2.2 in CCC is

Theorem 7.7.9 *If either of the following conditions holds for all $u \in (0, 1)$ then the mean-field condition* (7.7.12) *holds with $\gamma_n = t^{(n)}_{meet}$ and hence we have convergence of voter model densities to Wright-Fisher diffusions* (7.7.11).

Both conditions assume $\lim_{n\to\infty} \pi^{(n)}_{diag} = 0$.

(i) $\lim_{n\to\infty} t^{(n)}_{mix}/t^{(n)}_{meet} = 0$.
(ii) The $q^{(n)}$ Markov chains are reversible and satisfy

$$\lim_{n\to\infty} \frac{\log(e \vee t^{(n)}_{meet}\, \pi^{(n)}_{max}\, q^{(n)}_{max})}{g_n t^{(n)}_{meet}} = 0.$$

The last result can be applied to the torus in $d \geq 3$, but we have already analyzed that example, so we turn to some new ones.

Example 7.7.10 (Random regular graphs) When $r \geq 3$ these graphs satisfy the hypotheses of Theorem 6.3.1, so $t^n_{mix} \leq C \log n$. Formula (7.3.2) shows that the meeting time, is $\sim n(r-1)/(r-2)$. Thus we have $t^{(n)}_{mix}/t^{(n)}_{meet} \to 0$ as $n \to \infty$ and hence (i) of Theorem 7.7.9 holds.

Example 7.7.11 (Power law random graphs with $2 < \gamma < 3$) In this case

$$P(d(x) = k) \sim Ck^{-\gamma} \quad \text{with} P(d(x) = k) = 0 \text{for} k < 3.$$

We have assumed that the minimum degree is 3, so these graphs satisfy the hypotheses of Theorem 6.4.1 and hence have $t^{(n)}_{mix} \leq C \log n$. Formula (7.3.1) implies that

$$t_{meet} \sim \frac{1}{\pi(A)} \cdot \frac{1}{P_A(T_A \gg t_{mix})}.$$

Lemma 7.3.5 shows that $1/\pi(A) \approx n^\alpha$ where $\alpha = (2\gamma - 4)/(\gamma - 1)$. When $2 < \gamma < 3$ the power $\alpha > 0$ so $t^{(n)}_{mix}/t^{(n)}_{meet} \to 0$ and (i) in Theorem 7.7.9 holds. In Example 7.7.6 we showed that $\nu(\mathbf{1}) = \pi(A)$, so in Theorem 7.7.8, $\gamma_n \nu_n(\mathbf{1}) = 1/P_A(T_A \gg t_{mix})$. Since all vertices have degree ≥ 3 the last probability is bounded away from 0.

Why are these results true? The proof of Theorem 2.2 in CCC is somewhat sophisticated, and the details fill eight pages, so we will start by explaining the intuition behind why the result holds for the random regular graph. When (i) holds the lineages in the coalescent converge to equilibrium before they coalesce. Thus if the fraction of occupied sites is u at time $t - t^{(n)}_{mix}$ then finite-dimensional distributions of the state at time t are close to the voter model stationary distribution ν_u.

We computed in Example 7.7.5 that there are n vertices and $\sim nr/2$ edges.

$$\nu(\mathbf{1}) = 1/n \quad \text{and} \quad t^{(n)}_{meet} \sim n\frac{r-1}{r-2} \quad \text{so} \quad \nu(\mathbf{1})t^{(n)}_{meet} \to \frac{r-1}{r-2}.$$

If we run two coalescing random walks on the graph starting from the end points of each of these edges then the number of times they will not coalesce is

$$\frac{nr}{2} \cdot \frac{r-2}{r-1}.$$

The number of discordant edges is $(p_{01} + p_{10})nr/2$ so

$$\gamma_n \nu(\mathbf{1})(p_{01} + p_{10}) \to u(1-u).$$

Connection to Coalescence Times

Condition (i) should remind the reader of the results in Sect. 7.5 for reversible Markov chains on transitive graphs. There $m(Q) = t^Q_{meet}$, $\rho(Q) = t^Q_{mix}/m(Q)$, $\mathbf{C}$ is time to coalesce to one particle when we start with one particle at each site on the graph, and $Z_m = \text{exponential}(\binom{m}{2})$. Theorem 7.5.3 shows that the Wasserstein distance between $\mathbf{C}/m(Q)$ and $\sum_{m\geq 2} Z_m$ is bounded by $C|\rho(Q)\log\rho(Q)|^{1/6}$.

CCC consider a sequence of irreducible Q-matrices $(q^{(n)}, E_n)$ with stationary distributions $\pi^{(n)}$. For a given n they let $U_1, U_2, \ldots$ be i.i.d. with distribution $\pi^{(n)}$. Let $\widehat{X}^{U_1}_t, \ldots, \widehat{X}^{U_k}_t$ be coalescing random walks with the indicated initial positions that move independently until they hit, and when they hit they coalesce. For $1 \leq j < k$ define the coalescence times

$$\mathbf{C}^{(n)}_{k,j} = \inf\{t : |\{\widehat{X}^{U_1}_t, \ldots, \widehat{X}^{U_k}_t\}| = j\}.$$

On the complete graph, if we let $\gamma_n = |E_n|/2$ then

$$\frac{\mathbf{C}^{(n)}_{k,j}}{\gamma_n} \Rightarrow \sum_{m=j+1}^{k} Z_m. \tag{7.7.13}$$

The next result is Proposition 2.5 in CCC.

Theorem 7.7.12 *If the mean-field condition* (7.7.12) *holds then so does the convergence of coalescence times to Kingman's coalescent* (7.7.13). *In particular if either of the conditions in Theorem 7.7.9 hold then so does* (7.7.13) *with* $\gamma_n = t^{(n)}_{meet}$.

Proof. Let Z_j, $j \geq 2$, be independent exponential with rate $\binom{j}{2}$ and hence mean $1/\binom{j}{2}$. For $i < j$ let $S_{j,i} = Z_{i+1} + \cdots + Z_j$ be the time it takes for Kingman's coalescent to come down from j to i particles. If we let D_t be the pure death process in which $k \to k-1$ at rate $\binom{k}{2}$, i.e., the number of particles in Kingman's coalescent,

$$P_k(D_t = j) = P(S_{k,j} \leq t < S_{k,j-1}).$$

By the duality equation and the fact that the initial law of the voter model is μ_u, product measure with density u,

$$\begin{aligned} E^{(n)}_{\mu_u}(p_1(\xi_{\gamma_n t}))^k &= \sum_{j=1}^{k} u^j P^{(n)}\left(\left|\left\{\widehat{X}^{U_1}_{\gamma_n t}, \ldots \widehat{X}^{U_k}_{\gamma_n t}\right\}\right| = j\right) \\ &= \sum_{j=1}^{k} u^j P^{(n)}\left(\mathbf{C}^{(n)}_{k,j} \leq \gamma_n t < \mathbf{C}^{(n)}_{k,j-1}\right). \end{aligned} \tag{7.7.14}$$

On the other hand by assumption and the duality equation

$$\lim_{n\to\infty} E^{(n)}_{\mu_u}(p_1(\xi_{\gamma_n t})^k = E_u Y_t^k = E_k u^{D_t})$$
$$= \sum_{j=1}^{k} u^j P(S_{k,j} \le t < S_{k,j-1}). \tag{7.7.15}$$

Combining (7.7.14) and (7.7.15)

$$\sum_{j=1}^{k} u^j P^{(n)}\left(\mathbf{C}^{(n)}_{k,j} \le \gamma_n t < \mathbf{C}^{(n)}_{k,j-1}\right) = \sum_{j=1}^{k} u^j P(S_{k,j} \le t < S_{k,j-1}).$$

Since this holds for all u we must have for $1 \le j \le k$ that

$$\lim_{n\to\infty} P^{(n)}\left(\mathbf{C}^{(n)}_{k,j} \le \gamma_n t < \mathbf{C}^{(n)}_{k,j-1}\right) = P(S_{k,j} \le t < S_{k,j-1}).$$

Using dominated convergence

$$\lim_{n\to\infty} \int_0^\infty \lambda e^{-\lambda t} P^{(n)}\left(\mathbf{C}^{(n)}_{k,j} \le \gamma_n t < \mathbf{C}^{(n)}_{k,j-1}\right) dt$$
$$= \int_0^\infty \lambda e^{-\lambda t} P(S_{k,j} \le t < S_{k,j-1})\, dt.$$

This implies

$$\lim_{n\to\infty} E^{(n)}(e^{-\lambda \mathbf{C}^{(n)}_{k,j-1}}) - E^{(n)}(e^{-\lambda \mathbf{C}^{(n)}_{k,j}}) = E(e^{-\lambda S_{k,j-1}}) - E(e^{-\lambda S_{k,j}}).$$

Since $\mathbf{C}^{(n)}_{k,k} = 0 = S_{k,k}$ this gives the desired result.

Consider now the full coalescence time $\mathbf{C}^{(n)}_{\infty,j} = \inf\{t : |\{\widehat{X}^x_t : x \in E_n\}| = j\}$. Proposition 2.6 in CCC is

Theorem 7.7.13 *Let $\tau_1^{(n)}$ denote the first hitting time of 1 by the density process $p_1(\xi_{\gamma_n}\cdot)$ starting from product measure with density u, μ_u, and let τ_1^Y be the hitting time of 1 by the Wright-Fisher process starting from u. Then the following convergences are equivalent*

$$\tau_1^{(n)}/\gamma_n \Rightarrow \tau_1^Y \quad \textit{for all} u \in (0,1), \tag{7.7.16}$$

$$\frac{\mathbf{C}^{(n)}_{\infty,j}}{\gamma_n} \Rightarrow \sum_{m=j+1}^{\infty} Z_m. \tag{7.7.17}$$

CCC say that (7.7.16) does not follow immediately from the convergence of the density process to the Wright-Fisher diffusion since the hitting time of 1 is not a continuous function. However as we will show,

Lemma 7.7.14 *The hitting time of 1 is continuous a.e. with respect to the limiting Wright-Fisher diffusion, so Proposition 2.6 is indeed a corollary of Proposition 2.5.*

Proof. Let $\mathcal{L}f = (1/2)x(1-x)f''(x)$ be the generator of the standard Wright-Fisher process X_t. Let $\tau = \inf\{t : X \notin (0,1)$.

Claim: if $\mathcal{L}v = -1$, and $v(0) = v(1) = 0$, then $v(x) = E_x\tau$.

Proof of Claim. Let $\tau = \inf\{t : X_t \notin (0,1)\}$. Dynkin's formula implies that

$$v(X_t) - \int_0^t \mathcal{L}v(s)\,ds = v(X_t) + t \quad \text{is a martingale fort } < \tau.$$

Using the optional stopping theorem, it follows that $v(x) = E_x\tau$. □

Symmetry implies that $v'(1/2) = 0$. Since $v''(x) = 2/x(1-x)$ one can integrate to find the solution but let's skip that work and check that

$$v(x) = -2(x\log x + (1-x)\log(1-x)).$$

This v is 0 at $x = 0$ and $x = 1$. Differentiating gives

$$v'(x) = -2\left(\log x + 1 - \log(1-x) + (1-x)\frac{-1}{1-x}\right) = -2(\log x + \log(1-x)),$$
$$v''(x) = -2\left(\frac{1}{x} + \frac{1}{1-x}\right) = -\frac{2}{x(1-x)}.$$

The last equation implies $\mathcal{L}v = -1$.

Let $\tau_\epsilon = \inf\{t : X_t \notin (\epsilon, 1-\epsilon)$. By the Markov property if the starting point is in $(\epsilon, 1-\epsilon)$ then $\tau - \tau_\epsilon$ is the time to exit $(0,1)$ starting from where the particle exited $(\epsilon, 1-\epsilon)$. Using symmetry $E(\tau - \tau_\epsilon) \le v(\epsilon) \to 0$. Since $\epsilon \to \tau_\epsilon$ is increasing for small ϵ, i.e., this holds once the initial condition is in $(\epsilon, 1-\epsilon)$, we have $\tau_\epsilon \to \tau$ almost surely. This shows that τ is almost surely continuous with respect to P_x^{WF}. Using Theorem 3.2.10 in PTE5 the desired result follows.

Example 7.7.15 (Power law graphs with $2 < \gamma < 3$) In Sect. 7.4 we noted that Theorem 7.4.1 gave an upper bound on the coalescence time in this case that is the right order of magnitude. Using Theorems 7.7.12 and 7.7.13 with the convergence of voter model densities to the Wright-Fisher diffusion we can conclude that the coalescing random walk starting from all sites occupied when rescaled by the meeting time converges to Kingman's coalescent. This is not only a new result but the method seems applicable to a large number of examples.

References

Aldous D (1989a) Probability approximations via the Poisson clumping heuristic. Springer, New York

Aldous D (1989b) Stopping times and tightness. II Ann Probab 17:586–595

Aldous D, Fill J (2002) Reversible Markov chains and random walks on graphs. http://www.stat.berkeley.edu/~aldous/

Bramson M, Griffeath D (1980) Asymptotics for interacting particle systems on $\mathbb{Z}^d$. Probab Theory Rel Fields 53:183–196

Brown M (1983) Approximating IMRL distributions by exponential distributions with applications to first passage times. Ann Probab 11:419–427

Capannoli F, den Hollander F (2024) Interacting particle systems on random graphs. arXiv:2410.17766

Chen YT (2017) Meeting time of random walks on random regular graphs. Electron J Probab 54:1–18

Chen YT, Choi J, Cox JT (2016) On the convergence of densities of finite voter models to the Wright-Fisher diffusion. Ann Inst Henri Poincaré Prob Stat 52:286–322

Clifford P, Sudbury A (1973) A model for spatial conflict. Biometrika 60:581–588

Cooper C, Elsässer R, Ono H, Radzik T (2012) Coalescing random walks and voting on connected graphs. SIAM J Discrete Math 27:1748–1758

Cox JT (1989) Coalescing random walks and voter model consensus times on the torus $\mathbb{Z}^d$. Ann Probab 17:1333–1366

Cox JT, Greven A (1990) On the long term behavior of some finite particle systems. Probab Theory Rel Fields 85:195–237

Cox JT, Griffeath DS (1986) Diffusive clustering in the two dimensional voter model. Ann Probab 14:347–370

Dvoretsky A, Erdős P (1951) Some problems on random walk in space. In: Proceedings of the 2nd Berkeley symposium, pp 353–367

Ethier SN, Kurtz TG (1986) Markov processes: characterization and convergence. In: Wiley series in probability and statistics

Griffeath D (1979) Additive and cancellative interacting particle systems. Springer, Berlin

Hermon J, Li S, Yao D, Zhang L (2022) Mean-field behavior during the big bang regime for coalescing random walk. Ann Probab 50:1818–1884

Holley R, Liggett T (1975) Ergodic theorems for weakly interacting systems and the voter model. Ann Probab 4:195–228

Jansen S, Kurt N (2014) On notion(s) of duality for Markov processes. arXiv:1210.7193

Kelly FP (1977) The asymptotic behavior of an invasion process. J Appl Prob 14:584–590

Kesten H, van den Berg J (2000) Asymptotic density in a coalescing random walk model. Ann Probab 28:303–352

Kingman JFC (1982) The coalescent. Stoch Proc Appl 13:235–248

Oliveira RI (2012) On the coalescence time for reversible random walks. Trans Am Math Soc 364:2109–2128

Oliveira RI (2013) Mean-field conditions for coalescing random walks. Ann Probab 41:3420–3461

Sawyer S (1979) Results for the voter model. Ann Probab 7(3):447–463

Sood V, Redner S (2005) Voter model on heterogeneous graphs. Phys Rev Lett 94:178701

Suchecki K, Eguíluz VM, Miguel MS (2005) Conservation laws for the voter model in complex networks. Europhys Lett 69:228–234

Sudbury A (1976) The size of a region occupied by an invasion process. J Appl Prob 13:355–356

Varadhan SRS (2001) Probability theory, American math society, Providence, RI

Wakeley J (2009) Coalescent theory: an introduction. Roberts and Company, Greenwood Village, Colorado

Chapter 8
Coevolving Systems

This chapter is concerned with models in which the states of the vertices and the connections between them coevolve. This topic has been studied in the physics literature for many years. However, there are only a small number of rigorous results, which will be our focus here. We consider three types of systems.

Evolving Voter Models

During the 2010–2011 program on Complex Networks at SAMSI[1] Rick Durrett, James Gleeson, Alun Lloyd, Peter Mucha, Feng Shi, David Sivakoff, Josh Socloar, and Chris Varghese were part of a working group on *Dynamics On and Of Graphs*. Inspired by work of Holme and Newman (2006), which will be described later, they formulated a model in which on each step an oriented edge (x, y) is picked at random. If the two connected individuals have the same opinion nothing happens. If they hold different opinions then: with probability $1 - \alpha$ vertex x imitates the opinion of y, otherwise, i.e., with probability α, the link between them is broken and x makes a new connection to an individual chosen at random (i) from those with the same opinion, or (ii) from the network as a whole.

The two versions of the model are called (i) "rewire-to-same" and (ii) "rewire-to-random." The evolution of the system stops at time τ when there are no longer any discordant edges (that connect individuals with different opinions). At time τ, in each connected component of the graph all the individuals have the same opinion. For several months, participants in the SAMSI group thought that the two versions had the same behavior, but then we did some simulations.

Figure 8.1 gives simulation results starting from Erdős-Rényi graphs with N = 100,000 nodes and average degree $\lambda = 4$. Opinions are initially assigned randomly with the probability of opinion 1 given by $u = 0.5$, 0.25, 0.1, and 0.05. The figure shows the final fraction ρ of voters with the minority opinion from five realizations

[1] Statistics and Applied Mathematical Sciences Institute, an NSF funded institute in Research Triangle Park, which no longer receives funding.

R. Durrett, *Dynamics on Graphs*, Probability Theory and Stochastic Modelling 110,
https://doi.org/10.1007/978-3-032-21908-4_8

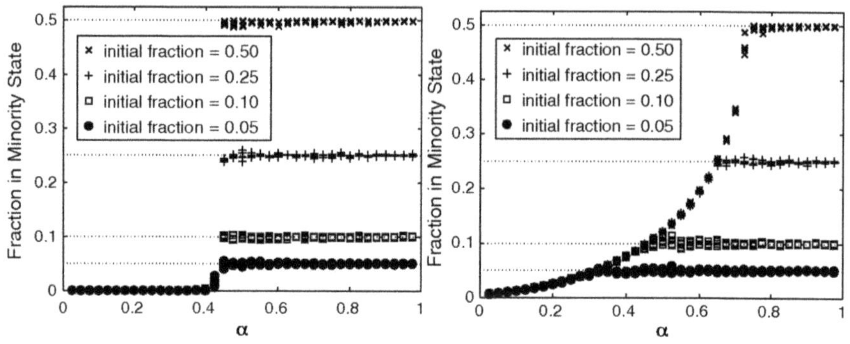

Fig. 8.1 Simulation results for rewire-to-same model (left panel) and rewire-to-random (right panel)

for each u. In case (i), there is a critical value α_c which does not depend on u, with $\rho \approx u$ for $\alpha > \alpha_c$ and $\rho \approx 0$ for $\alpha < \alpha_c$. In case (ii), the transition point $\alpha_c(u)$ depends on the initial density u. For $\alpha > \alpha_c(u)$, $\rho \approx u$, but for $\alpha < \alpha_c(u)$ we have $\rho(\alpha, u) = \rho(\alpha, 0.5)$. Since all of the $\rho(\alpha, u)$ agree with $\rho(\alpha, 0.5)$ when they are $< u$, we call the graph of $\rho(\alpha, 0.5)$ on $[0, \alpha_c(0.5)]$ the **universal curve**.

In Sect. 8.1.1 we will give heuristic arguments and numerical results from Durrett et al. (2012) in support of these conjectures. In Sect. 8.1.2 we will describe work of Basu and Sly (2017) that rigorously prove the existence of a phase transition between rapid disconnection and prolonged persistence (see Theorems 8.1.4 and 8.1.5). While the proofs of these results required an incredible amount of ingenuity they only work for dense Erdős-Rényi graphs and do not provide much insight into the exotic behavior shown by rewire-to-random.

The evolving voter model has been studied by many physicists. Even they think the model is hard, so further progress seems difficult. Readers may find more fertile directions for research on nonlinear versions surveyed by Kureh and Porter (2020), where the behavior of a node depends on how well it "fits in" to its neighborhood. Baldassarri et al. (2024) have investigated "opinion dynamics on dense dynamic graphs." Avena et al. (2025) have studied evolving voter models on random regular graphs with Volz-Meyers style edge swap dynamics. (See Sect. 8.2.1 for an explanation of this scheme.) A survey paper by den Hollander (2024) on the "Evolution of Discordance" can be consulted for more information about the area.

Evolving SIS Epidemics

Gross et al. (2006) introduced the following model: "we consider a network with a constant number of nodes N, and bidirectional links K. The nodes represent individuals that can be either susceptible (S) or infected (I).

- On every time step and for every link connecting an infected with a susceptible (an SI link) the susceptible becomes infected with a fixed probability p, and the infected recovers from the infection with probability γ, becoming susceptible again.
- In addition we allow susceptible individuals to rewire their links. With probability w for every SI link, the susceptible breaks its link to the infected and rewires to another randomly chosen susceptible. Double connections and self-connections are not allowed to form in this way."

Simulations suggested the phase diagram drawn in Fig. 8.2.

In the **Healthy** region the infection dies out. In the **Endemic** region there is positive probability of the infection surviving from a single infected individual and there is a unique equilibrium state, while in the **Bistable** region the epidemic always dies out starting from a single infected but there is a nontrivial stationary distribution. The authors observe that the **Oscillations** can only be observed in a relatively small parameter region so perhaps they are a finite size artifact. The clear and most striking observation is that the equilibrium density jumps from 0 to a positive level when increasing p moves the parameters from the **Healthy** to the **Bistable** region.

As usual physicists, who are free from the burdens of proof, can do more. Quoting Gross et al. "To study the effect of adaptive rewiring consider the threshold infection probability p^* that is necessary to maintain a static epidemic. On a random graph without rewiring ($w = 0$) the basic reproductive number $R_0 = p\langle k\rangle/r$, where $\langle k\rangle = 2K/N$ is the mean degree of nodes. Demanding that exactly one secondary infection is caused yields $p^* = \gamma/\langle k\rangle$. If rewiring is taken into account a single infected node will on average lose a constant fraction w of its links. Therefore the degree of such a

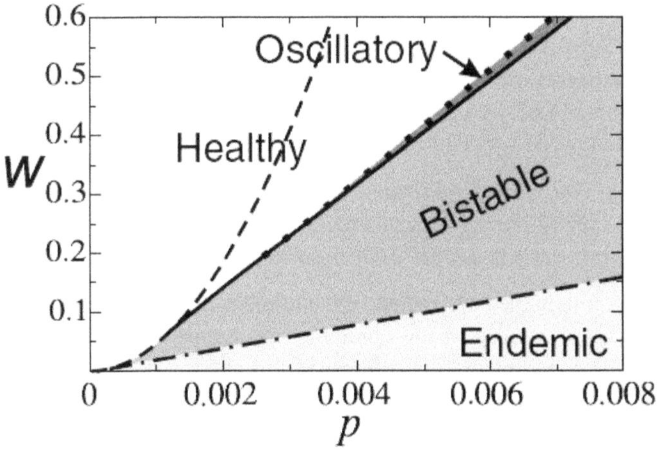

Fig. 8.2 Simulation with $N = 10^5$, $K = 10^6$ (average degree 20), and $\gamma = 0.002$. Dotted line is predicted critical value p^*

node can be written as $k(t) = \langle k \rangle \exp(-wt)$, where t is the time since infection. By averaging over the typical lifetime $1/\gamma$ of an infected node, we obtain the infection rate

$$p^* = \frac{w}{\langle k \rangle [1 - \exp(-w/\gamma)]}.$$

Note that this corresponds to $p^* = \gamma/\langle k \rangle$ for $w = 0$ but $p^* = w/\langle k \rangle$ for $w \gg \gamma$."

In Sect. 8.2.1 we give results obtained by Zanette (2007) using mean-field theory and the pair approximation. As far as we know there are no rigorous results for this model. In Sect. 8.2.2 we describe rigorous work of Chatterjee et al. (2022) on an SIS epidemic with model with **link inactivation**, i.e., susceptibles may drop their connections to infectious neighbors. This version of the model is much simpler to analyze since the graph does not change over time. At the end of the Sect. 8.2.2 we give references to other work and describe some open problems.

Evolving SIR Epidemics

In the summer of 2018, postdoc Matt Junge, graduate student Zoe Huang, and I worked with four students in the Duke math department's 8 week summer REU: Duke Opportunities in Math or DOMath. Three of the students continued to work with Matt and me in the Fall. The object of study was evoSIR, an epidemic model in which each infection lasts for exactly time 1 with the new feature that susceptibles who are neighbors of an infected break their connections at rate ρ and connect to a randomly chosen neighbor.

In order for the infection to be transmitted along an edge, it must occur before any rewiring and before time 1, so the transmission probability is

$$\tau_{f,r} = \frac{\lambda}{\lambda + \rho}(1 - e^{-(\lambda+\rho)}), \tag{8.0.1}$$

where f is for fixed time infections, and r is for rewiring. They also considered a much easier model delSIR in which connections were deleted instead. Their first result shows that evoSIR has the same critical value as delSIR.

Theorem 8.3.1. *The critical value for the fixed time epidemic with rewiring is given by $\mu_c = 1/\tau_{f,r}$. Moreover, in the subcritical regime the ratio of the expected epidemic size in delSIR to the size in evoSIR converges to 1.*

One can also compute the distribution of the number of secondary infections caused by an infected vertex early in the epidemic and hence the probability of a large epidemic. See Theorem 8.3.2. Since the delSIR model is essentially percolation, this probability is the same as the fraction of individuals infected in a large epidemic, and hence goes to 0 at the critical value $\mu_c = 1/\tau_r^f$. In evoSIR, as the simulations in Fig. 8.3 suggest the fraction of individuals infected is discontinuous at the critical value.

As the authors of Jiang et al. (2019) were finishing up the writing of their paper they learned of work of Britton et al. (2016), who studied a family of models called SIR-ω that interpolated between evoSIR and delSIR. See Sect. 8.3.3. Work of Leung et al.

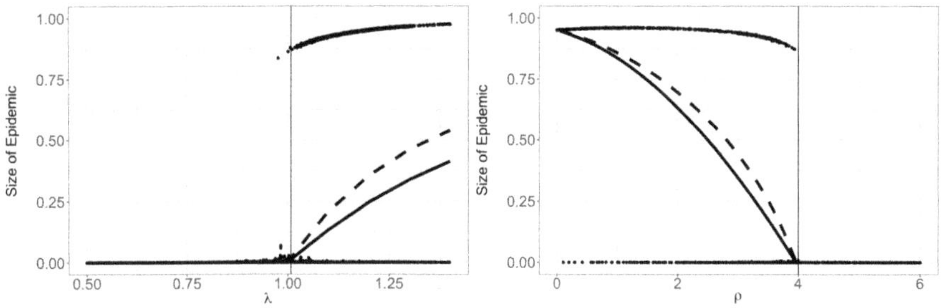

Fig. 8.3 Simulation of the fixed time evoSIR on an Erdős-Rényi graph with $\mu = 5$. In the left panel, the rewiring rate $\rho = 4$ and λ varies with $\lambda_c \approx 1.0084$ in agreement with Theorem 8.3.1. In the right panel, $\lambda = 1$ and ρ varies with $\rho_c \approx 4$. In both panels the top curve is the final size of the evoSIR epidemic, and the solid curve is the final size of the delSIR epidemic with the same parameters. The dashed line is an approximation developed in Jiang et al. (2019) that did not do a good job of predicting the final size in evoSIR

(2018) demonstrated the paradoxical fact that individual preventative measures (i.e., dropping connections to infected sites) may increase the final size of the epidemic. It is interesting to note that their simulations did not show a discontinuous phase transition. See Fig. 8.11.

Meanwhile back in Durham, Durrett and his graduate student Dong Yao used ideas of Janson et al. (2014) described in Sect. 4 to analyze the evolving SIR model on graphs generated by the configuration model. See Sect. 8.3.4. The trick is to grow the graph at the same time as you run the epidemic. Unfortunately their construction had an error: the algorithm resulted in rewiring connections between $I - I$ pairs, which should not occur. All was not lost: the formula was an upper bound on the true extent of the epidemic and a modification of the proof worked for the evoSI epidemic.

While the paper of Durrett and Yao (2022) was being refereed, work of Ball and Britton posted on the arXiv showed that the predictions of Durrett and Yao (2022) were not correct. See Fig. 8.14. Ball and Britton (2022) studied the SIR-ω model on Erdős-Rényi graphs. This is tractable because one can derive a limiting differential equation for S, I, i_E, and w, the fraction of susceptibles, infecteds, $I - S$ edges, and $S - S$ edges produced by rewiring. See (8.3.29). They were able to completely analyze the model except for one small gap, see (8.3.31), which was later filled by Chen et al. (2022).

Despite the large amount of work on this topic and the existence of a complete solution on Erdős-Rényi random graphs, much work remains to be done. The biggest open problem is to understand the behavior of evoSIR on graphs generated by the configuration model. While the proof of Durrett and Yao (2022) was not correct the answer may very well be correct. It gives the right answer in the Erdős-Rényi case, see Theorem 8.3.7.

8.1 Voter Models

Our starting point is the model of Holme and Newman (2006). They begin with a network of N vertices and M edges, where each vertex x has an opinion $\xi(x)$ from a set of G possible opinions and the number of people per opinion $\gamma_N = N/G$ stays bounded as N gets large. On each step of the process, a vertex x is picked at random. If its degree $d(x) = 0$, nothing happens. For $d(x) > 0$, (i) with probability α an edge attached to vertex x is selected and the other end of that edge is moved to a vertex chosen at random from those with opinion $\xi(x)$; (ii) otherwise (i.e., with probability $1 - \alpha$) a random neighbor y of x is selected and we set $\xi(x) = \xi(y)$. This process continues until the time τ at which there are no longer any edges connecting individuals with different opinions.

When $\alpha = 1$, only rewiring steps occur, so once all of the M edges have been touched, the graph has been disconnected into at least G components, each consisting of individuals who share the same opinion. Since none of the opinions have changed, all the components are small. By classical results for the coupon collector's problem, it requires $\tau \sim M \log M$ updates.

In contrast, for $\alpha = 0$ this system reduces to the voter model on a static graph. If we suppose that the initial graph is an Erdős-Rényi random graph in which each vertex has average degree $\lambda > 1$, then there is a "giant component" that contains a positive fraction of the vertices , μN. Results in Chap. 7 suggest that the voter model on the giant component will reach consensus in $O(N^2)$ steps of the vertices. The second largest component is small, having only $O(\log N)$ vertices, i.e., when N is large the size will be $\approx C_\lambda \log N$, where C_λ is a constant that depends on λ, so the end result is that one opinion has μN followers while all of the other groups are small.

Using simulation and finite size scaling, Holme and Newman showed that there is a critical value α_c so that for $\alpha > \alpha_c$ all of the opinions have a small number of followers at the end of the process, while for $\alpha < \alpha_c$ "a giant community of like-minded individuals forms." When the average degree $\lambda = 2M/N = 4$ and the number of individuals per opinion $\gamma_N \to 10$, this transition occurs at $\alpha_c \approx 0.46$.

8.1.1 Eight Guys at SAMSI

The behavior of our models for $\alpha > \alpha_c$ is easy to understand. As in the case of the Holme and Newman model, we expect consensus to be reached in $O(N \log N)$ steps. Let N_i be the number of vertices in state i. Since N_1 is a time change of a random walk the final fraction ρ with the minority opinion will be approximately equal to u while the number of jumps is $o(N^2)$.

Our first clue to the reason for a universal curve in the rewire-to-random model came from the next two figures (Fig. 8.4).

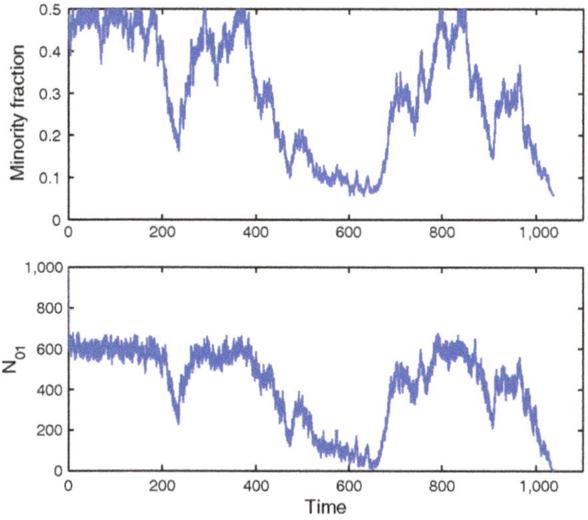

Fig. 8.4 Fraction of nodes with the minority opinion ($\min\{N_0, N_1\}/N$) and the number of discordant edges N_{01} versus time, for a simulation starting with an Erdős-Rényi graph with N = 10,000 and $\lambda = 4$. *Note* that the two statistics are strongly correlated

The situation becomes clearer when we plot the joint distribution in Fig. 8.5. In the left panel five simulations starting from $u = 0.2$, 0.35, 0.5, 0.65, and 0.8 are plotted. Note that the simulation comes quickly to a curve that then diffuses along it. In the right panel we do five simulations all starting at $u = 0.5$.

The fact that the values of the statistics N_{01}/N and N_{010}/N depend only on the fraction of 1's suggest that there is a one-dimensional family of quasi-stationary distributions indexed by the density. The system converges quickly to it and then diffuses along the curve until it hits an absorbing state. In rewire-to-random the axis is hit near 0.125 or 0.875. Thus the final fraction with the minority opinion $\rho \approx 0.125$, a value that agrees with the universal curve in Fig. 8.1 at $\alpha = 0.5$. In rewire-to-same the curve hits the axis at 0 or 1, so the minority fraction will be small.

The fact that the statistics N_{01}/N and N_{010}/N only depend upon the fraction of 1's $u = N_1/N$ density suggests that for each u there is a quasi-stationary. The next result, due to Cox and Greven (1990), shows that this is true for the voter model on the torus in dimensions $d \geq 3$. Here ν_p is the one parameter family of stationary distributions from Theorem 7.1.1.

Theorem 8.1.1 *If the voter model on the torus in $d \geq 3$ starts from product measure with density p then at time Nt it looks locally like $\nu_{\theta(t)}$ where the density θ_t changes according to the Wright-Fisher diffusion process*

$$d\theta_t = \sqrt{\beta_d \cdot 2\theta_t(1-\theta_t)}dB_t,$$

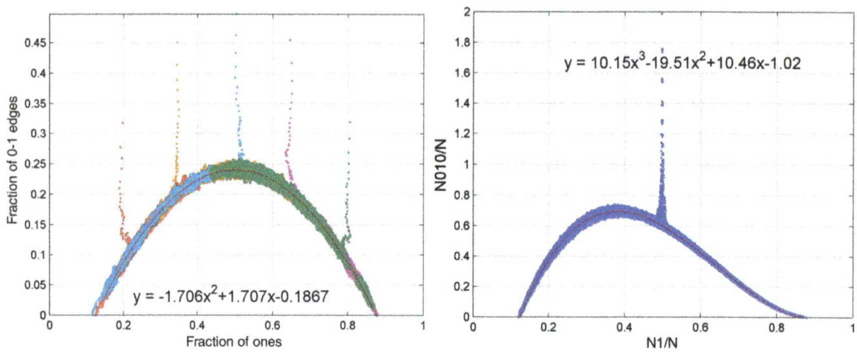

Fig. 8.5 Here the initial graph is the same but $\alpha = 0.5$. The left panel shows a plot of $(N_1/N, N_{10}/M)$. The right panel gives a plot of $(N_1/N, N_{010}/N)$

and β_d is the probability that two random walks starting from neighboring sites never hit.

The phenomena just described for $\alpha = 0.5$ also hold for other values of α. Figure 8.6 shows the arches that correspond to $\alpha = 0.1, 0.2, \dots, 0.7$. Numerical results show that the curves are well approximated by $c_\alpha u(1-u) - b_\alpha$. Let $(v(\alpha), 1 - v(\alpha))$ be the **support interval** where the arch has positive values. Simulations show that if $u < v(\alpha)$ then the simulated curve rapidly goes almost straight down and hits the axis where $N_{10} = 0$.

Conjecture 8.1.2 In the rewire-to-random model if $\alpha < \alpha_c(1/2)$ and $v(\alpha) < u \le 1/2$ then starting from product measure with density u of 1's, the evolving voter model converges rapidly to a quasi-stationary distribution $\nu_{\alpha,u}$. At time tN the evolving voter model looks locally like $\nu_{\alpha,\theta(t)}$ where the density changes according to a generalized Wright-Fisher diffusion process

$$d\theta_t = \sqrt{(1-\alpha)[c_\alpha \theta_t(1-\theta_t) - b_\alpha]}dB_t,$$

until θ_t reaches $v(\alpha)$ or $1 - v(\alpha)$.

If this conjecture is true then the universal curve in the right panel of Fig. 8.1 has $\rho(\alpha, 0.5) = v(\alpha)$ for $\alpha < \alpha_c(0.5)$. In the conjectured generalization of the Cox-Greven result the quantity under the square root is $(1-\alpha)N_{10}/M$ with $(1-\alpha) =$ the fraction of steps that are voter steps, since rewiring steps do not change the number of 1's.

Though the nature of the phase transition looks different in the rewire-to-same model, the underlying picture is the same. The right panel in Fig. 8.6 shows arches computed from simulations for the rewire-to-same model that correspond to the ones in the left panel for the rewire-to-random model. However, now all the arches have the same support interval, (0, 1). To be fair to physicists, we should note that Vazquez

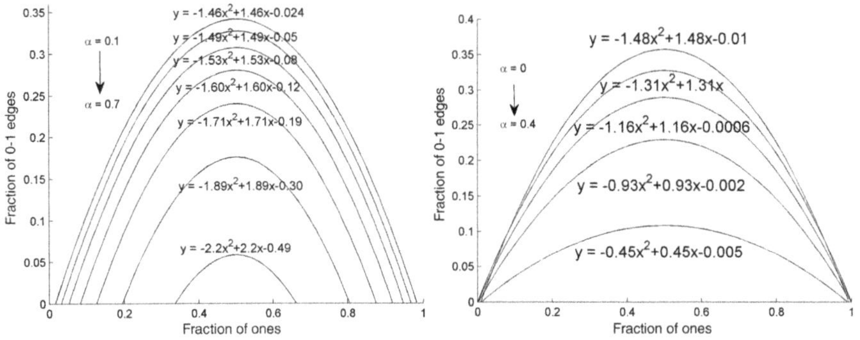

Fig. 8.6 This figure uses simulation data with $N = 10,000$, $\lambda = 4$. The left panel gives the observed arches for the rewire-to-random model. The right panel gives the observed arches for the rewire-to-same model

et al. (2008) noticed that in the rewire-to-same case: "For $p < p_c \approx 0.38$ the random walk has two stages. In the first and very short stage the random walk travels along the (magnetization) $m = 0$ axis to the point that correspond to the steady state." See their Fig. 8.4.

Conjecture 8.1.3 In the rewire-to-same model the behavior is as described in Conjecture 8.1.2 but now $b_\alpha = 0$, so α_c is independent of the initial density u, and for $\alpha < \alpha_c$, $\rho \approx 0$.

The simple formulas approximating the curves in Figs. 8.5 and 8.6 suggest that there might be a simple polynomial formulas. However, when Shi et al. (2013) studied the model with three or more opinions, their more accurate numerical results suggested the formulas were good approximation but not exact.

Numerical Results, Etc.

In addition to heuristic arguments, Durrett et al. (2012) used mean-field computations, the pair approximation, moment methods, etc. An excellent description of these methods and their application to epidemics on networks can be found in the excellent 2017 book by István Kiss, Joel Miller and Peter Simon which was cited in Chap. 4 (Fig. 8.7).

Mean-Field Approach. Let $u = N_1/N$ be the initial fraction of vertices in state 1. By considering all of the possible changes one arrives at the following differential equations.

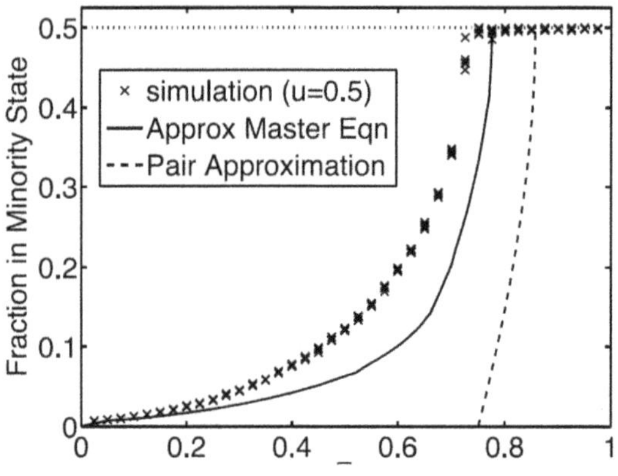

Fig. 8.7 Predictions of the ending minority fraction $\rho(\alpha)$ in the rewire-to-random case starting from product measure with density $u = 0.5$, from the pair approximation (dashed line) and approximate master equation (solid line), compared with simulation (x)

$$\begin{aligned}
\frac{dN_{10}}{dt} &= -(2-\alpha)N_{10} + (1-\alpha)[N_{100} - N_{010} + N_{110} - N_{101}] \\
\frac{1}{2}\frac{dN_{11}}{dt} &= (1-\alpha(1-u))N_{10} + (1-\alpha)[N_{101} - N_{011}] \\
\frac{1}{2}\frac{dN_{00}}{dt} &= (1-\alpha u)N_{10} + (1-\alpha)[N_{010} - N_{100}].
\end{aligned} \tag{8.1.1}$$

Here we have used the fact that in the initial phase modeled by these differential equations $N_1 \approx Nu$ and $N_0 \approx N(1-u)$. We have $N_{11} + 2N_{10} + N_{00} = M$ so the sum of the three differential equations is 0. Note that the derivatives of the two-vertex quantities N_{ij} involve two or three vertices. Derivatives of three-vertex quantities involve up to four vertices. To be able to solve the equations, we need some approximation to close the equations.

The Pair approximation is commonly used in physics and ecology (see Ellner (2001) and references therein). Intuitively, the mean-field approach assumes that the equilibrium state is a product measure, while the pair approximation assumes it is a Markov chain: $N_{100} = N_{10}N_{00}/N_0$. The calculations we are about to do are similar to the ones in Kimura and Hayakawa (2008). Omitting the first equation from (8.1.1), which is redundant:

$$\begin{aligned}
\frac{1}{2}\frac{dN_{11}}{dt} &= [1-\alpha(1-u)]N_{10} + (1-\alpha)[N_{101} - N_{011}] \\
\frac{1}{2}\frac{dN_{00}}{dt} &= (1-\alpha u)N_{10} + (1-\alpha)[N_{010} - N_{100}].
\end{aligned}$$

Using these two equations and the pair approximation:

$$\frac{[1-\alpha(1-u)]}{1-\alpha}N_{10} = \frac{N_{01}N_{11}}{uN} - \frac{N_{10}N_{01}}{(1-u)N}$$
$$\frac{(1-\alpha u)}{1-\alpha}N_{10} = \frac{N_{10}N_{00}}{(1-u)N} - \frac{N_{01}N_{10}}{uN}$$

which leads to the equations

$$\frac{N_{11}}{uN} - \frac{N_{10}}{(1-u)N} = 1 + \frac{\alpha u}{1-\alpha} \quad (8.1.2)$$
$$\frac{N_{00}}{(1-u)N} - \frac{N_{10}}{uN} = 1 + \frac{\alpha(1-u)}{1-\alpha}. \quad (8.1.3)$$

Adding uN times (8.1.2) to $(1-u)N$ times (8.1.3), we have

$$N_{11} + N_{00} - \left(\frac{u}{1-u} + \frac{1-u}{u}\right) N_{01} = \left[1 + \frac{(u^2+(1-u)^2)\alpha}{1-\alpha}\right] N. \quad (8.1.4)$$

Using (8.1.4) with $N_{11} + N_{00} = \lambda N - 2N_{01}$ we have

$$\lambda N - 2N_{01} - \left(\frac{u}{1-u} + \frac{1-u}{u}\right) N_{01} = \left[1 + \frac{(u^2+(1-u)^2)\alpha}{1-\alpha}\right] N. \quad (8.1.5)$$

A little algebra gives

$$2 + \frac{u}{1-u} + \frac{1-u}{u} = \frac{2u - 2u^2 + u^2 + 1 - 2u + u^2}{u(1-u)}$$
$$= \frac{1}{u(1-u)}. \quad (8.1.6)$$

Rearranging (8.1.5) and using (8.1.6) gives

$$\left(\lambda - 1 - \frac{(u^2+(1-u)^2)\alpha}{1-\alpha}\right) N = \frac{1}{u(1-u)} N_{01},$$

and solving we get the pair approximation for the arch:

$$\frac{N_{01}}{N} = u(1-u)\left(\lambda - 1 - \frac{(u^2+(1-u)^2)\alpha}{1-\alpha}\right).$$

Note that as α decreases from $\alpha_c(1/2) = (\lambda-1)/(\lambda-1/2)$ to $(\lambda-1)/\lambda$, the arch expands from a point to a curve that spans (0, 1).

The **approximate master equation (AME)** framework gives much better estimates. Here we follow the approach in Gleeson (2011). A more extensive discussion of the method can be found in Gleeson (2013). Let $\bar{S}_{k,m}(t)$ $[\bar{I}_{k,m}(t)]$ be the number of nodes at time t that are in state 0 (susceptible) [in state 1 (infected)], have degree k, and have m neighbors in state 1.

To describe the logic behind the AME, let x be a vertex and let y be one of its neighboring vertices. Three types of things can happen: (a) rewiring may break the connection between x and y or bring a new edge to connect to x, (b) x or y may influence the other by a voting step, or (c) the opinion of y may be changed by imitating one of its neighbors $z \neq x$. Exact equations can be written for the first two types of events in terms of $\bar{S}_{k,m}$ and $\bar{I}_{k,m}$ but the third type requires an approximation.

The AME for the susceptible sites is:

$$\begin{aligned}\frac{d}{dt}\bar{S}_{k,m} &= \alpha\{-(2-u)m\bar{S}_{k,m} + (1-u)(m+1)\bar{S}_{k,m+1} + (m+1)\bar{S}_{k+1,m+1})\} \\ &\quad + \alpha N_{01}[-2\bar{S}_{k,m} + \bar{S}_{k-1,m-1} + \bar{S}_{k-1,m}]/N \\ &\quad + (1-\alpha)[-m\bar{S}_{k,m} + (k-m)\bar{I}_{k,m}] \\ &\quad + (1-\alpha)[-\beta^S(k-m)\bar{S}_{k,m} + \beta^S(k-m+1)\bar{S}_{k,m-1} \\ &\qquad - \gamma^S m\bar{S}_{k,m} + \gamma^S(m+1)\bar{S}_{k,m+1}],\end{aligned} \tag{8.1.7}$$

where $\beta^S = N_{001}/N_{00}$ and $\gamma^S = 1 + N_{010}/N_{01}$. Here β^S gives the expected number of 1 neighbors of a 0–0 edge. In γ^S the ratio N_{010}/N_{01} gives the expected number of 0 neighbors of the 1 at the end of a 0–1 edge, and the $+1$ counts the 0 on the conditioning edge.

To obtain predictions from numerically solving the system to steady-state. Mathematica's NDSolve function was used starting with a Poisson degree distribution of mean degree $\lambda = 4$, vertices independently assigned the value 1 with probability u, and the equations cutoff at maximum degree $K = 15$. (The results are not appreciably affected by increasing K.) In Fig. 8.8, the AME gives a better approximation of the final minority fraction ρ than the PA.

Further Improvements in Numerical Methods. Böhme and Gross (2011) wrote equations for the probability distribution of the state of a vertex and its neighbors that they called q-fans and spiders. The equations are linear so one can solve for the equilibrium rather than solving the differential equations to get their limit.

Silk et al. (2014) took things further in what they called a "mathematical tripe jump" named after the event in the Olympics. One uses a heterogeneous moment expansion to generate an infinite set of ODEs, maps the ODE system to a PDE system by looking at generating functions, then solves the PDE by the method of characteristics. The results given in their Fig. 8.2 almost perfectly match simulation.

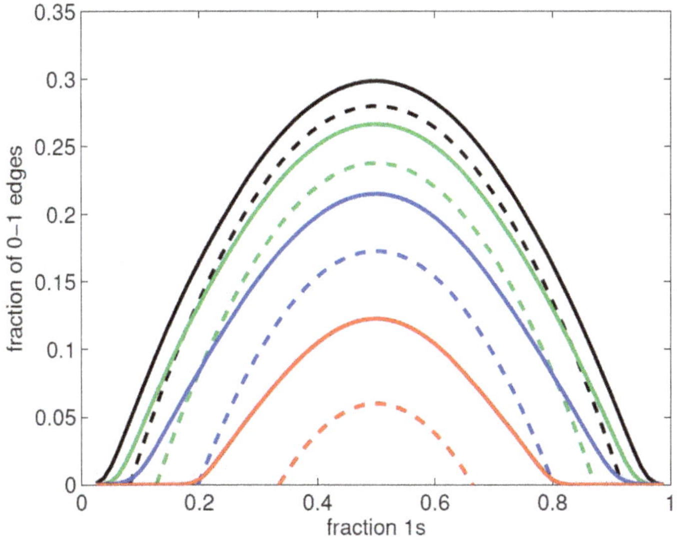

Fig. 8.8 Arches computed by approximate master equation (solid lines) versus simulation (dashes) for rewire-to-random model with $\alpha = 0.4, 0.5, 0.6, 0.7$. The curves decrease as α increases

8.1.2 Basu and Sly (2017)

These authors studied the evolving voter model on the dense Erdős-Rényi graph $ER(n, 1/2)$ in which edges between two vertices are independently present with probability 1/2. In order for there to be a transition in this version of the model, one must change the opinion update rate to $1 - \alpha = \beta/n$. Let τ be the time to reach an absorbing state in which there are no edges in the graph that connect two vertices with different opinions.

The first result holds for the rewire-to-random and for the rewire-to-same versions. It assumes that in the initial condition the states of vertices are independent and take the value 0 and 1 with equal probability. Let $N_*(t)$ be the number of edges with the minority opinion at time t, and for $1/2 > \epsilon > 0$ let $\tau_*(\epsilon) = \min\{t : N_*(t) \le \epsilon n\}$. They proved two main results

Theorem 8.1.4 *(Theorem 1.) Let* $1/2 > \epsilon' > 0$*. For both variants of the model there is a constant* $0 < \beta_0 < \beta_*(\epsilon')$ *so that the following hold*

(i) For all $\beta < \beta_0$ *and any* $\eta > 0$, $\{\tau < 6n^2, N_*(\tau) \ge 1/2 - \eta\}$ *holds with high probability.*
(ii) For all $\beta > \beta_*(\epsilon')$ *we have* $\tau_*(\epsilon') \le \tau$ *with high probability and*

$$\lim_{c \to 0} \liminf_{n \to \infty} P(\tau > cn^3) = 1.$$

In case (i) the frequency of the minority opinion does not change much by time τ so we are in the rapid disconnection regime. Conversely in case (ii) there is a substantial change in the minority frequency. Since voter model steps happen at rate β/n the number of steps to reach consensus is now $O(n^3)$.

Theorem 8.1.5 *(Theorem 2.) Let $\beta > 0$ be fixed. For the rewire-to-random model, there exists $\epsilon_*(\beta)$ so that $\tau < \tau_*(\epsilon)$ with high probability.*

The second result is only for rewire-to-random since the graph in the left half of Fig. 8.1 shows that it is false for rewire-to-same. Proving that the universal curve exists for rewire-to-random seems very difficult, as is the following

Open Problem. Prove that the critical value $\alpha_c(u)$ for rewire-to-same does not depend on the initial density u, and that for $\alpha < \alpha_c$ the ending minority fraction tends to 0 as $N \to \infty$.

Constructing the Process. Erdős-Rényi$(n, 1/2)$ has roughly $n^2/4$ edges so it will rapidly develop a large number of parallel edges, so we will need to carefully choose definitions to keep track of the state. Let n be a fixed positive integer. Let V be a fixed set with $|V| = n$, e.g., $V = \{1, 2, \dots, n\}$. Let $V^{(2)}$ be the set of unordered pairs of elements $\{x, y\}$ in V, with $x \neq y$, which are called **bonds**. Note that while we cannot prevent parallel edges there is no reason to allow self-loops.

Let $\mathcal{E}$ be the set of edges of the graph. We consider a discrete time Markov chain with state space

$$(\{0, 1\}^V, (V^{(2)})^{\mathcal{E}}).$$

Deviating from the notation of Basu and Sly, the first coordinate is a function $\xi_t : V \to \{0, 1\}$. $\xi_t(v)$ gives the opinion of v at time t but this is also written less formally as $v(t)$. The second coordinate is a function $Y_t : \mathcal{E} \to V^{(2)}$. $Y_t(e)$ gives the location of the edge e, which we can think of as little rods that are labeled and are moved around. The resulting multigraph is called $G(t)$. Let $N_i(t) = |\{v : X_t(v) = i\}|$ be the number of vertices in state $i - 1, 2$ and $N_*(t) = \min\{N_1(t), N_0(t)\}$ be the number of vertices with the minority opinion.

We take $G(0)$ to be distributed as $ER(n, 1/2)$ so each bond in $V^{(2)}$ contains one edge with probability 1/2 and 0 edges with probability 1/2. The dynamics we are about to describe will preserve the number of edges. The first step is to let $\mathcal{E}^\times$ be the set of discordant edges (the opinions are different at the two ends), pick one at random, and give it an orientation, say u, v). To decide which event will occur let $\zeta(t)$ be Bernoulli(β/n).

• If $\zeta(t) = 0$ we perform a rewiring step. To do this we pick w at random from $V - \{u\}$, and move (u, v) to (u, w), i.e., we set $Y_{t+1}(e) = (u, w)$.

• If $\zeta(t) = 1$ we perform a voter step in which v imitates u, ie. we set $X_t(v) = X_t(u)$.

Proof of Part (i) of Theorem 8.1.4. Most of Basu and Sly (2017) is devoted to the proof of part (ii) of Theorem 8.1.4, so following our usual *modus operandi* we will only prove the easy results. Let $D_{\max}(t)$ denote the maximum degree of a vertex in $G(t)$, which is the number of edges $e \in \mathcal{E}$ containing it, not the number of vertices that are connected to it by an edge.

Recall that τ is the first time there are no discordant edges. Let

$$\tau_1 = \min\{t : D_{\max}(t) \geq 8n\}, \qquad \tau_2 = \min\{t : N_*(t) \leq n/3\},$$

and $\tau_0 = \tau \wedge \tau_1 \wedge \tau_2$. Let $X_t = |\mathcal{E}_t^\times|$ be the number of discordant edges at time t

Lemma 8.1.6 *(Lemma 2.1.) There is a β_0 so that for all $\beta < \beta_0$ we have $\tau_0 \leq 6n^2$ with high probability.*

Proof. Let $\mathcal{F}_t$ be the filtration generated by the process up to time t. Observe that when an edge is rewired either X_t remains the same or decreases by 1. Conditional on $\mathcal{F}_t$ the probability that X_t decreases by a rewiring is at least $(N_* - 1)/(n-1)$. Also note that on a voter move X_t can increase by at most $D_{\max}(t)$. Hence we have for all $\lambda > 0$,

$$E(e^{\lambda X_{t+1}/n}|\mathcal{F}_t) \leq e^{\lambda X_t/n}\left[(1-\beta/n)\left(1+\frac{N_*-1}{n-1}(e^{-\lambda/n}-1)\right)+\frac{\beta}{n}e^{D_{\max}(t)/n}\right].$$

For large n on the event $\{t < \tau_0\}$ we have $(N_* - 1)/(n-1) \leq 1/4$. If we take λ sufficiently small so that $e^{8\lambda} \leq 1 + 9\lambda$ and $e^{-\lambda/n} - 1 \leq -\lambda/2n$ then on $\{t < \tau_0\}$ we have

$$E(e^{\lambda X_{t+1}/n}|\mathcal{F}_t) \leq e^{\lambda X_t/n}\left[(1-\beta/n)\left(1+\frac{1}{4}\cdot\left(-\frac{\lambda}{2n}\right)\right)+\frac{\beta}{n}(1+9\lambda)\right].$$

Doing some algebra the right-hand side is

$$\begin{aligned}
&= e^{\lambda X_t/n}\left[1-(1-\beta/n)\frac{\lambda}{8n}+\frac{\beta}{n}\cdot 9\lambda\right] \\
&\leq e^{\lambda X_t/n}\left[1-\frac{\lambda}{n}\left(\frac{1-\beta}{8}-9\beta\right)\right] \leq e^{\lambda X_t/n}\left[1-\frac{\lambda}{10n}\right].
\end{aligned}$$

when $\beta < \beta_0$. It follows from the above that

$$\begin{aligned}
P(\tau_0 > t|\mathcal{F}_0) &\leq E(\exp(\lambda X_t/n)1_{(\tau_0>t)}|\mathcal{F}_0) \\
&\leq e^{\lambda X_0/n}e^{-\lambda t/10n} \leq e^{\lambda n/2}e^{-\lambda t/10n},
\end{aligned}$$

since $X_0 \leq n^2/2$, and it follows that $P(\tau_0 > 6n^2) \leq \exp(-\lambda n/10)$.

Lemma 8.1.7 *(Lemma 2.2.) Recall τ_1 is the first time $D_{\max} \geq 8n$. With high probability $\tau_1 > 6n^2$.*

Proof. Let W_i be a sequence of i.i.d. random variables uniform on $\{1, \ldots, n\}$. Let $L_0 = 0$. For $i \geq 1$ let v_i be the root of the ith rewiring move and let

$$L_i = \min\{j > L_{j-1} : W_j \neq v_i\}.$$

$L_i - l_{i-1}$ are independent geometric($(n-1)/n$). It follows by a large deviations estimate that

$$L_{6n^2} \leq 13n^2/2 \quad \text{with exponentially high probability.}$$

To fill in the details note that if X is geometric(p) then

$$Ee^{\theta X} = \sum_{k=1}^{\infty}(1-p)^{k-1}pe^{\theta k} = \frac{pe^{\theta}}{1-(1-p)e^{\theta}}.$$

We are interested in the special case $p = (n-1)/n$. Formula (3.6.9) from the appendix tells us that is S_m is the sum of m geometric($(n-1)/n$)

$$\frac{1}{m}\log P(S_m \geq 13m/12) \leq -\left[\frac{13\theta}{12} - \theta - \log\left(\frac{1-1/n}{1-e^{\theta}/n}\right)\right].$$

If we take $\theta = \log 2$ then the right-hand side converges to $-(\log 2)/12$ as $n \to \infty$. Taking $m = 6n^2$ and using $\log 2 > 0.5$ it follows that for large n we have

$$P(L(6n^2) \leq 13n^2/2) \leq \exp(-n^2/24).$$

Let $N(v) = |\{i \leq 13n^2/2 : W_i = v\}|$. $N(v)$ is binomial($13n^2/2, 1/n$). Using Lemma A.1.6 with $m = 13n^2/2$, $p = 1/n$ and $y = 1/13$ (so $1 + y = 14/13$ we have

$$P(N(v) \geq 7n) \leq \exp\left(-\frac{13n}{2}\cdot\frac{1}{(13)^2}\cdot\frac{13}{2(14)}\right) = e^{-n/56}.$$

Since $D_{\max}(0) \leq n$ we have

$$P(\tau_1 \leq 6n^2) \leq ne^{-n/56} + P(L_{6n^2} \leq 13n^2/2),$$

which proves the desired result. □

Lemma 8.1.8 *(Lemma 2.3.) Recall τ_2 is the first time $N_*(t) \leq n/3$. With high probability $\tau_2 > 6n^2$.*

Proof. For $t \geq 1$ the number of relabeling moves up to time t, $RL(t)$, is dominated by a binomial($t, \beta/n$) random variable. On $\tau > t$ the change in the number of 0 vertices $N_0(t) - N_0(0)$ is a random walk Z_t that jumps by 1 or -1 with equal probability. The probability $\tau_2 \leq 6n^2 \wedge \tau$ is bounded by

$$P(N_*(0) \le 2n/5) + P(RL(6n^2) > 12\beta n) + P\left(\max_{i\le 12\beta n} |Z_i| \ge 3n/20\right).$$

Noticing that $P(N_*(0) \le 2n/5) \le 2P(\text{binomial}(n, 1/2) < 2n/5)$, and using Lemma A.1.6 with $m = n$, $p = 1/2$ and $y = 1/5$ we have

$$P(N_*(0) \le 2n/5) \le \exp(-(n/2)\cdot(1/25)\cdot 1/2) = \exp(-n/100).$$

Using Lemma A.1.8(d) again with $m = 6n^2$, $p = \beta/n$ and $y = 1$

$$P(RL(6n^2) > 12\beta n) \le \exp(-6\beta n/4).$$

Finally, using the Azuma-Hoeffding bound in Theorem A.2.1 with $m = 12\beta n$, $c_k \equiv 1$ and $L = 3n/20$ gives

$$P\left(\max_{i\le 12\beta n} |Z_i| \ge 3n/20\right) \le \exp(-(3n/20)^2/(2\cdot 12\beta n)) = \exp(-9n/9600\beta).$$

Combining the last four inequalities proves the desired result. In the last three inequalities the constants we have obtained are different from Basu and Sly. This is not a big deal but in the last instance their constant $e^{-2\beta n}$ has the β in the numerator when it should be in the denominator. □

Proof of part (i) of Theorem 8.1.4. Lemmas 8.1.6, 8.1.7, and 8.1.8 imply that with high probability $\tau \le 6n^2$. Using a random walk estimate as in the proof of Lemma 2.3 we conclude that the probability $N_*(t)$ drops below $(1/2 - \eta)n$ in $6n^2$ steps tends to 0 as $n \to \infty$. □

Proof of Theorem 8.1.5. For $0 < p < 1$ let $\mathcal{G}^*(n, p)$ be the collection of multigraphs where $N_1(G) = pn$ and the number of edges in G is $[12n^2/50, 13n^2/50]$, which has high probability under $ER(n, 1/2)$. Theorem 2 follows from

Theorem 8.1.9 *(Theorem 6.5) Let $\beta > 0$ be fixed and run the rewire-to-random dynamics starting from $G(0) \in \mathcal{G}^*(n, p)$. Let*

$$\tau = \min\{t : \mathcal{E}^\times(t) = \emptyset\} \quad \textit{and} \quad \tau_* = \tau_*(p/2) = \min\{t : N_*(t) \le np/2\}$$

Then there is a sufficiently small $p(\beta)$ so that for all $G(0) \in \mathcal{G}^*(n, p)$ we have $\tau < \tau_*$ whp.

Let S be the vertices with degree $\le 10n$. Let T be the vertices with degree $> 10n$. Noting that the degrees of vertices have mean $n/2$, and are almost independent, a simple calculation with Chebyshev's inequality shows that w.h.p. $|T| \le n/25$.

Let W_{SS} be the number of rewirings for edges with both end points in S and define W_{ST} and W_{TT} similarly. We will run the process for $10n^2$ steps. Let Y_{SS} be the number of edges at time $10n^2$ with both end points in S and define Y_{ST} and Y_{TT} similarly.

A Special Construction of Rewire-to-Random

- Let X_i and X'_i be two sequences of i.i.d. (shifted) geometric(β/n) taking values in $\{0, 1, 2, \ldots\}$ i.e., the number of failures before the first success.
- Let Z_i be i.i.d. Bernoulli(1/2).
- Let W_i be i.i.d. uniform on the vertices of the graph, which we will suppose are $\{1, 2, \ldots, n\}$.
- For each vertex v we will define counters $K_i(v)$, whose value at time i gives the number of rewiring updates with focal vertex v which are the number of updates with v as the focal vertex that are rewirings before it changes its opinion.
- We will define sequences L_i, L'_i, and T_i recursively. Let $L_0 = 0$, $T_0 = 0$ and $L'_0 = n$.

The main loop begins with choosing a discordant edge. Suppose the loop counter is i. Z_i is used to chosen the focal vertex for the edge. It is the end point with state Z_i. Let v be the chosen vertex.

Case 1. The reason for this construction and in particular the reason for two sequences x and X' can be seen in the proof of Lemma 6.3. If $v \in S$ and the opinion at v is 0, do the following: Set $L'_i = L'_{i-1}$.

If $K_{i-1}(v) > 0$ then define

$$T_i = \min\{k > T_{i-1} : W_k \neq v\},$$

which is the scheme we used previously to pick a vertex at random which is $\neq v$. Rewire the edge to (v, W_k), reduce $K_i(v)$ by 1, and set $L_i = L_{i-1}$.

If $K_{i-1}(v) = 0$ then change the opinion at v and set $L_i = L_{i-1} + 1$, and $K_i(v) = X(L_i)$ the new counter value. Since we did not use any uniformly chosen vertices we set $T_i = T_{i-1}$.

Case 2. If $v \notin S$ or the opinion at v is 1. Set $L_i = L_{i-1}$.

If $K_{i-1}(v) > 0$ then define

$$T_i = \min\{k > T_{i-1} : W_k \neq v\}.$$

Rewire the edge to (v, W_k), reduce $K_i(v)$ by 1, and set $L'_i = L'_{i-1}$.

If $K_{i-1}(v) = 0$ then change the opinion at v and set $L'_i = L'_{i-1} + 1$, and $K_i(v) = X'(L_i)$ the new counter value and set $T_i = T_{i-1}$.

Turning to the proof

Lemma 8.1.10 *(Lemma 6.2.) For a fixed p and $G(0) \in \mathcal{G}^*(n, p)$ the number of vertices of opinion 1 remains between $pn/2$ and $3pn/2$ for the first $10n^2$ steps whp.*

Proof. The number of voting steps is $\approx 10\beta n$ which is not enough to change the density. □

An element of the sequence X is said to be stubborn if it is at least $25n$.

Lemma 8.1.11 *(Lemma 6.3.) Let* $Y = |\{i \le L_{10n^2} : X_i > 25n\}$. *Then with high probability* $N_1(10n^2) \ge Y$.

Proof. We will be happy on $\mathcal{H} = \{\forall v \in G : |\{i \le T_{10n^2} : W_i = v\} \le 14n\}$. The sequence of W_i skips over some choices if they are equal to the current vertex but that happens only with probability $1/n$. An argument similar to that used in Lemma 2.2 shows that $P(T_{10n^2} \ge 11n^2)$ is exponentially small in n. A Chernoff bound shows that the probability a given v appears more than $14n$ times in the first $11n^2$ elements of the list is exponentially small.

On $\mathcal{H}$ if a stubborn X gets assigned to V then that point the vertex was in S and in state 1. Since the initial degree is $\le 10n$ and it will be chosen $\le 14n$ times, and the vertex does not change until after time $10n^2$, it follows from this reasoning that the vertices assigned to stubborn vertices are distinct. □

Let RL_{SS} be the number of relabelings that occur with both end points in S. Let R_{SS} be the number of times an edge with both end points in S was chosen and define R_{ST} and R_{TT} similarly. The proof is now completed by showing that if p is sufficiently small

Lemma 6.4. $RL_{SS} \le \beta n/20$ *whp.*

Lemma 6.5. $R_{SS} \le n^2/10$ *whp.*

Lemma 6.6. $R_{ST} \le 3n^2$ *whp.*

Lemma 6.7. $R_{SS} \le 6n^2$ *whp.*

We refer to Basu and Sly for the proofs of the four lemmas. Combining the last three lemmas implies that for a $p = p(\beta)$ sufficiently small and $G(0) \in \mathcal{G}^*(p)$ we have

$$R_{ss} + R_{ST} + R_{TT} < 10n^2 \quad \text{w.h.p.}$$

This implies that after $10n^2$ steps there are no disagreeing edges and hence $\tau \le 10n^2$, and the desired result follows from Lemma 8.1.10.

8.2 SIS Epidemics

8.2.1 *Rewiring to Avoid Infection*

In the introduction to this chapter we described the work of Gross et al. (2006), who introduced the model. Here we follow Zanette (2007), who analyzed the model using the tools physicists turn to first: mean-field and pair approximation calculations. He used a formulation different from Gross et al. (2006). Consider a population of N agents at the nodes of the network with M links.

- In the first sub-step an agent is chosen at random from the whole population. If it is infected it becomes healthy with probability γ.
- In the second substep a pair of linked agents are chosen. If both agents are susceptible or both infected nothing happens. Otherwise the S agent breaks the link with probability r. If it does not break the link then it becomes infected with probability λ.
- A time unit consists of N evolution steps. Thus each I-agent has a probability γ per time unit of becoming susceptible, so that the mean duration of the infected period is $1/\gamma$. Moreover each S-agent in contact with an I-agent becomes in turn infected with probability $(1-r)\lambda$ per time unit.

Let N_S, N_I be the number of susceptible and the number of infecteds

$$N_I + N_S = N, \qquad n_S = N_S/N, \quad \text{and} \quad n_I = N_I/N.$$

Let M_{II}, M_{SI}, M_{SS} be the number of edges of various types,

$$M_{II} + M_{SI} + M_{SS} = M \qquad m_{ij} = M_{IJ}/M.$$

Run time at rate $1/\gamma$. Let $\tilde{\lambda} = (1-r)\lambda/\gamma$, $z = 2M/N$ be the average degree, then $\tilde{r} = 2r/z\gamma$ since in the second step we are choosing edges instead of vertices.

Mean-Field Theory. In this approach we assume that the equilibrium state is always a product measure. We begin with the differential equation

$$n_I' = -n_I + \tilde{\lambda} m_{IS}, \tag{8.2.1}$$

which is exact. The $-n_I$ comes from $I \to S$. $\tilde{\lambda} m_{IS}$, from $IS \to II$. Introducing the mean-field approximation, $m_{IS} = 2n_I n_S = 2n_I(1-n_I)$ and the equation becomes

$$n_I' = -n_I + 2\tilde{\lambda} n_I(1-n_I) = n_I(2\tilde{\lambda} - 1 - n_I). \tag{8.2.2}$$

The critical value for survival (when $\tilde{r} = 0$) is $\tilde{\lambda}_c = 1/2$.

Pair Approximation. In this approximation we suppose that the state is always a Markov chain. In exact equations, derivatives of two site probabilities lead to three site probabilities. We write these in terms of two and one site probabilities using the Markov property. Let x, y, z be a path on the graph and $a, b, c \in \{0, 1\}$.

$$\begin{aligned}
P(\xi(x) = a, \xi(y) = b,\ \xi(z) = c) \\
&= P(\xi(x) = a, \xi(y) = b) P(\xi(z) = c | \xi(y) = b, \xi(x) = a) \\
&= P(\xi(x) = a, \xi(y) = b) P(\xi(z) = c | \xi(y) = b) \\
&= P(\xi(x) = a, \xi(y) = b) \frac{P(\xi(z) = c, \xi(y) = b)}{P(\xi(y) = b)} = \frac{m_{ab} m_{bc}}{n_b},
\end{aligned}$$

where in the second equality we have used the Markov property To be able to compare with other treatments we will include the third redundant equation

$$m'_{II} = -2m_{II} + \bar{\lambda} m_{IS} m_{SI}/n_S \tag{8.2.3}$$
$$m'_{IS} = 2m_{II} - m_{IS} - \bar{r} n_S m_{IS} + \bar{\lambda} m_{IS}(2m_{SS} - m_{IS})/n_S \tag{8.2.4}$$
$$m'_{SS} = m_{IS} + \bar{r} n_S m_{IS} - 2(\bar{\lambda} m_{IS} m_{SS})/n_S. \tag{8.2.5}$$

(8.2.3). $-2m_{II}$ from $II \to IS$; second term. $ISI \to III$ plus pair approximation.

(8.2.4). $2m_{II}$ from $II \to SI$. $-m_{IS}$ from $IS \to SS$. $-\bar{r} n_S m_{IS}$ from S breaking a connection to I and reconnecting to an S. Third term $ISS \to IIS$, 2 orientations for edge. Fourth term $ISI \to III$.

(8.2.5). From $m'_{II} + m'_{IS} + m'_{SS} = 0$.

Gross et al. (2006), see their (3) and (4), use the equations for m'_{II} and m'_{SS} but the notation is slightly different.

Equilibria satisfy the following equations

$$m_{IS} = \frac{n_I}{\tilde{\lambda}} \quad \text{and} \quad m_{II} = \frac{n_I^2}{2\tilde{\lambda}(1-n_I)} \tag{8.2.6}$$

by (8.2.1) and (8.2.3). Using (8.2.4) with $n_S = 1 - n_I$ and $2m_{SS} = 2 - 2m_{II} - 2m_{IS}$,

$$0 = 2m_{II} - m_{IS} - \tilde{r} m_{IS}(1-n_I) + \tilde{\lambda} m_{IS} \frac{2 - 2m_{II} - 3m_{IS}}{1-n_I}.$$

Multiplying by $(1-n_I)$

$$0 = 2m_{II}(1-n_I) - m_{IS}(1-n_I) - \tilde{r} m_{IS}(1-n_I)^2 + \tilde{\lambda} m_{IS}[2 - 2m_{II} - 3m_{IS}].$$

Substituting the values of m_{IS} and m_{II} from (8.2.6)

$$0 = \frac{2n_I^2}{2\tilde{\lambda}} - \frac{n_I}{\lambda}(1-n_I) - \tilde{r}\frac{n_I}{\lambda}(1-n_I)^2 + n_I\left[2 - 2\frac{n_I^2}{2\tilde{\lambda}(1-n_I)} - 3\frac{n_I}{\lambda}\right].$$

Factoring out an n_I from the first line and multiplying by $\tilde{\lambda}(1-n_I)$ we have

$$0 = n_I\Big\{n_I(1-n_I) - (1-n_I)^2 - \tilde{r}(1-n_I)^3 + [2\tilde{\lambda}(1-n_I) - n_I^2 - 3n_I(1-n_I)]\Big\}.$$

Combining the terms with no $\tilde{\lambda}$ or $\tilde{r}$ gives

$$n_I - n_I^2 - 1 + 2n_I - n_I^2 - n_I^2 - 3n_I + 3n_I^2 = -1.$$

Sorting terms by the powers of n_I

$$0 = n_I\left(2\tilde{\lambda} - 1 - \tilde{r} + (3\tilde{r} - 2\tilde{\lambda})n_I - 3\tilde{r}n_I^2 + \tilde{r}n_I^3\right), \tag{8.2.7}$$

which is formula (5) in Zanette (2007). When $\tilde{r} = 0$ this reduces to

$$0 = n_I(2\tilde{\lambda} - 1 - n_I),$$

which is the mean-field Eq. (8.2.2).

In addition to the trivial solution $n_I = 0$ the cubic has three solutions. One of them tends to infinity as $\tilde{r} \to 0$ and stays ≥ 1 for all $\tilde{r}$. The other two solutions $n_{I,2}^* < n_{I,1}^*$ are

$$1 - \sqrt{\frac{2\tilde{\lambda}}{3\tilde{r}}}\,[\cos(\alpha/3) \mp \sqrt{3}\sin(\alpha/3)] \quad \text{where} \quad \alpha = \arctan\sqrt{\frac{32\tilde{\lambda}^3}{27\tilde{r}} - 1}.$$

These solutions are real when $32\tilde{\lambda}^3/(27\tilde{r}) > 1$. That is, when $\tilde{r} < \tilde{r}_0 = (32/27)\tilde{\lambda}^3$. The results of these calculations are graphed in Fig. 8.9.

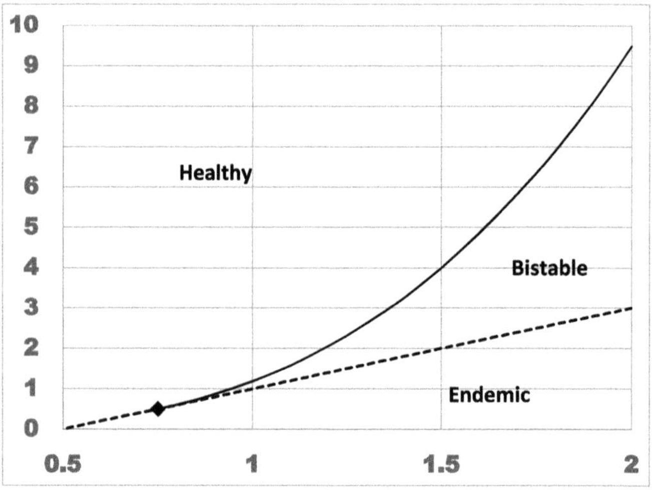

Fig. 8.9 Phase diagram as a function of $\tilde{\lambda}$ and $\tilde{r}$. $\tilde{\lambda}_c = 0.5$, the diamond marks (0.75, 0.5). The solid line where the equilibrium density drops discontinuously to 0 is $(32/27)\tilde{\lambda}^3$. The endemic region occurs when the smaller root is negative. The zero root occurs when the constant in the cubic (8.2.7) is 0 which is when $\tilde{r} = 2\tilde{\lambda} - 1$

Further Reading. da Silva et al. (2021) and Schapira and Valesin (2023) have considered contact processes on a dynamical d-regular graph where the graph evolves by degree-conserving Volz-Meyers edge swap dynamics, scaled so that in the limit as the number of nodes $n \to \infty$, each edge is involved in exchanges at rate v. The first paper shows that there is a threshold $\bar{\lambda}$ which depends on d and v so that for $\lambda > \bar{\lambda}$ the process survives exponentially long in n. They show that $\bar{\lambda} < \lambda_1(T_d)$. Combining this observation with a result of Mourrat and Valesin (2016) given in Theorem 5.3.4, they conclude that in some situations the process survives for $O(\log n)$ on the static graph and exponentially long on the evolving graph. The second paper completes the picture by showing that when $\lambda < \bar{\lambda}$, the system survives for time $\leq C \log n$. In addition they show that for fixed d, $v \to \lambda(d, v)$ is strictly decreasing.

8.2.2 Link Inactivation

Rewiring is difficult because the graph is constantly changing. A version of the evolving contact process that is somewhat easier to study is the model in which individuals temporarily suspend interactions with individuals who are infected (and this is arguably more sensible for applications than choosing a new connection at random who may be infected). To formulate the model it is convenient to replace each edge in the graph by two oriented edges, each of which may be either active or inactive. An active edge becomes inactive at rate α, and an inactive edge becomes active once its tail vertex becomes healthy. As usual in the contact process, infected vertices become healthy at rate 1, and infection is transmitted through active edges at rate λ.

To construct the process we use a graphical representation. Here we have written occupied instead of infected, and vacant instead of susceptible.

For each site x there is a Poisson process T_n^x with rate 1. At each arrival we put a dot ($\bullet$) at the site to indicate that any particle there will be killed.

For each oriented edge (x, y), there is a birth Poisson process with rate λ. At these arrivals we draw an arrow from x to y to indicate a birth will occur at y if it is vacant and x is occupied. We also have a deactivation Poisson process with rate α. At arrivals we draw an edge from x to $(x + y)/2$ and mark it with a $\times$. If x is occupied and y is vacant the edge will be deactivated until the next time it is vacant.

Chatterjee et al. (2022) have studied this model on infinite and finite graphs. Let $\mathcal{X}$ be the collection of finite sets of vertices. Since there is no monotonicity (see Fig. 8.10) there are potentially three different critical values

$$\begin{aligned}
\lambda_\alpha^- &= \inf\{\lambda : \exists \xi \in \mathcal{X}\ P_\xi(\xi_t \neq \emptyset \text{ for all } t \geq 0) > 0\}, \\
\lambda_{\alpha,w}^+ &= \sup\{\lambda : \forall \xi \in \mathcal{X}\ P_\xi(\xi_t \neq \emptyset \text{ for all } t \geq 0) = 0\}, \\
\lambda_\alpha^+ &= \sup\{\lambda : \forall \xi \in \mathcal{X}\ P_\xi(0 \in \xi_t \text{ infinitely often}) = 0\}.
\end{aligned}$$

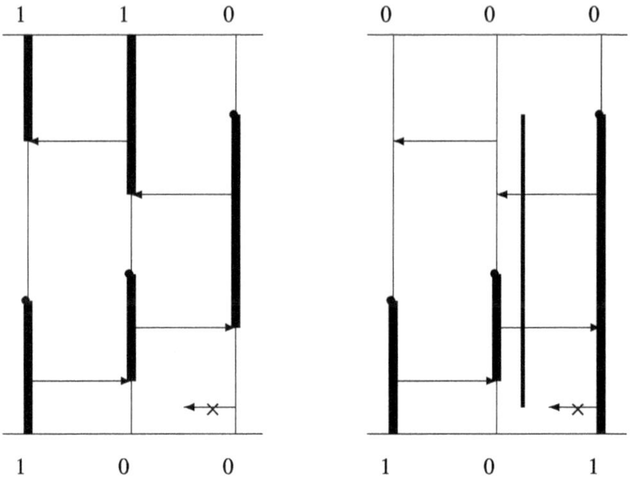

Fig. 8.10 An example of one graphical representation for the process on $\{1, 2, 3\}$ with two different initial conditions. Thick lines indicate occupied sites. On the left the deactivation edge has no effect since 3 is vacant. On the right, births from 3 to 2 are not allowed to cross the vertical line because $(3, 2)$ is deactivated. The example shows that this system is not attractive, i.e., $\xi_0 \supset \xi_0'$ does not guarantee that $\xi_t \supset \xi_t'$ for all $t \geq 0$

The last two definition brings in the issue of local versus global survival as in the contact process on trees.

On the integers with the usual nearest neighbors

Theorem 8.2.1 $1 + \alpha \leq \lambda_\alpha^- \leq \lambda_\alpha^+ \leq A + B\alpha$.

Ideas Behind the Proof**.** To prove the lower bound on the critical value look at $[\ell_t, r_t]$ where ℓ_t is the left-most and r_t is the right-most occupied site. Let L_n be the jump chain embedded in ℓ_t. In order for $L_{n+1} = L_n - 1$ the infected at L_n must try to infect $L_n - 1$ before that site avoids L_n and before L_n becomes healthy. When $\lambda < 1 + \alpha$ a downward jump has probability $< 1/2$. When L_n becomes healthy before infecting $L_n - 1$ (and this always happens when that site avoids L_n) then the process will jump to the right by ≥ 1. This shows that the left edge has positive drift. A similar argument shows that the right edge has negative drift, so survival is impossible.

The proof of the other direction involves a block construction. One defines good events that will have high probability if λ is large enough, and survival follows from a comparison with supercritical oriented percolation. Here the lack of monotonicity rears its ugly head again and the authors have to deal with the dependence between the events in different blocks. Since the details are somewhat complicated we refer the reader to Sect. 8.3 of the paper for details. □

Consider now the cycle $\mathbb{Z}$ mod n. Based on results for the contact process and a number of related systems, we expect the process on the cycle to survive for time $\leq C \log n$ below the critical value and for time $\geq \exp(cn)$ above it. However, since one cannot prove the existence of the critical value this goal is out of reach. Using the machinery developed to prove Theorem 8.2.1 the authors were able to prove

Theorem 8.2.2 *There are constants C, γ that depend on λ and α so that:*

(i) If $\lambda < 1 + \alpha$ then $P(\tau > C(\log n)^2) \to 0$.
(ii) If $\lambda > A + B\alpha$ (the constants in the previous result) then $P(\tau \le e^{\gamma n}) \to 0$.

Intuitively, the contact process with avoidance is monotone in the sense that if we consider it on a graph G and then restrict it to a subgraph H by declaring that all sites in $G - H$ are 0 then the process on G dominates the process restricted to H. For this reason, we expect that conclusion (ii) holds whenever there is a self-avoiding path of length n. The "inconvenient truth", to use a phrase of Al Gore, is that this is hard to prove rigorously, so this is one of the proofs in the book that you should not believe.

Perhaps the most surprising result in the paper is the result for the star graph. Suppose that initially all the sites are infected and all of the edges are active.

Theorem 8.2.3 *Let τ_{star} be the extinction time of the infection on the star graph with n vertices and let*

$$\Delta(\lambda, \alpha) = 2\left[(\lambda + \alpha + 1) - \sqrt{(\lambda + \alpha + 1)^2 - 4\alpha}\right]^{-1}.$$

Then there is an N so that for $n \ge N$

$$P\left(\frac{1}{K}(n/\log^4 n)^\Delta \le \tau_{star} \le Kn^\Delta\right) \to 1,$$

uniformly in n as $K \to \infty$.

Thus except for an annoying factor of $\log^4 n$ in the lower bound the contact process with avoidance survives for time $\approx n^\Delta$ where Δ depends on λ and α.

Further Reading. In the next section we will see a number of papers by Ball, Britton, and collaborators on SIR models with rewiring which include the possibility of dropping of edges

Linker and Reminik (2020) have proved theorems about a one-dimensional contact process in which edges open up at rate vp and close at rate $v(1 - p)$. For small enough v the process dies out, while for large v it behaves like a contact process with births at rate λp. For fixed v the infection dies out for small p, while if p is close to 1 then survival is possible for large enough λ. Some of the results hold for general vertex-transitive regular graphs. This paper is closely related to Reminik's (2008) paper, which was written while he was a Ph.D. student at Cornell. The model is a site version of link inactivation. Sites can be in state $1 =$ occupied, $0 =$ habitable but vacant, and $-1 =$ uninhabitable. The fact that transitions $1, 0 \to -1$ occur at rate

α make it possible to do a fairly complete analysis, including proving a complete convergence theorem.

Jacobsen et al. (2016) have proved ODE limits for stochastic epidemic models on dynamic multilayer networks, as Janson et al. (2014) did for the Miller-Volz equations in Sect. 4.5.

Open Problems. Three interesting papers for those who want to try their hands at proving results from the physics literature. Szabó-Solticzky et al. (2016) study the occurrence of oscillatory behavior in a paper in *Journal of Mathematical Biology*. Tunc et al. (2013), writing in the *Journal of Statistical Physics*, developed a mean-field description and found two distinct regimes: slow network dynamics where the adaptation mechanism simply reduces the number of contacts, and fast dynamics where more efficient adaptation reduces the spread of the disease by targeting dangerous connections. Finally a 13 page paper by Guo et al. (2013), and published in *Physica Review E*, also finds two regimes, and concludes that adaptation of the topology suppresses the infection, while it promotes the network evolving towards a topology that exhibits assortative mating, modularity and a binomial-like degree distribution.

8.3 SIR Epidemics

As discussed in Chap. 4, in the SIR model, individuals are in one of three states: $S =$ susceptible, $I =$ infected, $R =$ removed (cannot be infected). $S - I$ edges become $I - I$ at rate λ, i.e., after a time T with an exponential(λ) distribution: $P(T > t) = e^{-\lambda t}$. As in Sects. 4.3 and 4.4 we will consider two versions of the model that differ in the length of time individuals remain infected. In the first, infections always last for time 1. In the second, infection times are exponential(1) distributed. Once individuals leave the infected state, they enter the removed state.

8.3.1 *Volz and Meyers (2007, 2009)*

We begin by discussing a paper not mentioned in the summary at the beginning of the chapter, but which we have mentioned many times due to its unusual update rule. To motivate their work the authors note that homogeneously mixing models ignore three aspects of epidemics

(i) A given individual has contact with only a small number of other individuals in the population at any one time, and contacts that can result in disease transmission are usually short and repeated events.
(ii) The number and frequency of contacts between individuals can be very heterogeneous.
(iii) The numbers and identities of an individual's contacts will change as time goes by.

The first two points have been addressed by static network models. The authors say that the main motivation for their paper is to address the third point by introducing the **neighbor exchange** (NE) model. In this model the individual's number of contacts is fixed while the composition of those contacts changes at a specified rate. The dynamics in which individuals change partners: connections $A - B$ and $C - D$ become $A - D$ and $B - C$ is a little unusual (except for square dancers) but as the reader will see it leads to a system that it is easier to analyze than most evolving networks.

Using notation that should be familiar from Sect. 4.4 the edge (ego,alter) is reassigned to (ego,alter′) at rate ρ (for rewire) or equivalently (for the resulting ODE) the two edges are swapped at rate $\rho/2$. Using notation that was introduced earlier, let $\mathcal{A}_{X,Y}$ be the set of arcs with ego in X and alter in Y, let $\mathcal{A}_X$ be the set of arcs with ego in X and let $|\mathcal{A}|$ be the number of arcs in the graph (which is twice the number of edges). Define the fraction of arcs in $\mathcal{A}_{X,Y}$ and in $\mathcal{A}_X$ by

$$M_{X,Y} = \frac{|\mathcal{A}_{X,Y}|}{|\mathcal{A}|} \quad \text{and}$$
$$M_X = \frac{|\mathcal{A}_X|}{|\mathcal{A}|}.$$

We replace the rate r at which infected individuals infect susceptible neighbors used in VM by β, and the recovery rate μ used in VM by our favorite letter γ. Let S, I, and R be the fraction of susceptible, infected, and removed nodes at time t and let

$$p_I = \frac{M_{S,I}}{M_S} \quad \text{and}$$
$$p_S = \frac{M_{S,S}}{M_S}. \tag{8.3.1}$$

Letting $g(x)$ be the probability generating function of the degree distribution p_k and $\theta(t)$ the probability an edge has not transmitted an infectious contact (which is defined precisely right after (4.4.1), we have

$$S = g(\theta),$$
$$d\theta/dt = -\beta p_I \theta, \tag{8.3.2}$$

where p_I and p_S satisfy

$$dp_I/dt = \beta p_S p_I \theta \frac{g''(\theta)}{g'(\theta)} - \beta p_I(1 - p_I) - \gamma p_I, \tag{8.3.3}$$

$$dp_S/dt = \beta p_S p_I \left(1 - \theta \frac{g''(\theta)}{g'(\theta)}\right). \tag{8.3.4}$$

This agrees with the formulas in Table 3 in Volz and Meyers (2007) except for the unfortunate typo that $g''(\theta)$ has lost its derivatives in the formulas for dp_I/dt and dp_S/dt. To confirm that this is indeed a typo look at formulas (2.14) and (2.15) in VM and the result below.

Repeating the reasoning used to derive (8.3.3) and (8.3.4) on the neighbor exchange system leads to

$$dp_I/dt = \beta p_S p_I \theta \frac{g''(\theta)}{g'(\theta)} - \beta p_I(1-p_I) - \gamma p_I + \rho(M_I - p_I) \tag{8.3.5}$$

$$dp_S/dt = \beta p_S p_I \left(1 - \theta \frac{g''(\theta)}{g'(\theta)}\right) + \rho \left[\frac{\theta g'(\theta)}{g'(1)} - p_S\right] \tag{8.3.6}$$

$$dM_I/dt = -\gamma M_I + \beta p_I \left[\frac{\theta^2 g''(\theta) + \theta g'(\theta)}{g'(1)}\right]. \tag{8.3.7}$$

The additional terms in the first two equations come from rewiring. The third equation is needed since M_I, the fraction of arcs with an infected ego, cannot be expressed in terms of the other three variables.

Simulations of the fraction of individuals that have been infected given in Fig. 8.2 of their paper match the solution of the differential equation, except for a time shift, which seems likely to be due to the fact that in the initial phase the infects are a branching process, so we have

$$I(t) \approx W e^{\alpha t} = e^{\alpha t - S},$$

where W is random and $S = -\log(W)/\alpha$ is a random time shift. Note that in agreement with this computation, the stochastic curves are translates in time of the deterministic curve.

The 2007 paper closes with an application of the NE model to data on syphilis among Atlanta adolescents, so we will now move on to the 2009 paper that computes the reproductive ratio R_0 for the model. This quantity introduced in Sect. 4.1 gives the number of secondary infections caused by one infected individual in an otherwise susceptible population. Since they are dealing with epidemics on graphs generated by the configuration model they instead look at what they call $R_\star$ which gives the number of secondary infections caused by an infected individual early in the epidemic but who is not the first infected, since as we learned in Chap. 4 this is the quantity which gives the condition for the epidemic to be supercritical $R_\star > 1$.

The probability of transmitting the infection to a susceptible neighbor is

$$\tau = \frac{\beta}{\beta + \gamma + \rho},$$

since infection must occur before recovery and rewiring. After a change of notation this is their (3.4). Some mental gymnastics with generating functions leads them to conclude in (3.11) that

$$R_\star = \tau\left(\frac{\rho}{\gamma} + \frac{\gamma+\rho}{\gamma}\cdot\frac{g''(1)}{g'(1)}\right).$$

Using this result they compute epidemic thresholds and branching process approximation to the initial phase of the epidemic. We refer the reader to the paper for details.

8.3.2 *Jiang et al. (2019)*

Here we prove the two results mentioned in the introduction to the chapter. The object of study was SIR dynamics in which each infection lasts for exactly time 1 with the new feature that susceptibles who are neighbors of an infected break their connections at rate ρ and connect to a randomly chosen neighbor. As explained in Sect. 4.3, the SIR model with fixed infection times is simple because each edge will be $S - I$ (or $I - S$) only once. When that happens the infection will be transmitted to the other end with probability

$$\tau_f = P(T \le 1) = 1 - e^{-\lambda}, \tag{8.3.8}$$

and the events for different edges are independent. Here the 'f' in the subscript is for "fixed time." Due to the last observation, we can delete edges with probability $e^{-\lambda}$ and the connected components of the resulting graph will give the epidemic sizes when one member of the cluster is infected. In Chap. 4 we proved

Theorem 4.3.1. *If the original graph is Erdős-Rényi with mean degree μ, then the reduced graph is Erdős-Rényi with mean degree $\mu\tau_f$. So, a large epidemic occurs with positive probability if $\mu\tau_f > 1$. If z_0 is the fixed point smaller than 1 of the generating function*

$$G(z) = \exp(-\mu\tau_f(1-z)), \tag{8.3.9}$$

then $1 - z_0$ gives both the limiting probability an infected individual will start a large epidemic, and the fraction of individuals who will become infected when a large epidemic occurs.

We now introduce rewiring of $S - I$ edges at rate ρ, i.e., susceptibles break their connection with infected individuals and rewire to an individual chosen at random. In order for the infection to be transmitted along an edge, it must occur before any rewiring and before time 1. To compute this probability, note that (i) the probability that infection occurs before rewiring is $\lambda/(\lambda+\rho)$ and (ii) the minimum of two independent exponentials with rates λ and ρ is an exponential with rate $\lambda+\rho$, so the transmission probability is

$$\tau_{f,r} = \frac{\lambda}{\lambda + \rho}(1 - e^{-(\lambda+\rho)}). \tag{8.3.10}$$

Here the second 'r' subscript is for "rewire." Our first result shows that evoSIR has the same critical value as delSIR in which edges are deleted at rate ρ.

Theorem 8.3.1 *The critical value for the fixed time epidemic with rewiring is given by the solution of* $\mu\tau_{f,r} = 1$*. Moreover, in the subcritical regime then the ratio of the expected epidemic size in delSIR to the size in evoSIR converges to 1.*

***Sketch of Proof*.** The formula for the critical value is easily seen to be correct for the delSIR since, by the reasoning above, there is a large epidemic if and only if the reduced graph in which edges are retained with probability $\tau_{f,r}$ has a giant component. It is clear than the delSIR model has a larger ($\geq$) critical value than evoSIR. Thus, we only have to prove the reverse inequality.

To compare the two evolutions we first run delSIR until it dies out. Once we do this we randomly rewire the edges deleted in delSIR. The number of sites infected in our subcritical delSIR is $\leq C \log n$ with high probability. This gives a trivial upper bound on the number of edges deleted in delSIR and hence rewired in evoSIR. A rewired edge will only produce a new infected if it is rewired to a vertex that is infected or will later become infected.

If we let Y be the number of new infections then $P(Y = 1) \leq (C \log^3 n)/n$ and $P(Y \geq 2) \leq (C \log^6 n)/n^2$. (An extra $\log n$ comes from worry about vertices of large degree.) The second possibility will not happen for any possible starting point and hence can be ignored. It remains then to estimate the possible effects of one extra infection. The proof is not difficult but it does take a few pages to write out the details, so we refer the reader to Jiang et al. (2019) for details. □

When n is large, the degree distribution, which is binomial$(n - 1, \mu/n)$, is approximately Poisson with mean μ. Due to Poisson thinning, the number of new infections directly caused by one I is asymptotically Poisson with mean $\mu\tau_{f,r}$, and hence has limiting generating function

$$\hat{G}(z) = \exp(-\mu\tau_{f,r}(1 - z)). \tag{8.3.11}$$

Theorem 8.3.2 *If* z_0 *is the fixed point* < 1 *of* $\hat{G}(z)$*, then* $1 - z_0$ *gives the probability of a large delSIR or evoSIR epidemic.*

***Sketch of Proof*.** Again this is obvious for delSIR, since it is just percolation. It remains to show that the probability of a large epidemic in evoSIR is not larger. To do this apply the method of proof of Theorem 1.3.2 to delSIR. Run the exploration process to time $r = \beta \log n$. Let $F_0 = \{A_r = 0\}$, $F_1 = \{0 < A_r < \gamma \log n\}$ and $F_2 = \{A_r \geq \gamma \log n\}$. The probability of a large delSIR cluster $P(B_d) = (1 - o(1)P(F_2)$. A large evoSIR cluster is likely in case F_2, very unlikely in case F_0 and case F_1 has small probability. For more details see Sect. 8.3 of Jiang et al. (2019). □

The delSIR model is essentially percolation, so $1 - z_0$ is also the fraction of individuals infected in a large epidemic. This proportion goes to 0 at the critical value $\mu_c = 1/\tau_r^f = 1$. In evoSIR, as the simulations in Fig. 8.3 suggest, it is discontinuous at the critical value.

8.3.3 Britton et al. (2016–2018)

As the authors of Jiang et al. (2019) were finishing up the writing of their paper, they learned of papers by Britton and collaborators that study epidemics on evolving graphs with exponential infection times. Britton et al. (2016) studied a one-parameter family of models (SIR-ω) that interpolates between delSIR and evoSIR.

SIR-ω Epidemic. In this model, an infected individual infects each neighbor at rate λ, and recovers at rate γ. A susceptible individual drops its connection to an infected individual at rate ω. The edge is rewired with probability α and dropped with probability $1 - \alpha$. Since evoSIR ($\alpha = 1$) and delSIR ($\alpha = 0$) have the same critical values and survival probability it follows that this holds for all $0 \leq \alpha \leq 1$.

These epidemics take place on graphs generated by a configuration model with n vertices and degree distribution D, $\mathbb{CM}(n, D)$, as defined in Chap. 2. Britton et al. (2016) studied the initial phase of the epidemic starting with one infected individual at vertex x (chosen uniformly at random from all vertices) using a branching process approximation. Let Z_m^n be the number of vertices at distance m from x in the graph. For any fixed $k \in \mathbb{N}$, $\{Z_m^n, 0 \leq m \leq k\}$ converges to the following two-phase branching process $\{Z_m, 0 \leq m \leq k\}$. The number of children in the first generation has the distribution D while subsequent generations have the distribution $D^* - 1$ where D^* has the *size-biased* degree distribution

$$P(D^* = j) = \frac{jp_j}{m_1}, \quad j \geq 0.$$

Here $p_j = P(D = j)$ and $m_1 = E(D)$ is the mean of D. $D^* - 1$ has mean

$$E(D^* - 1) = \sum_j (j - 1)\frac{jp_j}{m_1} = \frac{m_2 - m_1}{m_1},$$

where $m_j = E(D^j)$

In the **SIR-ω** model, the probability that an infection will cross an $S-I$ edge is

$$\tau = \frac{\lambda}{\lambda+\gamma+\omega}.$$

Thus we get another two-phase branching process $\bar{Z}_k$ defined as follows. $\bar{Z}_0 = 1$, $\bar{Z}_1 = \text{binomial}(D, \tau)$ and future generations have offspring distribution $\text{binomial}(D^* - 1, \tau)$. One can see from this description that the limiting branching process $\bar{Z}_m$ will have positive survival probability if

$$1 < \tau E(D^* - 1) = \frac{(m_2 - m_1)\tau}{m_1}.$$

Correspondingly, there will be a large epidemic in the SIR-ω model if

$$R_0 = \frac{\lambda}{\lambda+\gamma+\omega} \cdot \frac{m_2 - m_1}{m_1} > 1. \tag{8.3.12}$$

Thus for fixed values of γ and ω

$$\lambda_c = (\gamma+\omega)\frac{m_1}{m_2 - 2m_1}. \tag{8.3.13}$$

When $\gamma = 0$ (no recovery) and $\omega = \rho$, which is the SI-ω model in our notation,

$$\lambda_c = \frac{\rho m_1}{m_2 - 2m_1}. \tag{8.3.14}$$

The critical values of delSI and evoSI only depend on the ratio ρ/λ, so it is natural to define a parameter (this α is different from the α used in the definition of the SIR-ω model)

$$\alpha = \rho m_1/\lambda \tag{8.3.15}$$

that has $\alpha_c = m_2 - 2m_1$. We will only consider the del and evo endpoints of the one parameter family of models SIR-ω, so after the discussion of previous work is completed, there should be no confusion between our α and theirs.

Work of Leung, Ball, Sirl, and Britton (2018) demonstrated the paradoxical fact that individual preventative measures may lead to a larger final size of the epidemic. They proved this rigorously for SI epidemics on the configuration model with two degrees and conducted simulation studies for many social networks. Note that in Fig. 8.11 (taken from their Fig. 8.1) the final size increases with the rewiring rate when it is small. This simulation does not suggest that the phase transition was discontinuous.

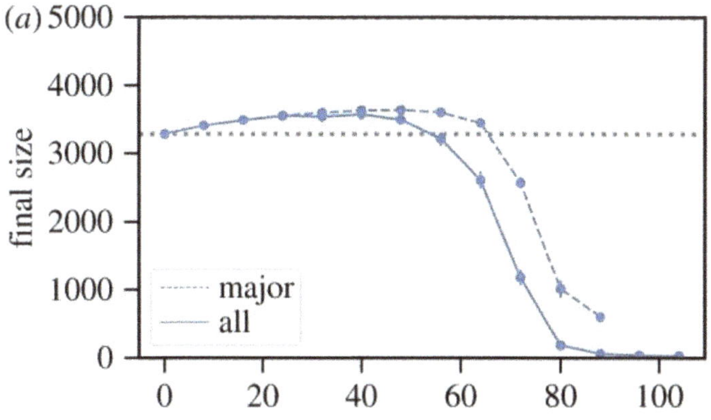

Fig. 8.11 Social distancing can lead to an increase in the final epidemic size in the configuration model. The x-axis indicates the rewiring rate. The horizontal line is the final size when $\omega = 0$

8.3.4 Durrett and Yao (2020)

Here we are referring to the never published paper on the arXiv. These authors studied the evolving SIR and SI epidemics but, for reasons that will become clear as the story unfolds, we will concentrate on the evoSI and delSI models. Let I_t be the number of infected vertices at time t. For evoSI or delSI epidemics the final size is

$$I_\infty = \lim_{t\to\infty} I_t,$$

which exists since $I(t)$ is increasing. We say a **large outbreak** occurs if a positive fraction of the population becomes infected. To be precise, this means that there is an $\epsilon > 0$ so that

$$\limsup_{n\to\infty} P_1(I_\infty/n > \epsilon) > 0, \tag{8.3.16}$$

where P_1 is the process starting with one randomly chosen vertex infected.

Our first result identifies λ_c. Let Z_m be the number of vertices at distance m from x, then as $n \to \infty$, the process Z_m converges to the *two-phase branching process* in which $Z_0 = 1$, the number of children in the first generation has distribution D with $P(D = k) = p_k$, and the number of offspring in subsequent generations has distribution $D^* - 1$, where deviating from the notation used earlier $P(D^* = k) = kp_k/\mu$. Though the term is slightly inaccurate, it is convenient to call the distribution of $D^* - 1$ the size-biased degree distribution. One can see from this description that the limiting branching process will have a giant component if

$$1 < E(D^* - 1) = \frac{m_2}{m_1} - 1 = \frac{m_2 - m_1}{m_1}.$$

Before moving on, we will pause to compute the generating function of $D^* - 1$. Let G be the generating function of D,

$$G(z) = \mathbb{E}z^D = \sum_{j=0}^{\infty} d_j z^j. \tag{8.3.17}$$

The distribution of $D^* - 1$ satisfies $P(D^* - 1 = j - 1) = jd_j/m_1$, so

$$G^*(z) = Ez^{(D^*-1)} = \sum_{j=1}^{\infty} \frac{jd_j}{m_1} z^{j-1} = \frac{G'(z)}{m_1}.$$

If we consider the delSI model then the probability the infection will cross an $S - I$ edge is

$$\tau = \lambda/(\lambda + \rho).$$

If we independently keep edges with probability τ and delete them with probability $1 - \tau$, then there will be a large epidemic in the delSI model if

$$\frac{\lambda}{\lambda + \rho} \cdot \frac{m_2 - m_1}{m_1} > 1. \tag{8.3.18}$$

Rearranging gives $\lambda(m_2 - m_1) > (\lambda + \rho)m_1$, which holds if

$$\lambda > \rho m_1/(m_2 - 2m_1).$$

As noted earlier, the critical values of delSI and evoSI only depend on the ratio ρ/λ, so it is natural to define a parameter $\alpha = \rho m_1/\lambda$. This α is NOT the one used by Britton et al.

Theorem 8.3.3 *Suppose $ED^3 < \infty$.*

(i) The critical values of delSI and evoSI are equal to $\hat{\alpha}_c = m_2 - 2m_1$.
(ii) In addition, when $\alpha < \alpha_c$ the probability of a large epidemic is the same in the two models.

The assumption $ED^3 < \infty$ is natural because it implies that the size-biased distribution has finite variance. The proof of Theorem 8.3.3 is based on two proofs from Jiang et al. that we discussed in Sect. 8.3.2. Here and in what follows, formulas are sometimes easier to evaluate if we use the "factorial moments"

$$\mu_k = E[D(D-1)\cdots(D-k+1)],$$

since these can be computed from the kth derivative of the generating function. To translate between the two notations:

$$\mu_1 = m_1, \qquad \mu_2 = m_2 - m_1, \qquad \mu_3 = m_3 - 3m_2 + 2m_1.$$

In particular, $\alpha_c = m_2 - 2m_1 = \mu_2 - \mu_1$.

Limiting Sizes of delSI and evoSI

To compute the probability of a large epidemic in delSI we note that generating function of degree distribution for the reduced graph is

$$\begin{aligned} G_0(z) &= \sum_{j=0}^{\infty}\sum_{k=j}^{\infty} d_k \binom{k}{j} \tau^j (1-\tau)^{k-j} z^j \\ &= \sum_{k=0}^{\infty} d_k \sum_{j=0}^{k} \binom{k}{j} (\tau z)^j (1-\tau)^{k-j} = G(z\tau + (1-\tau)). \end{aligned}$$

Similarly, in the reduced graph the offspring distribution in the second phase of the branching process has generating function

$$G_1(z) = G^*(z\tau + (1-\tau)). \tag{8.3.19}$$

To compute the size of the delSI epidemic now, note that the probability the homogeneous branching process started by a first generation particle dies out is the smallest solution of $G_1(\zeta) = \zeta$, therefore the size of the giant component in delSI is $1 - G_0(\zeta)$.

The computation of the limiting size of a large epidemic in the evoSI model is much more complicated. Recall that G is the generating function of the degree distribution D as defined in (8.3.17). Consider the following function

$$f(w) = \log\left(\frac{m_1 w}{G'(w) + \alpha(1-w)G(w)}\right) + \frac{\alpha}{2}(w-1)^2. \tag{8.3.20}$$

This function may look mysterious, but as we will explain below, it allows us to compute the limiting size of the epidemic. Ultimately that recipe turned out to be incorrect but it provided a mechanism to identify when the phase transition in evoSI was discontinuous.

Theorem 8.3.4 *Suppose $\alpha < \alpha_c$ and let*

$$\sigma = \sup\{w : 0 < w < 1, f(w) = 0\} \ \textit{with} \ \sup(\emptyset) = 0, \tag{8.3.21}$$

$$\nu = 1 - \exp\left(-\frac{\alpha}{2}(\sigma-1)^2\right) G(\sigma). \tag{8.3.22}$$

If we suppose

($\star$) *if* $0 < \sigma < 1$ *there is a* $\delta > 0$ *so that* $f < 0$ *on* $(\sigma - \delta, \sigma)$

then for any $\epsilon > 0$,

$$\lim_{n\to\infty} P_1(I_\infty/n < \nu + \epsilon) = \lim_{\eta\to 0}\liminf_{n\to\infty} P_1(I_\infty/n > \nu - \epsilon | I_\infty/n > \eta) = 1.$$

Using Theorem 8.3.4, we get the following almost necessary and sufficient conditions for a discontinuous phase transition in the final epidemic size. We say "almost" since the case $\Delta = 0$ is left out.

Theorem 8.3.5 *Assume* $ED^3 < \infty$. *Consider the evoSI model on the configuration model* $\mathbb{CM}(n, D)$. *Let*

$$\Delta = -\frac{\mu_3}{\mu_1} + 3(\mu_2 - \mu_1).$$

If $\Delta > 0$ *then there is a discontinuous phase transition. For some* $\epsilon_0 > 0$ *and some* $\delta > 0$,

$$\lim_{\eta\to 0}\liminf_{n\to\infty} P_1(I_\infty/n > \epsilon_0 | I_\infty/n > \eta) = 1 \quad \textit{for all } \alpha_c - \delta < \alpha < \alpha_c. \tag{8.3.23}$$

If $\Delta < 0$, *then there a continuous phase transition. For any* $\epsilon > 0$, *there exists some* $\delta > 0$, *so that*

$$\lim_{n\to\infty} P_1(I_\infty/n > \epsilon) = 0 \quad \textit{for } \alpha_c - \delta < \alpha < \alpha_c. \tag{8.3.24}$$

Example 8.3.6 (Erdős-Rényi graph with mean degree μ) For $D = \text{Poisson}(\mu)$, $\mu_k = \mu^k$, so the critical value $\alpha_c = \mu_2 - \mu_1 = \mu^2 - \mu$, which is positive if $\mu > 1$. $\Delta = 2\mu^3 - 3\mu^2$ so the phase transition for evoSI is discontinuous if $\mu > 3/2$. The example in Fig. 8.12 has $\lambda = 1.4$ and, as predicted, the phase transition is continuous. In Fig. 8.13, $\lambda = 3$, so the phase transition is discontinuous.

Ideas that Go Into the Proofs. We follow the approach that Janson, Luczak, and Windridge used in Sect. 4.5 to prove the Miller-Volz equations. Specifically (i) we run the epidemic at the same time as we grow the graph, and (ii) we change time in the process (see the discussion before Lemma 4.5.3) by multiplying all the transition rates by

$$(X_t - 1)/(\lambda X_{I,t}),$$

where X_t is the total number of half-edges and $X_{I,t}$ is the number of infected half-edges. Since the set of infected half-edges makes new connections at rate $\lambda X_{I,t}$ and connections to a given half-edge with probability $1/(X_t - 1)$, the time change means that connections to a given half-edge occur at rate 1.

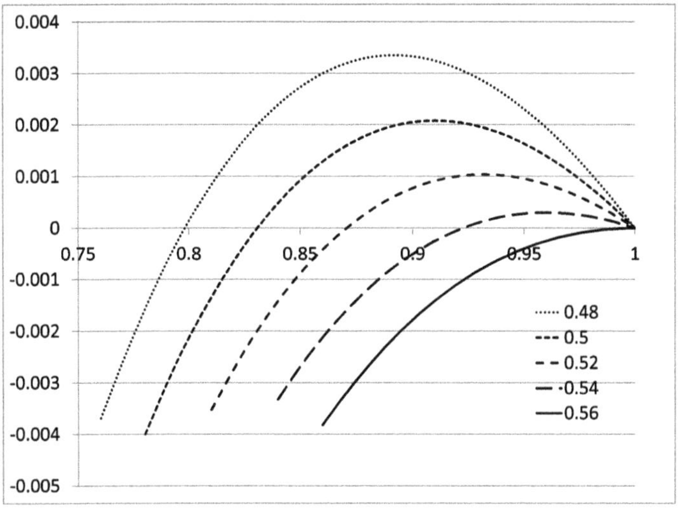

Fig. 8.12 The behavior of $f(w)$ for the Erdős-Rényi graph with mean 1.4, which has $\alpha_c = \mu^2 - \mu = .56$. Notice that as α increases to 0.56 the intersection with the x axis tends to 1, so the transition is continuous

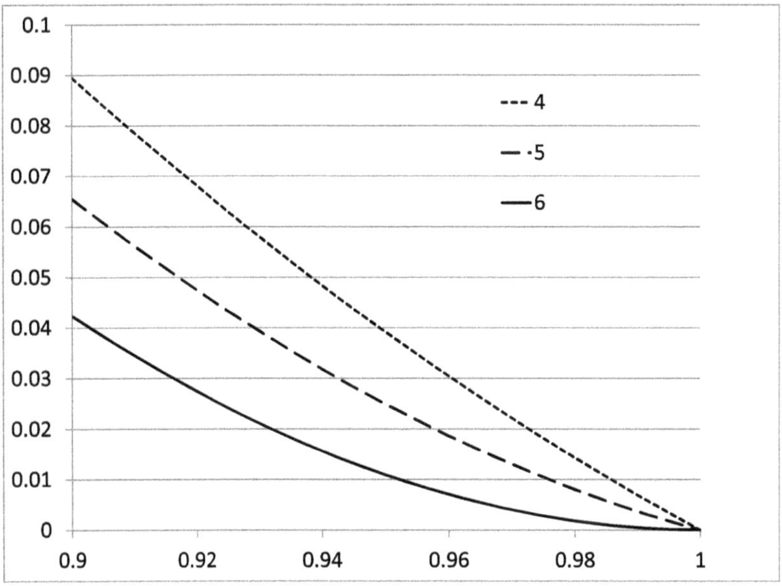

Fig. 8.13 The behavior of $f(w)$ for w near 1 for the Erdős-Rényi graph with mean 3, which has $\Delta > 0$. $\alpha_c = \mu^2 - \mu = 6$. Notice that when $\alpha \le \alpha_c$, $f(w) > 0$ for $w \in [0.9, 1)$, so σ is bounded away from 1

In evoSI the dynamics of $S_{t,k}$, the number of susceptible vertices with k half-edges is given by

$$dS_{t,k} = -\left(\lambda X_{I,t}\frac{kS_{t,k}}{X_t - 1}\right)dt + \left(1_{\{k\geq 1\}}\rho X_{I,t}\frac{S_{t,k-1}}{n}\right)dt - \left(\rho X_{I,t}\frac{S_{t,k}}{n}\right)dt + dM_{t,k},$$

where $M_{t,k}$ is a martingale. Using $\bar{S}_{t,k}$ for the time changed process the new dynamics are

$$d\left(\frac{\overline{S}_{t,k}}{n}\right) = -\left(k\frac{\overline{S}_{t,k}}{n}\right)dt + \left(1_{(k\geq 1)}\frac{\rho}{\lambda}\frac{\overline{X}_t - 1}{n}\frac{\overline{S}_{t,k-1}}{n}\right)dt$$
$$-\left(\frac{\rho}{\lambda}\frac{\overline{X}_t - 1}{n}\frac{\overline{S}_{t,k}}{n}\right)dt + d\left(\frac{\overline{M}_{t,k}}{n}\right). \tag{8.3.25}$$

Note the important fact that the number of infected half-edges no longer appears in the equation.

Let γ_n be the first time there are no infected half-edges, $w = \exp(-z)$ and $m_1 = ED$. Most of the work in the proof is to show

$$\sup_{0\leq z\leq\gamma_n}\left|\frac{\overline{X}_z}{n} - m_1 w^2\right| \to 0$$
$$\sup_{0\leq z\leq\gamma_n}\left|\frac{\sum_k \overline{S}_{z,k}}{n} - F_0(t)\right| \to 0 \tag{8.3.26}$$
$$\sup_{0\leq z\leq\gamma_n}\left|\frac{\sum_k k\overline{S}_{z,k}}{n} - F_1(t)\right| \to 0.$$

If a_S is the initial fraction of susceptibles then

$$F_0(t) = a_S \exp(-(\alpha/2)(w-1)^2)G(w), \tag{8.3.27}$$
$$F_1(t) = a_S \exp(-(\alpha/2)(w-1)^2)w(G'(w) + \alpha(1-w)G(w)).$$

From the results above, we see that

$$\frac{\overline{X}_z}{\overline{X}_{S,z}} \to \frac{m_1 w}{\exp(-(\alpha/2)(w-1)^2)\cdot(G'(w) + \alpha(1-w)G(w))}. \tag{8.3.28}$$

The logarithm of the right-hand side is $f(w)$, so when $\alpha < \alpha_c$ the largest root $\sigma < 1$ of $f(w) = 0$ gives the time $z = -\log(\sigma)$ at which the infection dies out in the time-changed process.

Theorem Lost (and Regained)

On my birthday (August 17) in 2020 Ball and Britton put their paper *Epidemics on networks with preventative rewiring* on the arXiv. Figure 8.5 from their paper, reproduced here as Fig. 8.14, made it painfully clear that our result was wrong.

About three months later, we received two referees reports on the paper, one with 160 comments. From these reports we learned that while the 30 pages of calculations needed to carry out the proof were mostly correct, our construction was flawed. In it $I - I$ connections could be broken and rewired, but this should not take place.

Almost immediately we realized our proof was half-right. Our computation provided an upper bound on the amount of the infection. The younger author embarked on a search for lower bounds which would show that the criterion in Theorem 8.3.5 was accurate. With the covid epidemic happening around him, Dong toiled in his apartment and met with me via zoom until he found a solution. The paper was resubmitted in July, 2021, at about the time he completed his Ph.D. at Duke. The final version accepted one year later.

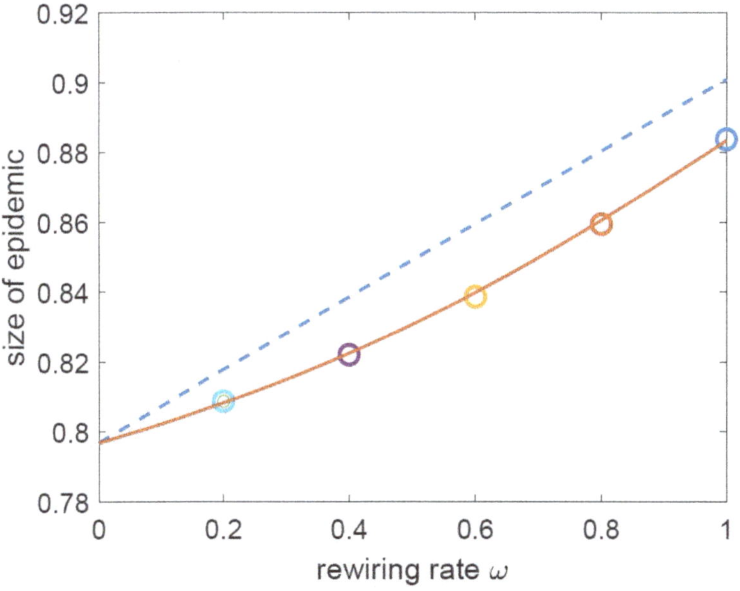

Fig. 8.14 Plots of our formula (dashed curve) versus their formula (solid curve) for evoSI when $\mu = 2$, $\lambda = 1$, and $\alpha = 1$ (evoSI). Circles give mean sizes of 1000 simulated major epidemics in a population of size 5000

8.3.5 Ball and Britton (2022)

These authors analyzed the evoSIR and evoSI epidemics on Erdős-Rényi graphs. Having just spent a long time on the second model we will restrict our attention to the first. The authors use a construction which exploits properties that are special to the Erdős-Rényi case. See Sect. 2.3 of their paper and also Britton and O'Neill (2002) for earlier examples of the use of this construction.

To prove results about the epidemics on an Erdős-Rényi random graph with mean degree μ they first consider a tree in which each vertex has a Poisson(μ) number of descendants and develop a branching process approximation for the SIR-ω epidemic in which infections (births) cross an edge with probability $\lambda/(\lambda+\gamma+\omega)$. Let $I(t)$ be the total number of individuals, $I_E(t)$ be the number of infectious edges (in state $I-E$), and $T(t)$ be the total progeny in the branching process on the tree. Let $I^n(t)$, $I_E^n(t)$ and $T^n(t)$ be the corresponding quantities for an Erdős-Rényi($n, \mu/n$) random graph on n vertices where initially a randomly chosen vertex is infected. They show in their Theorem 2.1 that if $t_n = \inf\{t : T(t) \geq \log n\}$ then the two systems can be defined on the same space so that

$$\sup_{0\leq t\leq t_n} |(I^n(t), I_E^n(t), T^n(t)) - (I(t), I_E(t), T(t))| \to_p 0$$

as $n \to \infty$.

Let $S^n(t)$ be the number of susceptibles at time t and $W^n(t)$ be the number of susceptible-susceptible edges created by rewiring by time t and let $X^n(t) = (S^n(t), I^n(t), I_E^n(t), W^n(t))$. Let $x(t) = (s(t), i(t), i_E(t), w(t))$ be the solution of the ODE

$$\begin{aligned}
\frac{ds}{dt} &= -\lambda i_E, \\
\frac{di}{dt} &= -\gamma i + \lambda i_E, \\
\frac{di_E}{dt} &= -\lambda i_E + \lambda\mu i_E s - \lambda\frac{i_E^2}{2} + 2\lambda i_E\frac{w}{s} - \omega i_E(1-\alpha+\alpha(1-i)), \\
\frac{dw}{dt} &= w\alpha i_E s - 2\lambda i_E\frac{w}{s}.
\end{aligned} \tag{8.3.29}$$

Here α is the probability an edge is rewired as in the definition of SIR-ω model. Theorem 2.2 in their paper, which is proved by using results of Darling and Norris (2008), shows that for any $t_0 > 0$

$$\sup_{0\leq t\leq t_0} |X^n(t)/n - x(t)| \to_p 0$$

as $n \to \infty$, provided that $I^n(t)/n \to i(0) > 0$.

It is interesting to note, see their Sect. 8.3, that the ODE in (8.3.29) is closely related to the "pair approximation" for the SIR-ω model. To explain the phrase in quotes, we note that "mean-field equations" come from pretending that the states of sites are independent; the pair approximation from assuming it is a Markov chain. In practice, this approach means that probabilities involving three sites are reduced to probabilities involving 1 and 2 sites using a conditional independence property. For the details of the computation see Chap. 7 in Andersson and Britton (2000).

Letting $T^n = n - S^n(\infty)$ (which is the final size of the epidemic) they make Conjecture 2.1 that one can interchange two limits $n \to \infty$ and $t \to \infty$ to conclude

$$T^n/n \to 1 - s(\infty).$$

To formulate a result that is independent of the validity of the conjecture they let

$$x^\epsilon(t) = (s^\epsilon(t), i^\epsilon(t), i_E(t), w^\epsilon(t))$$

be the solution to the ODE when

$$x^\epsilon(0) = (1-\epsilon, \epsilon, L^{-1}\epsilon, 0) \quad \text{where} \quad L = \frac{\lambda}{\lambda(\mu-1)-\omega}.$$

Letting $\tau_{SIR} = 1 - \lim_{\epsilon\downarrow 0} s^\epsilon(\infty)$, their Theorem 2.3 states

$$\lim_{\lambda\downarrow\lambda_c} \tau_{SIR} \begin{cases} = 0 & \text{if } \gamma > \omega(2\alpha-1) \text{or } \mu < 2\omega\alpha/[\omega(2\alpha-1)-\gamma], \\ > 0 & \text{if } \gamma < \omega(2\alpha-1) \text{and } \mu > 2\omega\alpha/[\omega(2\alpha-1)-\gamma]. \end{cases} \tag{8.3.30}$$

When $\alpha = 1$, $\omega = \rho$ and $\gamma = 1$ then we have a discontinuous phase transition if

$$\rho > 1 \quad \text{and} \quad \mu > \frac{2\rho}{\rho - 1}.$$

For the case of the delSIR model ($\alpha = 0$, $\omega = \rho$ and $\gamma = 1$) studied in Jiang et al. (2019), the transition is always continuous. This follows from (8.3.30) since $\gamma = 1 > -\omega$. From the last calculation we see that the phase transition is always continuous if $\alpha < 1/2$.

Theorem 2.4 in Ball and Britton (2022) gives results for the epidemic starting from a single infected individual. If we let τ^1_{SIR} be the limiting size of an epidemic conditionally on it being large then

$$\lim_{\lambda\downarrow\lambda_c} \tau^1_{SIR} \begin{cases} = 0 & \text{if } \gamma > \omega(3\alpha-1) \text{or } \mu < 3\omega\alpha/[\omega(3\alpha-1)-\gamma], \\ > 0 & \text{if } \gamma < \omega(2\alpha-1) \text{and } \mu > 2\omega\alpha/[\omega(2\alpha-1)-\gamma]. \end{cases} \tag{8.3.31}$$

See their paper for a precise statement. Remark 2.4 in their paper states the conjecture that the 3's in the first condition should be 2's. Recently Chen et al. (2022) have proved that this conjecture is correct.

To close this discussion we will state the result from Yao and Durrett (2020) for evoSIR and discuss its relation to (8.3.31)

Theorem 8.3.7 *Assume $ED^3 < \infty$. Consider the evoSIR model on the configuration model $\mathbb{CM}(n, D)$. Let*

$$\Delta = -\frac{\mu_3}{\mu_1} + \frac{2\rho}{(1+\rho)}(\mu_2 - \mu_1).$$

If $\Delta > 0$ then we have a discontinuous phase transition, i.e. for some $\epsilon_0 > 0$ and $\delta > 0$

$$\liminf_{\eta \to 0} \liminf_{n \to \infty} P_1(R_\infty/n > \epsilon_0 | R_\infty/n > \eta) > 0 \quad \text{for all } \alpha_c - \delta < \alpha < \alpha_c.$$

If $\Delta < 0$ then we have a continuous phase transition, i.e. for any $\epsilon > 0$ there is a $\delta > 0$ so that

$$\lim_{n \to \infty} P_1(R_\infty/n > \epsilon) = 0. \quad \text{for all } \alpha_c - \delta < \alpha < \alpha_c.$$

To make the connection we first transform the last result. In the Poisson case $\mu_i = \mu^i$ so the limit is 0 if

$$0 > -\frac{\mu^3}{\mu} + \frac{2\rho}{1+\rho}(\mu^2 - \mu).$$

Multiplying by μ and doing some algebra this is

$$0 > \frac{\rho - 1}{1+\rho}\mu^3 - \frac{2\rho}{1+\rho}\mu^2.$$

This is valid if $\rho < 1$ or $\mu < 2\rho/(\rho - 1)$.

To connect with (8.3.31) (with 3 replaced by 2) note that Yao and Durrett (2020) only consider the case $\alpha = 1$, take $\gamma = 1$, and use ρ instead of ω so the limit is 0 if

$$1 > \rho \quad \text{or} \quad \mu < \frac{2\rho}{\rho - 1}.$$

References

Andersson H, Britton T (2000) Stochastic epidemic models and their statistical analysis. Springer, New York

Avena L, Baldasso R, Bazra RS, den Hollander F, Quattropani M (2025) The voter model on random regular graphs with random rewiring. arXiv: 2501.08703

Baldassarri S, Braunsteins P, den Hollander F, Mandjes M (2024) Opinion dynamics on dense random graphs

Ball F, Britton T (2022) Epidemics on networks with preventative rewiring. Rand Struct Alg 61:250–297

Basu R, Sly A (2017) Evolving voter model and dense random graphs. Ann Appl Probab 27:1235–1288

Böhme GA, Gross T (2011) Analytical calculation of fragmentation transitions in adaptive networks. Phys Rev E 83:035101(R)

Britton T, O'Neill PD (2002) Bayesian inference for stochastic epidemics in populations with random social structure. Scand J Stat 29:375–390

Britton T, Juher D, Saldana J (2016) A network model with preventive rewiring: comparative of the initial phase. Bull Math Biol 78:2427–2454. Erratum: 79:1687–1689

Chatterjee S, Sivakoff D, Wascher M (2022) The effect of avoiding known infected neighbors on the persistence of a recurring infection process. Electron J Probab 27:1–40

Chen W, Hou Y, Yao D (2022) SIR epidemics on evolving Erdős-Rényi graphs. arXiv: 2208.11923

Cox JT, Greven A (1990) On the long term behavior of some finite particle systems. Probab Theory Rel Fields 85:195–237

da Silva GLB, Oliveira RI, Valesin D (2021) The contact process over a dynamical d-regular graph. arXiv: 2111.11757

Darling RWR, Norris JR (2008) Differential equation approximations for Markov chains. Probab Surv 5:37–79

Durrett R, Yao D (2022) Susceptible-infected epidemics on evolving graphs. Electron J Probab 27:110

Durrett R, Gleeson J, Lloyd A, Mucha P, Shi F, Sivakoff D, Socloar J, Varghese C (2012) Graph fission in an evolving voter model. Proc Natl Acad Sci 109:3682–3687

Ellner SP (2001) Pair approximation for lattice models with multiple interaction scales. J Theor Biol 210:435–447

Gleeson JP (2011) High-accuracy approximation of binary-state dynamics on networks. Phys Rev Lett 107:068701

Gleeson JP (2013) Binary-state dynamics on complex networks: pair approximation and beyond. Phys Rev X 3:021004

Gross T, D'Lima CD, Blasius B (2006) Epidemic dynamics on an adaptive network. Phys Rev Lett 96:208701

Guo D, Trajanovski S, van de Bovenkamp R, Wang H, van Mieghen P (2013) Epidemic threshold and topological structure of susceptible-infected-susceptible epidemics in adaptive networks. Phys Rev E 88:042802

den Hollander F (2024) Evolution of discordance. arXiv: 2410.17808

Holme P, Newman MEJ (2006) Nonequilibrium phase transition in the coevolution of networks and opinions. Phys Rev E 74:056108

Jacobsen KA, Burch MG, Tien JH, Rempala GA (2016) The large graph limit of a stochastic epidemic model on a dynamic multilayer network. J Biol Dyn 12:746–768

Janson S, Luczak M, Windridge P (2014) Law of large numbers for the SIR epidemic on a random graph with given degrees. Rand Struct Alg 45:726–763

Jiang Y, Kassem R, York G, Junge M, Durrett R (2019) SIR epidemics on evolving graphs. arXiv: 1901.06568

Kimura D, Hayakawa Y (2008) Coevolutionary networks with homophily and heterophily. Phys Rev E 78:016103

Kureh YH, Porter MA (2020) Fitting in and breaking up: a nonlinear version of coevolving voter models. Phys Rev E 101:062303

Leung KY, Ball F, Sirl D, Britton T (2018) Individual preventive social distancing during an epidemic may have negative population-level outcomes. J Roy Soc Interf 15:20180296

Linker A, Remenik D (2020) The contact process with dynamic edges on $\mathbb{Z}$. Electron J Probab 25:20

Mourrat J-C, Valesin D (2016) Phase transitions of the contact process on random regular graphs. Electron J Probab 21:31

Remenik D (2008) The contact process in a dynamic random environment. Ann Appl Probab 18:2392–2420

Schapira B, Valesin D (2023) The contact process on dynamic regular graphs: monotonicity and subcritical phase. arXiv: 2309.17040

Shi B, Mucha P, Durrett R (2013) Multi-opinion coevolving voter model with infinitely many phase transitions. Phys Rev E 88:062818

Silk H, Demirel G, Homer M, Gross T (2014) Exploring the adaptive voter model dynamics with a mathematical triple jump. New J Phys 16:093051

Szabó-Solticzky A, Barthouze L, Kiss IZ, Simon PL (2016) Oscillating epidemics in a dynamic network model: stochastic and mean-field analysis. J Math Biol 72:1153–1176

Tuunc I, Shkarayev MS, Shaw LD (2013) Epidemics in adaptive social networks with temporary link deactivation. J Statist Phys 151:355–366

Vazquez F, Eguíluz VM, San Miguel MS (2008) Generic absorbing transition in coevolution dynamics. Phys Rev Lett 100:108702

Volz E, Meyers LA (2007) Susceptible-infected recovered epidemics in dynamic contact networks. Proc Roy Soc B 274:2925–2933

Volz E, Meyers LA (2009) Epidemic threshold in dynamic contact networks. J Roy Soc Interf 6:233–241

Yao D, Durrett R (2020) Epidemics on evolving graphs. arXiv: 2003.08541

Zanette DH (2007) Coevolution of agents and networks in an epidemic model. arXiv: 0707.1249

Appendix A
Large Deviations

A.1 Chernoff's Theorem

This material is from Sect. 2.7 of PTE5, which can be consulted for the missing details. Let $X_1, X_2, \ldots$ be i.i.d. and let $S_m = X_1 + \cdots + X_m$. To make the results we develop here easier to use on concrete examples we will systematically avoid the use of the letter n. Suppose that the **moment-generating function**

(H1) $$\varphi(\theta) = E\exp(\theta X_i) < \infty \quad \text{for some } \theta > 0$$

and the mean $\mu = EX_i > -\infty$ exists. Let $a > \mu$ and $\gamma_m = P(S_m > ma)$.

Lemma A.1.1 *If $k, \ell \geq 0$ then $\gamma_{k+\ell} \geq \gamma_k \gamma_\ell$, so as $m \to \infty$, $\gamma_m/m \to \sup_{k\geq 1} \gamma_k/k$ and hence*

$$P(S_m \geq ma) \leq e^{m\gamma(a)}. \tag{A.1.1}$$

Proof Using arithmetic and independence we have

$$P(S_{k+\ell} \geq (k+\ell)a) \geq P(S_k \geq ka, S_\ell \geq \ell a) = P(S_k \geq ka) \cdot P(S_\ell \geq \ell a).$$

For the proof of the limit result see Lemma 2.7.1 in PTE5. The inequality follows from the formula for the limit. □

To calculate $\gamma(a)$ we use the **exponential Chebyshev inequality**. If $\theta > 0$ then

$$e^{\theta ma} P(S_m \geq ma) \leq E\exp(\theta S_m) = \phi(\theta)^m.$$

Letting $\kappa(\theta) = \log\phi(\theta)$ we have

$$P(S_m \geq ma) \leq \exp(-m[a\theta - \kappa(\theta)]). \tag{A.1.2}$$

This is an upper bound for all $\theta > 0$, so it is natural to optimize it by finding the maximum of $\theta a - \kappa(\theta)$.

R. Durrett, *Dynamics on Graphs*, Probability Theory and Stochastic Modelling 110,
https://doi.org/10.1007/978-3-032-21908-4

Suppose for the moment that

(H1′) $\varphi(\theta) = E\exp(\theta X_i) < \infty$ for all $\theta > 0$.

In this case if $\theta > 0$

$$\frac{d}{d\theta}\{\theta a - \log\varphi(\theta)\} = a - \varphi'(\theta)/\varphi(\theta)$$

so (assuming things are nice) the maximum occurs when $a = \varphi'(\theta)/\varphi(\theta)$. To show that things are nice when (H1′) holds we begin by defining

$$F_\theta(x) = \frac{1}{\varphi(\theta)}\int_{-\infty}^{x} e^{\theta y}\,dF(y) \quad \text{for } \theta > 0.$$

It is easy to see that if the moment generating function is finite on $(\theta - \delta, \theta + \delta)$ F_θ is a distribution function with mean

$$\int x\,dF_\theta(x) = \frac{1}{\varphi(\theta)}\int_{-\infty}^{\infty} xe^{\theta x}\,dF(x) = \frac{\varphi'(\theta)}{\varphi(\theta)}$$

$$\phi''(\theta) = \int_{-\infty}^{\infty} x^2 e^{\theta x}\,dF(x)$$

$$\frac{d}{d\theta}\frac{\varphi'(\theta)}{\varphi(\theta)} = \frac{\varphi''(\theta)}{\varphi(\theta)} - \left(\frac{\varphi'(\theta)}{\varphi(\theta)}\right)^2 = \int x^2\,dF_\theta(x) - \left(\int x\,dF_\theta(x)\right)^2 \geq 0$$

since the last expression is the variance of F_θ. If we assume

(H2) the distribution F is not a point mass at μ

then $\varphi'(\theta)/\varphi(\theta)$ is strictly increasing on $\{\theta > 0\}$. Since we have $\varphi'(0)/\varphi(0) = \mu$, this shows that for each $a > \mu$ there is at most one $\theta_a \geq 0$ that solves $a = \varphi'(\theta_a)/\varphi(\theta_a)$.

Theorem A.1.2 *Let* $\theta_+ = \sup\{\theta : \phi(\theta) < \infty\}$. *Suppose in addition to (H1) and (H2) that there is a* $\theta_a \in (0, \theta_+)$ *so that* $a = \varphi'(\theta_a)/\varphi(\theta_a)$. *Then, as* $n \to \infty$,

$$n^{-1}\log P(S_n \geq na) \to -a\theta_a + \log\phi(\theta_a).$$

Proof Using (A.1.2) we see that the right-hand side is an upper bound on the left-hand side for all n, and hence lim sup $LHS \leq RHS$. To prove lim inf $LHS \leq RHS$, the other inequality, pick $\lambda \in (\theta_a, \theta_+)$, let $X_1^\lambda, X_2^\lambda, \ldots$ be i.i.d. with distribution F_λ and let $S_n^\lambda = X_1^\lambda + \cdots + X_n^\lambda$. Writing dF/dF_λ for the Radon-Nikodym derivative of the associated measures, it is immediate from the definition that $dF/dF_\lambda = e^{-\lambda x}\varphi(\lambda)$. If we let F_λ^n and F^n denote the distributions of S_n^λ and S_n, then

Lemma A.1.3 $\dfrac{dF^n}{dF_\lambda^n} = e^{-\lambda x}\varphi(\lambda)^n$.

We leave the proof to the reader. If $\nu > a$, then the lemma and monotonicity imply (∗)

$$P(S_n \geq na) \geq \int_{na}^{n\nu} e^{-\lambda x}\varphi(\lambda)^n dF_\lambda^n(x) \geq \varphi(\lambda)^n e^{-\lambda n\nu}(F_\lambda^n(n\nu) - F_\lambda^n(na)).$$

F_λ has mean $\varphi'(\lambda)/\varphi(\lambda)$, so if we have $a < \varphi'(\lambda)/\varphi(\lambda) < \nu$, then the weak law of large numbers implies

$$F_\lambda^n(n\nu) - F_\lambda^n(na) \to 1 \text{ as } n \to \infty.$$

From the last conclusion and (∗) it follows that

$$\liminf_{n\to\infty} n^{-1} \log P(S_n > na) \geq -\lambda\nu + \log\phi(\lambda).$$

Since $\lambda > \theta_a$ and $\nu > a$ are arbitrary, the proof is complete. □

To explain the assumption $\theta < \theta_+$ we consider two examples

Example A.1.4 Suppose $F(0) = 0$. If $x_0 = \sup\{x : F(x) < 1\} < \infty$ then $\phi'(\theta)/\phi(\theta) \to x_0$ as $\theta \to \infty$. If $y > x_0$ then $P(S_m \leq My) = 1$ for all m. If $F(x_0) < 1$ then

$$\frac{1}{m}\log P(S_m \geq mx_0) = \log(1 - F(x_0))$$

so we have large deviations results for all levels.

Example A.1.5 (Perverted exponential) Consider a density function with $f(x) = Cx^{-\alpha}e^{-x}$ for $x \geq 1$ where $\alpha > 2$ and C_α is chosen so that it integrates to 1. Then $\phi(\theta) < \infty$ when $\theta \leq 1$ and $\phi'(\theta)/\phi(\theta) \to \phi'(1)/\phi(1) \equiv a_0 < \infty$ as $\theta \uparrow 1$. Theorem 2.7.10 in PTE5 shows that $\gamma(a)$ is linear for $a > a_0$.

Binomial large deviations

If $P(X = 1) = p$ and $P(X = 0) = 1 - p$ then

$$\varphi(\theta) = E\exp(\theta X) = (1 - p) + pe^\theta.$$

$\varphi'(\theta) = pe^\theta$ so in order to solve $a = \phi'(\theta_a)/\phi(\theta_a)$ we want

$$\begin{aligned} a(1-p) + ape^{\theta_a} &= pe^{\theta_a} \\ a(1-p) &= p(1-a)e^{\theta_a} \\ \theta_a &= \log(a/p) + \log((1-p)/(1-a)). \end{aligned}$$

This means

$$\varphi(\theta_a) = (1-p) + p \cdot a(1-p)/p(1-a) = (1-p)(1 + a/(1-a)) = \frac{1-p}{1-a}$$

and hence we have

$$\begin{aligned} -a\theta_a + \log\varphi(\theta_a) &= -a\log(a/p) - a\log((1-p)/(1-a)) + \log((1-p)/(1-a)) \\ &= -a\log(a/p) - (1-a)\log((1-a)/(1-p)) \equiv -H(a). \end{aligned}$$

Lemma A.1.6 *If $X = binomial(m, p)$ then for $0 < b < p < c < 1$*

(a) $P(X \le mb) \le \exp(-H(b)m)$.
(b) $P(X \ge mc) \le \exp(-H(c)m)$.

Remark A.1.7 To see why this is the answer, note that from the definition of the Binomial distribution and Stirling's formula, $n! \sim n^n e^{-n}\sqrt{2\pi n}$,

$$\begin{aligned} P(X = ma) &= \binom{m}{ma} p^{ma}(1-p)^{m(1-a)} \\ &\approx \frac{m^m}{(ma)^{ma}(m(1-a))^{m(1-a)}} p^{ma}(1-p)^{m(1-a)}. \end{aligned}$$

Now cancel the m^m's in the fraction and take $(1/n)\log$ of both sides.

Proof Suppose $a > p$. Then for any $\theta > 0$

$$e^{na\theta} P(X \ge na) \le ((1-p) + pe^\theta)^n,$$

which we can rewrite as

$$P(X \ge na) \le \left[\exp\{-\theta a + \log((1-p) + pe^\theta)\}\right]^n.$$

To optimize we differentiate the term in set braces with respect to θ and set

$$0 = -a + \frac{pe^\theta}{(1-p) + pe^\theta}.$$

Solving we have $e^\theta = (a(1-p))/(p(1-a))$. Using

$$((1-p) + pe^\theta) = pe^\theta/a = (1-p)/(1-a)$$

and plugging in we have

$$= \exp\left(-a \log\left(\frac{a(1-p)}{p(1-a)}\right) + \log((1-p)/(1-a))\right)$$
$$= \exp(-H(a)).$$

For $a < p$ the first equality is valid for $\theta < 0$ but the remaining calculations are the same. □

In many cases the following simplification is useful

Lemma A.1.8 *If $X = binomial(n, p)$ then*

(a) $P(X \le n(p - z)) \le \exp(-nz^2/2p)$.
(b) $P(X \ge n(p + z)) \le \exp(-nz^2/2(p + z))$.

Taking $z = py$ this becomes

(c) $P(X \le np(1 - y)) \le \exp(-npy^2/2)$.
(d) $P(X \ge np(1 + y)) \le \exp(-npy^2/2(1 + y))$.

Proof We begin with (b). The function defined in Lemma A.1.6 has $H(p) = 0$ and

$$H'(a) = \log(a/p) - \log((1-a)/(1-p))$$

so $H'(p) = 0$. If $p < a \le 1$, Taylor's theorem implies that there is a $y \in [p, a]$ so that $H'(a) = H''(y)(a - p)^2/2$. Differentiating again $H''(a) = 1/a(1 - a)$, which is minimized at $a = 1/2$. Therefore if $a \le 1/2$, $H(a) \ge (a - p)^2/2a$, while if $a \ge 1/2$, $H(a) \ge 2(a - p)^2 \ge (a - p)^2/2a$. Substituting $a = p + z$ gives the second result. The argument for (a) is almost identical, but when we substitute $a = p - z$ we can let $a = p$ in the denominator. □

Four more concrete examples

To prepare for the computations, we recall some important formulas:

$$\kappa(\theta) = \log\phi(\theta) \quad \kappa'(\theta) = \phi'(\theta)/\phi(\theta) \quad \theta_a \text{ solves } \kappa'(\theta_a) = a$$
$$\gamma(a) = \lim_{n\to\infty} (1/n) \log P(S_n \ge na) = -a\theta_a + \kappa(\theta_a).$$

Example A.1.9 (Normal distribution) Without loss of generality we can assume the mean is 0. If the variance is σ^2, the density function is $(2\pi\sigma^2)^{-1/2}e^{-x^2/2\sigma^2}$ so

$$\phi(\theta) = \int (2\pi\sigma^2)^{-1/2}e^{-x^2/2\sigma^2}e^{\theta x} = \exp(\theta^2/2\sigma^2)$$

and it follows that for $a > 0$

$$\kappa(\theta) = \sigma^2\theta^2/2 \qquad \kappa'(\theta) = \sigma^2\theta \qquad \theta_a = a/\sigma^2$$
$$\gamma(a) = -a\theta_a + \kappa(\theta_a) = -a^2/2\sigma^2 + a^2/\sigma^2 = a^2/2\sigma^2.$$

One can check the last result by observing that S_n has a normal distribution with mean 0 and variance $n\sigma^2$, and then using the formula for the tail of the distribution function for the normal. See Theorem 1.2.6 in PTE5.

Example A.1.10 **Exponential(1)** has density function $f(x) = e^{-x}$ for $x \geq 0$ so

$$\phi(\theta) = \int_0^\infty e^{-x} e^{\theta x}\, dx = 1/(1-\theta) \qquad \text{for } \theta < 1$$

and it follows that

$$\kappa(\theta) = -\log(1-\theta) \qquad \kappa'(\theta) = 1/(1-\theta) \qquad \theta_a = 1 - 1/a$$
$$\gamma(a) = -a\theta_a + \kappa(\theta_a) = -a + 1 + \log a \quad \text{for } a > 1.$$

To extend to a general λ, note that if X_i are exponential(1) then $Y_i = X_i/\lambda$ is exponential(λ) so if $a > 1$

$$P(Y_1 + \cdots + Y_m > ma/\lambda) = P(X_1 + \cdots + X_m > am).$$

Letting $T_m = X_1 + \cdots + X_m$ and $T_m^\lambda = Y_1 + \cdots + Y_m$

$$\frac{1}{m} \log P\left(T_m^\lambda > \frac{ma}{\lambda}\right) = \frac{1}{m} \log P(T_m > ma). \tag{A.1.3}$$

This simple argument does not work for discrete random variables, so we have to compute for a general parameter.

Example A.1.11 **Poisson(λ)** has $P(X = k) = e^{-\lambda}\lambda^k/k!$ for $k = 0, 1, 2, \ldots$ so

$$\phi(\theta) = \sum_{k=0}^\infty e^{-\lambda} \frac{\lambda^k}{k!} e^{\theta k} = \exp(-\lambda(1 - e^\theta)).$$

Replacing e^θ by z leads to the familiar formula for the generating function.

If $X = \text{Poission}(\lambda)$ and $Y = \text{Poission}(\mu)$ are independent then $X + Y = \text{Poission}$ $(\lambda + \mu)$. From this it follows that

$$P(S_m^\lambda = k) = e^{-\lambda m}(\lambda m)^k/k!$$

where we have added a superscript to indicate the parameter. Using Stirling's formula

$$k! \sim k^k e^{-k}\sqrt{2\pi k} \tag{A.1.4}$$

we have

$$P(S_m^\lambda = k) \sim e^{-\lambda m + k} \left(\frac{\lambda m}{k}\right)^k / \sqrt{2\pi k}$$

and if $k/\lambda m \to a$ we have a scaling result similar to (A.1.3)

$$\frac{1}{\lambda m}\log P(S_m^\lambda = k) = -\frac{\lambda m - k}{\lambda m} - \frac{k}{\lambda m}\log\left(\frac{k}{\lambda}\right) + o(1)$$
$$\to a - 1 - a\log a = \lim_{m\to\infty}\frac{1}{m}\log P(S_m^1 = k).$$

To convert this into an asymptotic for $\geq k$ we note that if $j/(\lambda m) \geq a > 1$ we have

$$P(S_n^\lambda = j+1) = \frac{\lambda m}{j+1}P(S_n^\lambda = j) \leq \frac{1}{a}P(S_n^\lambda = j).$$

From this we see that (ignoring annoying rounding)

$$P(S_m^\lambda \geq \lambda m) \leq \frac{1}{1-1/a}P(S_m^\lambda = \lambda m)$$

The dominant contribution is from values near λm, so $\leq$ is actually $\sim$.

Example A.1.12 **Geometric**(p) has $P(X = k) = (1-p)^{k-1}p$ for $k = 1, 2, \ldots$ so if $(1-p)e^\theta < 1$ then

$$\phi(\theta) = \sum_{k=1}^\infty (1-p)^{k-1}pe^\theta = \frac{pe^\theta}{1-(1-p)e^\theta}$$

and it follows that $\kappa(\theta) = \theta + \log p - \log(1-(1-p)e^\theta)$,

$$\kappa'(\theta) = 1 + \frac{(1-p)e^\theta}{1-(1-p)e^\theta}.$$

Setting $a = \kappa'(\theta)$ leads to $(a-1)[1-(1-p)e^\theta] = (1-p)e^\theta$ so

$$a - 1 = a[(1-p)e^\theta] \quad\text{and}\quad \theta_a = \log\left(\frac{a-1}{a(1-p)}\right).$$

At this point it takes courage to compute $\gamma(a)$, but some nice things happen,

$$\kappa(\theta_a) = \theta_a + \log p + \log[1-(1-p)e^{\theta_a}]$$

looks ugly but $(1-1-p)e^{\theta_a} = 1-(a-1)/a = 1/a$ so

$$\gamma(a) = -a\theta_a + \kappa(\theta_a) = -(a-1)\theta_a + \log(ap).$$

To check this result, we note that the sum of m geometric(p) has a **negative binomial distribution**

$$P(S_m = k) = \binom{k-1}{m-1} p^m (1-p)^{k-m} \qquad \text{for } k \geq m.$$

To see this note that in order for $S_m = k$ the mth success must come at time k and there must be exactly $k-1$ successes on trials $1, 2, \ldots, k-1$. If $k = am$ with $a > 1$,

$$P(S_m = am) = \binom{am-1}{m-1} p^m (1-p)^{m(a-1)}.$$

Dropping the -1's, using Stirling's formula, ignoring the square roots, and noticing that the exponentials cancel out when it is applied to a binomial coefficient,

$$\begin{aligned} P(S_m = am) &\approx \frac{(am)^{am}}{m^m (m(a-1))^{m(a-1)}} \cdot p^m (1-p)^{m(a-1)} \\ &= (ap)^m \cdot \left(\frac{a(1-p)}{a-1} \right)^{m(a-1)} = (ap)^m \exp(-\theta_a m(a-1)). \end{aligned}$$

A.2 Azuma-Hoeffding inequality

To quote Wikipedia, "In probability theory, the Azuma-Hoeffding inequality (named after Kazuoki Azuma and Wassily Hoeffding) gives a concentration result for the values of martingales that have bounded differences."

Theorem A.2.1 *If X_k is a supermartingale with $|X_k - X_{k-1}| \leq c_k$ and X_0 is constant then*

$$P(X_m - X_0 \geq x) \leq \exp\left(\frac{-x^2}{2\sum_{k=1}^m c_k^2} \right).$$

This immediately implies a result for submartingales and combining the two we have

Theorem A.2.2 (Azuma-Hoeffding inequality) *If M_k is a martingale and $|M_k - M_{k-1}| \leq c_k$ and M_0 is constant then*

$$P(|M_m - M_0| \geq x) \leq 2\exp\left(\frac{-x^2}{2\sum_{k=1}^m c_k^2} \right).$$

Proof It suffices to prove Theorem A.2.1. Let $\mathcal{F}_k = \sigma(X_0, \ldots X_k)$, and let $Y_k = X_k - X_{k-1}$ for $1 \leq k \leq m$. Then $\sum_{k=1}^{\ell} Y_k = X_\ell - X_0$ and $E(Y_{k+1}|\mathcal{F}_k) \leq 0$. -Let $\lambda > 0$

$$P(X_m - X_0 \geq x) = P\left(\exp\left\{ \lambda \sum_{k=1}^m Y_i \right\} \geq e^{\lambda x} \right) \leq e^{-\lambda x} E\left(\exp\left\{ \lambda \sum_{k=1}^m Y_i \right\} \right)$$

by Markov's inequality. Since $-c_i \leq Y_i \leq c_i$ and $e^{\lambda x}$ is convex

$$e^{\lambda Y_i} \leq \frac{1 - Y_i/c_i}{2} e^{-\lambda c_i} + \frac{1 + Y_i/c_i}{2} e^{\lambda c_i} = \cosh(\lambda c_i) + (Y_i/c_i)\sinh(\lambda c_i),$$

which implies

$$E(e^{\lambda Y_i} | \mathcal{F}_{i-1}) \leq \cosh(\lambda c_i). \tag{A.2.1}$$

From this it follows that

$$\begin{aligned} E\left(\exp\left\{\lambda \sum_{k=1}^{n} Y_i\right\}\right) &= E\left(E(e^{\lambda Y_n} | \mathcal{F}_{n-1}) \cdot \exp\left\{\lambda \sum_{i=1}^{n-1} Y_i\right\}\right) \\ &\leq \cosh(\lambda v_n) E\left(\exp\left\{\lambda \sum_{i=1}^{n-1} Y_i\right\}\right). \end{aligned}$$

Using induction gives

$$E\left(\exp\left\{\lambda \sum_{k=1}^{n} Y_i\right\}\right) \leq \prod_{i=1}^{n} \cosh(\lambda c_i).$$

The right-hand side of the last inequality is

$$\begin{aligned} \prod_{i=1}^{n} \cosh(\lambda c_i) &= \prod_{i=1}^{n} \sum_{m=0}^{\infty} \frac{(\lambda c_i)^{2m}}{(2m)!} \\ &\leq \prod_{i=1}^{n} \sum_{m=0}^{\infty} \frac{(\lambda c_i)^{2m}}{2^m \cdot m!} = \exp\left(\frac{\lambda^2}{2} \sum_{i=1}^{n} c_i^2\right) \end{aligned}$$

where in the second step we have removed the odd integers from the factorial. Taking $\lambda = x / \sum_{i=1}^{n} c_i^2$ we have

$$e^{\lambda x} = \exp\left(\frac{\lambda^2}{2} \sum_{i=1}^{n} c_i^2\right),$$

which gives the desired result. □

Bibliography

Albert R, Barabási AL (2002) Statistical mechanics of complex networks. Rev Mod Phys 74:47–97
Alon N, Spencer S (2000) The probabilistic method, 2nd edn. Wiley
Azuma K (1967) Weighted sums of certain dependent random variables. Tôhoku Math J 19(3):357–367
Barrat A, Barthelemy M, Vespignani A (2008) Dynamical process on complex networks. Cambridge University Press
Bollobás B (1985) Random graphs. Academic Press, London
Castellano C, Fortunato S, Loreto V (2009) Statistical physics of social dynamics. Rev Mod Phys 81:591–646
Chernoff H (1952) A measure of asymptotic efficiency for tests of a hypothesis based on the sum of observations. Ann Math Stat 23(4):493–507
Chung F, Lu L (2006) Complex graphs and networks. American Mathematical Society, Providence, RI
Cohen R, Havlin S (2010) Complex networks: structure, robustness and function. Cambridge University Press
Durrett R (2019) Probability: theory and examples, 5th edn. Cambridge University Press
Easley D, Kleinberg (2010) Networks, crowds and markets. Cambridge University Press
Feller W (1968) An introduction to probability theory and its applications, vol I, 3rd edn. Wiley
Feller W (1971) An introduction to probability theory and its applications, vol II, 2nd edn. Wiley
Frieze A, Karonski M (2022) Introduction to random graphs. Cambridge University Press
Grimmett G (2010) Probability on graphs. IMS Textbooks. Cambridge University Press
Hoeffding W (1963) Probability inequalities for sums of bounded random variables. J Am Stat Assoc 58:13–30
Janson S, Luczak T, Rucinski A (2000) Random graphs. Wiley-Interscience, New York
Kiss IZ, Miller JC, Simon PL (2017) Mathematics of epidemics on networks. Interdisciplinary Applied Mathematics. Springer, p 46
Lyons R, Peres Y (2016) Probability on trees and networks. Cambridge University Press
Newman MEJ (2010) Networks: an introduction. Oxford University Press
Pastor-Satoras R, Castellano C, van Mieghan P, Vespignani A (2015) Epidemic processes in complex networks. Rev Mod Phys 87:925–979
Penrose M (2003) Random geometric graphs. Oxford University Press
van der Hofstad R (2017) Stochastic processes on random graphs. In: Lecture notes for the 47th summer school in probability Saint-Flour

R. Durrett, *Dynamics on Graphs*, Probability Theory and Stochastic Modelling 110,
https://doi.org/10.1007/978-3-032-21908-4

van der Hofstad, Remco (2023) Random graphs and complex networks, vol 2. Cambridge Series in Statistical and Probabilistic Mathematics. Cambridge University Press

Varadhan SRS (2001) Probability theory. American Mathematical Society

Index

R. Durrett, *Dynamics on Graphs*, Probability Theory and Stochastic Modelling 110,
https://doi.org/10.1007/978-3-032-21908-4

www.ingramcontent.com/pod-product-compliance
Ingram Content Group UK Ltd.
Pitfield, Milton Keynes, MK11 3LW, UK
UKHW021840081026
476418UK00003B/74
* 9 7 8 3 0 3 2 2 1 9 0 7 7 *